STUDENT SOLUTIONS MANUAL

for Yoshiwara, Yoshiwara, and Drooyan's

Elementary Algebra

EQUATIONS & GRAPHS

Gale Hughes

Brooks/Cole
Thomson Learning.

Pacific Grove • Albany • Belmont • Boston • Cincinnati • Johannesburg • London • Madrid • Melbourne
Mexico City • New York • Scottsdale • Singapore • Tokyo • Toronto

Project Development Editor: *Michelle Paolucci*
Marketing Team: *Leah Thomson and Debra Johnston*
Senior Editorial Assistant: *Erin Wickersham*
Production Coordinator: *Dorothy Bell*
Cover Design: *Vernon T. Boes*
Printing and Binding: *Webcom Limited*

For more information, contact:
BROOKS/COLE
511 Forest Lodge Road
Pacific Grove, CA 93950 USA
www.brookscole.com

Printed in Canada

10 9 8 7 6 5 4 3 2 1

ISBN 0-534-36261-3

Table of Contents

How to Be Successful in Your Math Class

The key to success in a math class (as in most endeavors) is persistence. You cannot learn mathematics in one great rush the night before the exam; but you can master it in small chunks a little at a time. You should plan to study math for at least one hour every night. Don't give up until you have a good grasp of the lesson and can work the problems on your own. If you get behind in a math class, it is very difficult to catch up.

1. **Attend class every day.**
   Studies have shown that success in math classes is correlated strongly with attendance. If you must miss a class, find out beforehand what the class will cover. Read the lesson and complete the assignment anyway, just as if you had attended.
   a. Use class time wisely. This is your best opportunity to learn the material.
   b. Take notes. Learn to summarize what the instructor says, not just what he or she writes on the board.
   c. Don't be afraid to ask questions when you don't follow the lesson.

2. **Read the text book.**
   Reading a math book is not like reading a novel. You will need to read the material more than once to understand and retain it.
   a. Read the new material *before* it will be covered in class.
   b. Read with a pencil in hand so that you can make notes to yourself, underline important points, or put question marks in the margins.
   c. Read the section again after it has been covered in class.

3. **Look over your handouts and class notes.**
   The sooner you can review your notes after class, the better. People forget most of what they hear very quickly, and reviewing your notes will help you retain the new information.
   a. Look for points where your notes reinforce the material in the textbook.
   b. Try to fill in any steps or information you may have missed in class.
   c. Write a sentence or two summarizing the main points of the lesson.

4. **Do the homework problems.**
   Most of your learning takes place when you work problems. If you do some of your work in a study group or tutoring center, you will have someone to consult as soon as you hit a snag.
   a. If you get stuck on a problem, refer to the textbook or your notes for help.
   b. Call a classmate on the phone and try to figure out together the problems you had trouble with.

   c. Mark any problems you can't get, but don't stop! Skip those problems for now, and continue on to the end of the assignment.

5. Get help right away.
Mathematics builds upon earlier material, so if you don't understand today's lesson you will have even more trouble tomorrow or the next day.
   a. Make a list of points you don't understand and problems you need help with.
   b. Ask your instructor or a tutor for help *today* -- don't put it off!
   c. Fill in your notes with the answers to your questions, and make sure you can work all the problems that gave you trouble.

6. Prepare for exams.
In addition to keeping up with daily work, you must prepare specifically for exams. Always study 100% of the material the exam will cover. If you omit some topics, you won't be sure which problems you should work on during the exam!
   a. Begin studying for the exam a week ahead of time, so that you will have a chance to get help on any topics you are unsure about.
   b. Make a check-list or outline of the material the exam will cover, and review each topic until you have mastered it.
   c. Have a classmate or tutor make up a sample exam (or make one yourself), and practice working problems under exam conditions.

Math Myths

Here are some false statements about learning mathematics.

1. Some people have "math minds."
2. Math is mainly a lot of memorization.
3. It doesn't matter how much you study math -- either you get it or you don't.
4. Since nobody else is asking questions, they must all understand everything.
5. I'm the only one in the class who couldn't get this problem.
6. Everyone will think I'm dumb if I ask this question.
7. If you understand what the teacher is doing at the board, then you are okay; you don't need to work a lot of problems.
8. If you can't work a problem on the first try, you won't be able to get it without help.
9. I don't need to read the book to pass this course -- the tests will only cover what we do in class.
10. If I get behind I can catch up later -- I can get a tutor to show me the quick way to do this.

CHAPTER 1

Homework 1.1

1a. Yes. Higher degrees result in higher salaries for all groups.

b. African-Americans have higher salaries on average than Asians in all educational levels except Master's degree.

c. No high school diploma: $\frac{16,487}{26,115} \approx 0.63 = 63\%$

High school diploma: $\frac{21,121}{27,376} \approx 0.77 = 77\%$

Bachelor's degree: $\frac{33,817}{44,426} \approx 0.76 = 76\%$

Master's degree: $\frac{41,431}{52,787} \approx 0.78 = 78\%$

Doctorate: $\frac{46,873}{59,348} \approx 0.79 = 79\%$

Professional degree: $\frac{41,029}{77,877} \approx 0.53 = 53\%$

d. Anglo: $\frac{44,426 - 27,376}{27,376} = \frac{17,050}{27,376} \approx 0.623 = 62.3\%$

Latino: $\frac{33,817 - 21,121}{21,121} = \frac{12,696}{21,121} \approx 0.601 = 60.1\%$

African-American: $\frac{34,290 - 22,040}{22,040} = \frac{12,250}{22,040} \approx 0.556 = 55.6\%$

Asian: $\frac{33,758 - 21,608}{21,608} = \frac{12,150}{21,608} \approx 0.562 = 56.2\%$

3a.

National Origin	US Population in 1990
Chinese	$1.041 \times 806 + 806 \approx 1645$
Filipino	$0.816 \times 775 + 775 \approx 1407.4$
Japanese	$0.209 \times 701 + 701 \approx 847.5$
Asian Indian	$1.256 \times 361 + 361 \approx 814.4$
Korean	$1.263 \times 353 + 353 \approx 798.8$
Vietnamese	$1.348 \times 262 + 262 \approx 615.2$

b.

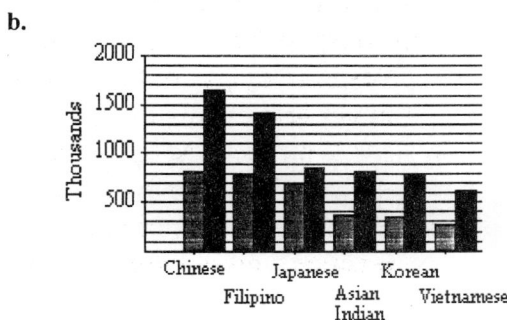

5a. 9 items **b.** 150

c.

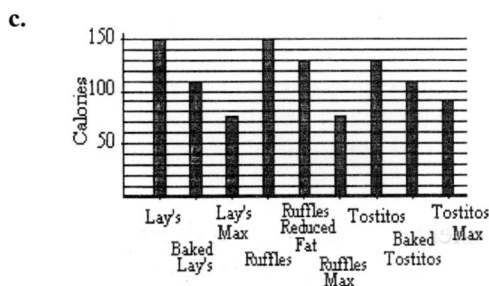

d. 10 **e.**

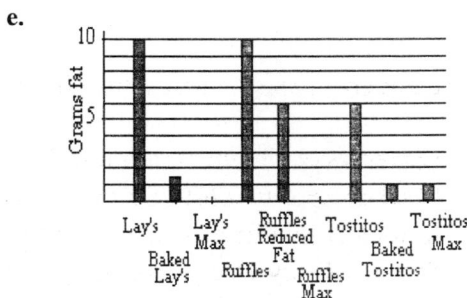

7a.

Quiz Score	0	1	2	3	4	5	6	7	8	9	10
Number of Students	1	2	1	4	7	9	10	8	3	2	3

b. Mode = 6 **c.** 50 students **d.** Median = 6

e. Total $= 0 \times 1 + 1 \times 2 + 2 \times 1 + 3 \times 4 + 4 \times 7 + 5 \times 9 + 6 \times 10 + 7 \times 8 + 8 \times 3 + 9 \times 2 + 10 \times 3$

$= 0 + 2 + 2 + 12 + 28 + 45 + 60 + 56 + 24 + 18 + 30$

$= 277$

f. Mean $= \dfrac{277}{50} = 5.54$

Homework 1.2

1a. The year; The number of US bicycle commuters in millions

b-c.

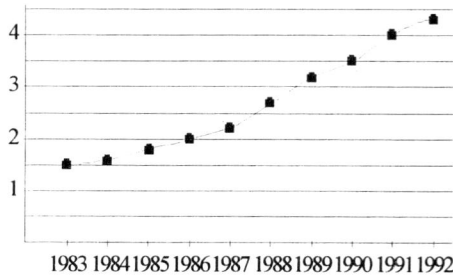

d. The graph is a smooth curve because the number of bicycle commuters changes smoothly.

3a. 12 tick marks

c.

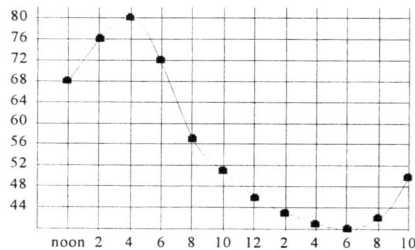

b. The high is 80° F and the low is 40° F.

d. The graph is a smooth curve because the temperature changes smoothly.

e. The temperature is about 72° F at 1 pm and about 45° F at 9 am.

f. About 12:30 pm and again at about 6:15 pm

g. 6 pm to 8 pm had the greatest temperature change. We can see that this is the greatest change in a two-hour period by noting that it corresponds to the steepest section of the graph, that is, the two-hour segment with the largest vertical change.

5a.

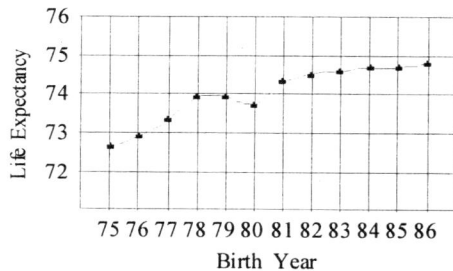

Birth Year

The graph is a smooth curve because the life expectancy changes smoothly.

b. Life expectancy dropped from 1979 to 1980.

c. The increase was greatest from 1977-1978 and from 1980-1981.

d. Life expectancy remained constant from 1978-1979 and from 1984-1985. The graph is horizontal between these intervals.

7a. Number of years with account; Amount in account

c. About 13.7 yr

d. About $2550 - 2350 = \$200$

b. About $3000

e. About $5600 - 5200 = \$400$

9a. Number of minutes after served; Temperature of soup

c. About 70° F

d. About $180 - 125 = 55°$ F

b. About 3.7 min

e. About 60° F

2

11a.

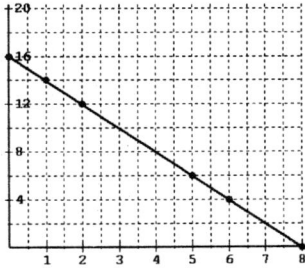

b. This is a straight line.

13a.

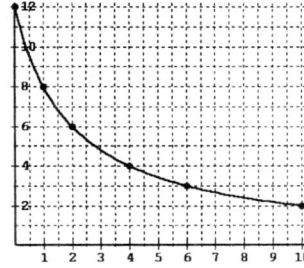

b. This is not a straight line.

15. Graph C shows an increasing speed for a time, then instantly going to zero.

17a. I. Graph C
  II. Graph D
  III. Graph A

b. Graph B: Delbert walks toward school, but before he gets there he returns home where he stays for a time. He then walks to school and stays there.

Homework 1.3

1.

Ridgecrest	70	75	82	86	90	R
Calculation	$70+12$	$75+12$	$82+12$	$86+12$	$90+12$	$R+12$
Sunnyvale	82	87	94	98	102	$R+12$

a. Add 12 to Ridgecrest's temperature.
b. Temp in Sunnyvale = Temp in Ridgecrest + 12
c. $S = R + 12$

d.

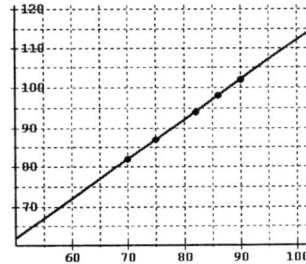

3.

Miles Driven	40	60	95	120	145	170	d
Calculation	$200-40$	$200-60$	$200-95$	$200-120$	$200-145$	$200-170$	$200-d$
Remaining	160	140	105	80	55	30	$200-d$

a. Subtract the miles driven from 200.
b. Miles remaining = 200 − miles driven
c. $r = 200 - d$

d.

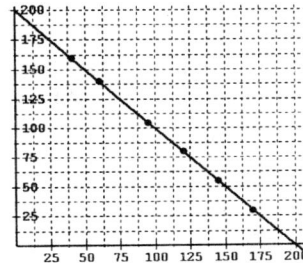

3

5.

Total Bill	24	30	33	45	54	81	b
Calculation	$\frac{24}{3}$	$\frac{30}{3}$	$\frac{33}{3}$	$\frac{45}{3}$	$\frac{54}{3}$	$\frac{81}{3}$	$\frac{b}{3}$
Milton's Share	8	10	11	15	18	27	$\frac{b}{3}$

a. Divide the total bill by 3.

b. Milton's share $= \dfrac{\text{total bill}}{3}$

c. $s = \dfrac{b}{3}$

d.

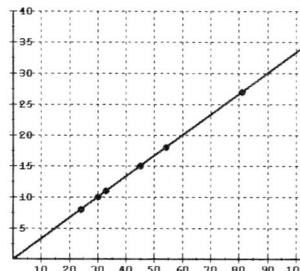

7.

m	g
2	5
3	6
5	8
10	13
12	15
16	19
18	21
m	$m + 3$

$g = m + 3$

9.

t	w
0	20
2	18
4	16
5	15
6	14
10	10
12	8
t	$20 - t$

$w = 20 - t$

11.

b	x
0	0
2	1
4	2
5	2.5
6	3
8	4
9	4.5
b	$\frac{b}{2}$

$x = \dfrac{b}{2}$

13.

z	3	6	8	12	15	18	20	z
r	2	4	$\frac{16}{3}$	8	10	12	$\frac{40}{3}$	$\frac{2}{3}z$

$r = \dfrac{2}{3}z$

15. $W = 1.2 \times n$ (other columns possible)

n	0	5	10	15	20
W	0	6	12	18	24

17. $M = \dfrac{3}{2} \times x$ (other columns possible)

x	0	2	4	6	8
M	0	3	6	9	12

19a.

x	0	10	30	40	60	70
y	70	60	40	30	10	0

b. $y = 70 - x$

4

Homework 1.4

1. Addition
   sum of
   increased by
   more than
   exceeded by
   total

3. Multiplication
   times
   twice
   (fraction) of
   product of

5. "Product of" means multiplication.
$$4y$$

7. "Twice" means two times.
$$2b$$

9. "115% of" means 1.15 times.
$$1.15g$$

11. "Decreased by" means subtraction.
$$t - 5$$

13. "Quotient of" means division.
$$\frac{7}{w}$$

15. The cost of the light bulb is unknown.
   Cost of light bulb: b
   "Times" indicates multiplication.
$$3b$$

17. The savings account balance is unknown.
   Saving account balance: s
   "Three-fifths of" indicates $\frac{3}{5}$ times.
$$\frac{3}{5}s$$

19. The price of a pizza is unknown.
   Price of a pizza: p
   "Divided by" indicates division.
$$\frac{p}{6}$$

21. The weight of copper is unknown.
   Weight of copper: w
   "Divided by" indicates division.
$$\frac{w}{16}$$

23. The rebate and the sale price are unknown.
   Rebate: r
   Sale price: p
   "Deducted from" indicates subtraction.
$$p - r$$

25. The base and the height are unknown.
   Base: b
   Height: h
   "Product" indicates multiplication.
$$bh$$

27. The heights of the roof and the tree are unknown.
   Height of the tree: t
   Height of the roof: r
   "Difference between" indicates subtraction.
$$t - r$$

29a. $p = 1.25w$
$$p = 1.25(28) = \$35$$

b.

w	12	16	20	30	36	40
p	15	20	25	37.50	45	50

31a. $d = rt$
$d = 180t$

b. $d = 180(2) = 360$ mi
$d = 180(3.5) = 630$ mi
$d = 180(12) = 2160$ mi

33a. $P = R - C$
$P = 6000 - C$

b. $P = 6000 - 800 = \$5200$
$P = 6000 - 1000 = \$5000$
$P = 6000 - 2500 = \$3500$

35a. $A = \dfrac{S}{n}$
$A = \dfrac{540}{n}$

b. $A = \frac{540}{20} = 27$
$A = \frac{540}{25} = 21.6$
$A = \frac{540}{30} = 18$

37. Area $= lw = x(2) = 2x$

39. Perimeter $= b + b + b + b = 4b$

41. Area $= \frac{1}{2}bh = \frac{1}{2}v(4) = 2v$

43. (algebraic) expression

45. product; factors

47. difference

49. To <u>evaluate</u> an expression is to substitute a specific value for each of the variables.

51a.

x	0	5	15	20	25	30
y	15	20	30	35	40	45

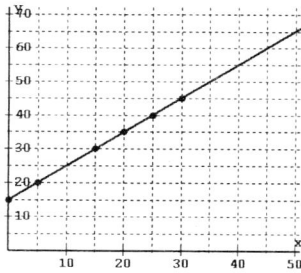

b. $y = x + 15$

c. $\left.\begin{array}{l} \text{For } x = 40: \ y = 40 + 15 = 55 \\ \text{For } x = 50: \ y = 50 + 15 = 65 \end{array}\right\}$ Yes

53a.

x	0	500	1000	2000	2500	3000
y	0	15	30	60	75	90

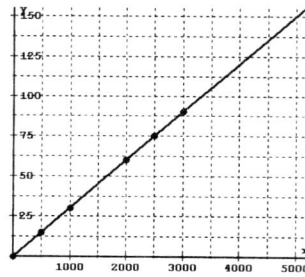

b. $y = 0.03x$

c. $\left.\begin{array}{l} \text{For } x = 4500: \ y = 0.03(4500) = 135 \\ \text{For } x = 5000: \ y = 0.03(5000) = 150 \end{array}\right\}$ Yes

Homework 1.5

1. graph

3. horizontal

5. ordered pair

7. solution

9. To graph an equation:
Step 1 Make a table of values
Step 2 Choose scales for the axes.
Step 3 Plot the points and connect them with a smooth curve.

11a. $D = T - 5$ (other columns possible)

T	5	10	15	20	50
D	0	5	10	15	45

b.

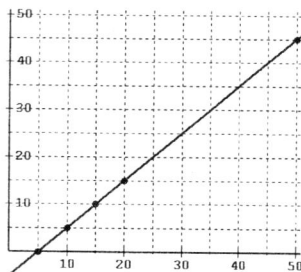

13a. $r = \dfrac{w}{2}$ (other columns possible)

w	1	2	4	6	10
r	0.5	1	2	3	5

b.

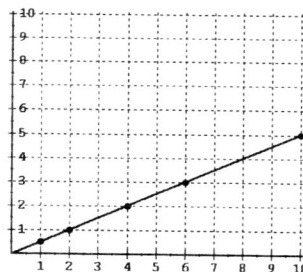

15a. $d = 1000 - l$ (other columns possible)

l	0	300	500	800	1000
d	1000	700	500	200	0

b.

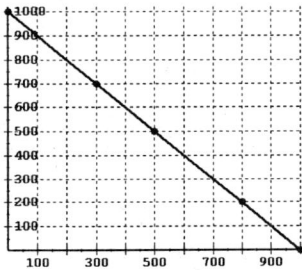

17a. $S = \dfrac{3600}{m}$ (other columns possible)

m	10	20	30	40	100
S	360	180	120	90	36

b.

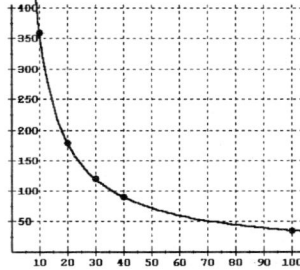

19. $y = \frac{3}{4}x$

a. For $(8, 6)$: $y = \frac{3}{4}(8) = 6 \Rightarrow$ yes

b. For $(12, 16)$: $y = \frac{3}{4}(12) = 9 \Rightarrow$ no

c. For $(2, 3)$: $y = \frac{3}{4}(2) = \frac{3}{2} \Rightarrow$ no

d. For $\left(6, \frac{9}{2}\right)$: $y = \frac{3}{4}(6) = \frac{9}{2} \Rightarrow$ yes

21. $w = z - 1.8$

a. For $(10, 8.8)$: $w = 10 - 1.8 = 8.2 \Rightarrow$ no

b. For $(6, 7.8)$: $w = 6 - 1.8 = 4.2 \Rightarrow$ no

c. For $\left(2, \frac{1}{5}\right)$: $w = 2 - 1.8 = 0.2 = \frac{1}{5} \Rightarrow$ yes

d. For $(9.2, 7.4)$: $w = 9.2 - 1.8 = 7.4 \Rightarrow$ yes

23a. No **b.** Yes **c.** No **d.** No

25a. Yes **b.** Yes **c.** No **d.** No

27a. The first variable is x, and the second variable is y.

b.

x	0	2	5	6	10	12	16
y	16	14	11	10	6	4	0

(other last two columns possible)

c. $y = 16 - x$

29a. The first variable is h, and the second variable is d.

b.

h	3	6	7.5	9	10.5	12	15
d	1	2	2.5	3	3.5	4	5

(other last three columns possible)

c. $d = \dfrac{h}{3}$

31. Equation (f) (other columns possible)

x	0	5	10	15
y	15	20	25	30

33. Equations (c) and (e) (other columns possible)

x	0	1	2	3
y	0	0.2	0.4	0.6

35a-c.

x	y
4	0.5
8	1
10	1.25
16	2

All the tables are the same because all the equations are equivalent.

Midchapter 1 Review

1. The heights of the bars represent the frequencies of each range.

2. A <u>variable</u> (usually represented by a letter) is a numerical quantity that changes over time or in different situations.

3. An <u>algebraic expression</u> is any meaningful combination of numbers, variables, and operation symbols. An <u>equation</u> is a statement that two expressions are equal.

4. <u>Factors</u> are quantities that are multiplied. <u>Terms</u> are quantities that are added.

5a. $w - p$ **b.** $56 - 5.6 = 50.4$

6a. $\dfrac{m}{q}$ **b.** $\dfrac{12}{192} = 0.0625$

7a. Distance: $d = rt$
b. Profit: $P = R - C$
c. Interest: $I = Prt$
d. Percentage or Part: $P = rW$
e. Average value: $A = \dfrac{S}{n}$

8. Perimeter of a Rectangle: $P = 2l + 2w$
Area of a Rectangle: $A = lw$
Area of a Circle: $A = \pi r^2$
Circumference of a Circle: $C = 2\pi r$

9. The first or independent variable is displayed on the horizontal axis and the second or dependent variable is displayed on the vertical axis.

10. An <u>ordered pair</u> whose <u>coordinates</u> make the equation true is called a <u>solution</u> of the equation.

11a. Great Britain; Israel   **b.** Sweden; Israel

c.   **d.** Sweden; Japan

Country	Percentage
Canada	$\frac{63}{295} \approx 0.214 = 21.4\%$
France	$\frac{63}{577} \approx 0.109 = 10.9\%$
Great Britain	$\frac{120}{659} \approx 0.182 = 18.2\%$
Israel	$\frac{9}{120} = 0.075 = 7.5\%$
Japan	$\frac{23}{500} = 0.046 = 4.6\%$
Mexico	$\frac{71}{500} = 0.142 = 14.2\%$
South Africa	$\frac{100}{400} = 0.250 = 25.0\%$
Sweden	$\frac{141}{349} \approx 0.404 = 40.4\%$
United States	$\frac{51}{435} \approx 0.117 = 11.7\%$

12a. 88%; 1.4%   **b.** Pakistan; Sudan and Pakistan   **c.** Jamaica and Thailand
d. Yemen Arab Rep; Sri Lanka   **e.** Yes. Countries with a high birth rate have a low female literacy rate, and countries with a low birth rate have a high female literacy rate.

13a. 9 states   **b.** 13%   **c.** Mode = 14%; 10 states
d. 3 states   **e.** 27 states
f. Median = 14%;

$$\text{Mean} = (8 \times 1 + 9 \times 2 + 10 \times 2 + 11 \times 5 + 12 \times 8 + 13 \times 6$$
$$+ 14 \times 10 + 15 \times 3 + 16 \times 9 + 17 \times 4 + 18 \times 1) \div 51$$
$$= (8 + 18 + 20 + 55 + 96 + 78 + 140 + 45 + 144 + 68 + 18) \div 51$$
$$= \frac{690}{51} \approx 13.5\%$$

14a. II   **b.** IV   **c.** I   **d.** For III: A soccer ball is kicked into the air, bounces several times and then rolls to a stop.

15a. 13,000 yr **b.** 0.3 or 30% **c.** About 5700 yr

16.

Interest	1150	1000	750	620	480
Calculation	$1200 - 1150$	$1200 - 1000$	$1200 - 750$	$1200 - 620$	$1200 - 480$
Principal	50	200	450	580	720

a. Subtract the interest from 1200.
b. Principle $= 1200 -$ interest
c. $P = 1200 - I$

d.

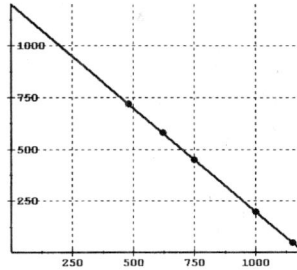

17a. About \$15,000; About \$11,000
b. 1986 to 1987; 1980 to 1981
c. $13,200 - 9500 = \$3700;$
$33,000 - 24,000 = \$9000;$
$39,500 - 27,000 = \$12,500$

d.

Year	Percent
1970	$\dfrac{9,567}{13,264} \approx 0.721 = 72.1\%$
1975	$\dfrac{13,542}{17,477} \approx 0.775 = 77.5\%$
1980	$\dfrac{19,469}{24,311} \approx 0.801 = 80.1\%$
1985	$\dfrac{23,853}{32,822} \approx 0.727 = 72.7\%$
1990	$\dfrac{26,653}{39,238} \approx 0.679 = 67.9\%$

18a.

x	16	32	40	56
y	2	4	5	7

b. $y = \dfrac{x}{8}$

19. The temperature is unknown.
Temperature: t
"6° hotter" indicates more (addition).
$t + 6$

20. The cost of gasoline is unknown.
Cost of gasoline: c
"Divided three ways" indicates division.
$\dfrac{c}{3}$

21. The total bill is unknown.
Total: t
"8% of" indicates 0.08 times.
$0.08t$

22. The height of the triangle is unknown.
Height: h
"5 inches less than" indicates subtraction.
$h - 5$ (Note: Order of terms are reversed.)

23a. $d = rt$
$d = r(3)$
$d = 3r$
b. $d = 3(6) = 18$ mi
$d = 3(20) = 60$ mi

24a. $I = Prt$
$I = (1500)r(2)$
$I = 3000r$
b. $I = 3000(0.065) = \$195$
$I = 3000(0.083) = \$249$

9

25a. $n = 2s$

b.

s	5	10	20	50
n	10	20	40	100

(other columns possible)

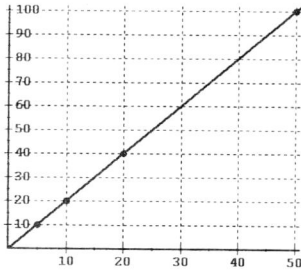

26a. $y = 8 - x$ (other columns possible)

x	1	2	4
y	7	6	4

Graph III

b. $y = x + 8$ (other columns possible)

x	1	2	4
y	9	10	12

Graph II

c. $y = 8x$ (other columns possible)

x	1	2	4
y	8	16	32

Graph I

Homework 1.6

1.

t	q
2	11
4	13
6	15
9	18
21	30
30	39

To find a value for q, add 9 to the value of t.
To find a value for t, subtract 9 from the value of q.

3.

n	p
0	0
2	10
4	20
5	25
7	35
11	55

To find a value for p, multiply the value of n by 5.
To find a value for n, divide the value of p by 5.

5. For $x = 10$:

$$x - 4 = 6$$
$$10 - 4 \overset{?}{=} 6$$
$$6 = 6 \quad \text{True}$$

Therefore, it is a solution.

7. For $y = 24$:

$$4y = 28$$
$$4(24) \overset{?}{=} 28$$
$$96 = 28 \quad \text{False}$$

Therefore, it is not a solution.

9. For $z = 19$:

$$\frac{0}{z} = 0$$
$$\frac{0}{19} \overset{?}{=} 0$$
$$0 = 0 \quad \text{True}$$

Therefore, it is a solution.

11.

$$x - 3 = 11$$
$$\underline{+3 \qquad +3}$$
$$x \quad = 14$$

Because 3 is subtracted from x, add 3 to both sides.

Check $x = 14$:
$$x - 3 = 11$$
$$14 - 3 \overset{?}{=} 11$$
$$11 = 11 \quad \text{True}$$

13.

$$10.6 = 7.8 + y$$
$$\underline{-7.8 \quad -7.8}$$
$$2.8 = \qquad y$$

Because 7.8 is added to y, subtract 7.8 from both sides.

Check $y = 2.8$:
$$10.6 = 7.8 + y$$
$$10.6 \overset{?}{=} 7.8 + 2.8$$
$$10.6 = 10.6 \quad \text{True}$$

15.

$$3y = 108$$
$$\frac{3y}{3} = \frac{108}{3}$$
$$y = 36$$

Because y is multiplied by 3, divide both sides by 3.

Check $y = 36$:
$$3y = 108$$
$$3(36) \overset{?}{=} 108$$
$$108 = 108 \quad \text{True}$$

17.

$42 = 3.5b$

$\dfrac{42}{3.5} = \dfrac{3.5b}{3.5}$

$12 = b$

Because b is multiplied by 3.5, divide both sides by 3.5.

Check $b = 12$: $42 = 3.5b$

$42 \overset{?}{=} 3.5(12)$

$42 = 42$ $\quad$ True

19.

$2.6 = \dfrac{a}{1.5}$

$1.5(2.6) = 1.5\left(\dfrac{a}{1.5}\right)$

$3.9 = a$

Because a is divided by 1.5, multiply both sides by 1.5.

Check $a = 3.9$: $2.6 = \dfrac{a}{1.5}$

$2.6 \overset{?}{=} \dfrac{3.9}{1.5}$

$2.6 = 2.6$ $\quad$ True

21.

$x - 4 = 0$

$\underline{+4 \quad\quad +4}$

$x \quad = \quad 4$

Because 4 is subtracted from x, add 4 to both sides.

Check $x = 4$: $x - 4 = 0$

$4 - 4 \overset{?}{=} 0$

$0 = 0$ $\quad$ True

23.

$34x = 212$

$\dfrac{34x}{34} = \dfrac{212}{34}$

$x = \dfrac{106}{17}$

Because x is multiplied by 34, divide both sides by 34.

Check $x = \dfrac{106}{17}$: $\quad 34x = 212$

$34\left(\dfrac{106}{17}\right) \overset{?}{=} 212$

$212 = 212$ $\quad$ True

25.

$6z = 20$

$\dfrac{6z}{6} = \dfrac{20}{6}$

$z = \dfrac{10}{3}$

Because z is multiplied by 6, divide both sides by 6.

Check $z = \dfrac{10}{3}$: $\quad 6z = 20$

$6\left(\dfrac{10}{3}\right) \overset{?}{=} 20$

$20 = 20$ $\quad$ True

27.

$9 = k + 9$

$\underline{-9 \quad\quad -9}$

$0 = k$

Because 9 is added to k, subtract 9 from both sides.

Check $k = 0$: $9 = k + 9$

$9 \overset{?}{=} 0 + 9$

$9 = 9$ $\quad$ True

29.

$I = Prt$

$75 = P(0.03)(1)$

$75 = 0.03P$ $\quad$ Divide both sides by 0.03

$\dfrac{75}{0.03} = \dfrac{0.03P}{0.03}$

$2500 = P$ $\quad$ Clive loaned his brother \$2500.

31.

$A = \dfrac{S}{n}$

$38.25 = \dfrac{S}{8}$ $\quad$ Multiply both sides by 8

$8(38.25) = 8\left(\dfrac{S}{8}\right)$

$306 = S$ $\quad$ Andy had 306 total points.

33.

$d = rt$

$234 = 13 \cdot t$ $\quad$ Divide both sides by 13

$\dfrac{234}{13} = \dfrac{13t}{13}$

$18 = t$ $\quad$ It takes 18 hr.

35.

$A = lw$

$400 = l \cdot 16$ $\quad$ Divide both sides by 16

$\dfrac{400}{16} = \dfrac{16l}{16}$

$25 = l$ $\quad$ The roll is 25 ft long.

37. An <u>equation</u> is a statement that two quantities are equal.

39. A <u>given number is a solution</u> if when the given number is substituted for the variable, the equation becomes a true statement.

41. A <u>trial and error method</u> for solving an equation involves substituting different values for the variable. When a value gives a true statement, that value is a solution to the equation.

In Problems 43 and 45, use Fact 1: The sum of the angles in a triangle is 180°.

43.
$$x + 80 + 35 = 180$$
$$x + 115 = 180$$
$$\underline{-115 \quad -115}$$
$$x \quad = \quad 65°$$

45.
$$x + 90 + 37 = 180$$
$$x + 127 = 180$$
$$\underline{-127 \quad -127}$$
$$x \quad = \quad 53°$$

In Problems 47 and 49, use Fact 2: A straight angle has a measure of 180°.

47.
$$x + 135 = 180$$
$$\underline{-135 \quad -135}$$
$$x \quad = \quad 45°$$

49.
$$x + 18 = 180$$
$$\underline{-18 \quad -18}$$
$$x \quad = \quad 162°$$

Homework 1.7

1. <u>Evaluating an expression</u> involves substituting a value for the variable and performing the indicated operations. In $B = 1.08P$, if $P = 100$, we can substitute 100 for P and evaluate the expression $1.08(100)$ to find the value of B. <u>Solving an equation</u> involves finding the value(s) for a variable to make the equation true. In $B = 1.08P$, if $B = 216$, we can solve the equation $216 = 1.08P$ to find the value of P.

3. *Step 2* Find a quantity that can be described in two different ways, and write an equation using the variable to model the problem.

Step 3 Solve the equation and answer the question.

5a. Price of used car: u
Price of new car: n
$$u = n - 3400$$

b.
$$u = 14,500 - 3400$$
$$u = 11,100$$
A used car costs $11,100.

c.
$$9200 = n - 3400$$
$$\underline{+3400 \qquad +3400}$$
$$12600 = n$$
A new car costs $12,600.

7a. Number of games won: w
Number of games played: p
$$w = 0.60p$$

b.
$$w = 0.60(120)$$
$$w = 72$$
They won 72 games.

c.
$$96 = 0.60p$$
$$\frac{96}{0.60} = \frac{0.60p}{0.60}$$
$$160 = p$$
They played 160 games.

9a. Profit: P
Selling price: S
$$P = 0.18S$$

b.
$$P = 0.18(60)$$
$$P = 10.8$$
The profit is $10.80.

c.
$$7.20 = 0.18S$$
$$\frac{7.20}{0.18} = \frac{0.18S}{0.18}$$
$$40 = S$$
The selling price is $40.

11. Price mother paid: p
$$89 = p - 26$$
$$\underline{+26 \qquad +26}$$
$$115 = p$$
Her mother paid $115.

13. Emily's monthly income: I
$$360 = 0.40I$$
$$\frac{360}{0.40} = \frac{0.40I}{0.40}$$
$$900 = I$$
She makes $900 per month.

15a.
$$d = rt$$
$$200 = r \cdot 19.32$$
$$\frac{200}{19.32} = \frac{19.32r}{19.32}$$
$$10.35 \approx r$$
About 10.35 meters per sec

b.
$$d = rt$$
$$100 = r \cdot 9.84$$
$$\frac{100}{9.84} = \frac{9.84r}{9.84}$$
$$10.16 \approx r$$
About 10.16 meters per sec

c. Johnson was faster.

17. $x + 7 = 26$

19. $\dfrac{x}{7} = 26$

21. $\dfrac{x}{26} = 7$

23a.

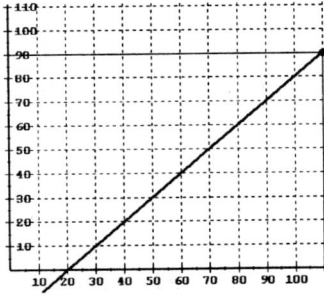

$p = 110$
$$90 = p - 20$$
$$\underline{+\,20 \qquad +\,20}$$
$$110 = p$$

b.

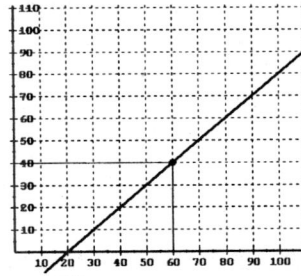

$p = 60$
$$p - 20 = \quad 40$$
$$\underline{+\,20 \qquad +\,20}$$
$$p \quad = \quad 60$$

25a.

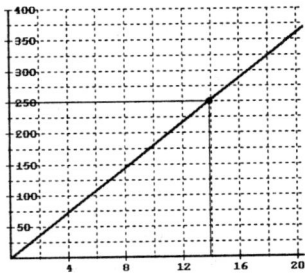

$g \approx 13.9$
$$250 = 18g$$
$$\frac{250}{18} = \frac{18g}{18}$$
$$13.9 \approx g$$

b.

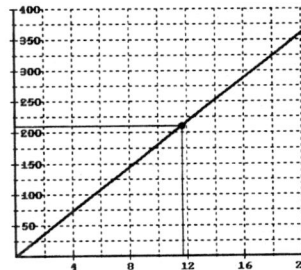

$g \approx 11.5$
$$18g = 210$$
$$\frac{18g}{18} = \frac{210}{18}$$
$$g \approx 11.7$$

27a. $(50, 4)$

b. When the price is $50, the tax is $4.

29a. $(4, 16)$

b. When the height is 4 inches, the weight is 16 ounces.

31. Since she ate all but three apples, three apples are left.

33. The perimeter is 38 cm.

35. The bus made six stops.

Homework 1.8

1. If a, b, and c are any numbers,
then $(a + b) + c = a + (b + c)$
and $(a \cdot b) \cdot c = a \cdot (b \cdot c)$

3. <u>Parentheses and brackets</u> are used as grouping symbols to show which part of an expression to simplify first.

5. False. Multiplication and division are done left to right.

7a. $(20 - 2) \cdot 8 + 1$ **b.** $20 - 2 \cdot (8 + 1)$ **c.** $20 - 2 \cdot 8 + 1$ (No change needed)

9. $\dfrac{(5 + 7) \cdot 4}{10 - 8}$

11. First multiply 32 times 12, and divide that result by 4. Subtract that amount from 825, and then add 2.

13. Incorrect to subtract 9 minus 3.
$$\begin{aligned}
(5 + 4) - 3(8 - 3 \cdot 2) &\quad \text{Add inside parentheses} \\
= 9 - 3(8 - 3 \cdot 2) &\quad \text{Multiply inside parentheses} \\
= 9 - 3(8 - 6) &\quad \text{Subtract inside parentheses} \\
= 9 - 3(2) &\quad \text{Multiply before subtract} \\
= 9 - 6 \\
= 3
\end{aligned}$$

15.
$$\begin{aligned}
2 + 4(3) &\quad \text{Multiply before add} \\
= 2 + 12 \\
= 14
\end{aligned}$$

17.
$$\begin{aligned}
15 - \frac{3}{4}(16) &\quad \text{Multiply before subtract} \\
= 15 - 12 \\
= 3
\end{aligned}$$

19.
$$\begin{aligned}
6 \div \frac{1}{4} \cdot 3 &\quad \text{Multiply and divide left to right} \\
= 6 \cdot 4 \cdot 3 \\
= 72
\end{aligned}$$

21.
$$\begin{aligned}
3 + 3(2 + 3) &\quad \text{Add inside parentheses} \\
= 3 + 3(5) &\quad \text{Multiply before add} \\
= 3 + 15 \\
= 18
\end{aligned}$$

23.
$$\begin{aligned}
\frac{1}{3} \cdot 12 - 3\left(\frac{5}{6}\right) &\quad \text{Multiply before subtract} \\
= 4 - \frac{5}{2} &\quad \text{LCD is 2} \\
= \frac{8}{2} - \frac{5}{2} \\
= \frac{3}{2}
\end{aligned}$$

25.
$$\begin{aligned}
2 + 3 \cdot 8 - 6 + 3 &\quad \text{Multiply first} \\
= 2 + 24 - 6 + 3 &\quad \text{Add and subtract in order} \\
= 26 - 6 + 3 &\quad \text{from left to right} \\
= 20 + 3 \\
= 23
\end{aligned}$$

27.
$$\begin{aligned}
\frac{3(8)}{12} - \frac{6 + 4}{5} &\quad \text{Simplify above fraction bar} \\
= \frac{24}{12} - \frac{10}{5} &\quad \text{Divide} \\
= 2 - 2 &\quad \text{Subtract} \\
= 0
\end{aligned}$$

29.
$$\begin{aligned}
28 + 6 \div 2 - 2(5 + 3 \cdot 2) &\quad \text{Multiply inside parentheses} \\
= 28 + 6 \div 2 - 2(5 + 6) &\quad \text{Add inside parentheses} \\
= 28 + 6 \div 2 - 2(11) &\quad \text{Multiply and divide} \\
= 28 + 3 - 22 &\quad \text{Add and subtract in order} \\
= 31 - 22 &\quad \text{from left to right} \\
= 9
\end{aligned}$$

31. $3[3(3+2)-8]-17$ Add inside parentheses
$= 3[3(5)-8]-17$ Multiply inside brackets
$= 3[15-8]-17$ Subtract inside brackets
$= 3[7]-17$ Multiply before subtract
$= 21-17$
$= 4$

33. $\dfrac{3(3)+5}{6-2(2)} + \dfrac{2(5)-4}{9-4-2}$ Simplify above and below fraction bar

$= \dfrac{9+5}{6-4} + \dfrac{10-4}{5-2}$ Add and subtract above and below fraction bar

$= \dfrac{14}{2} + \dfrac{6}{3}$ Divide

$= 7+2$ Add

$= 9$

35. $7[15-24 \div 2] - 9[5(4+2)-4(6+1)]$ Add inside parentheses
$= 7[15-24 \div 2] - 9[5(6)-4(7)]$ Divide and multiply inside brackets
$= 7[15-12] - 9[30-28]$ Subtract inside brackets
$= 7[3] - 9[2]$ Multiply before subtract
$= 21 - 18$
$= 3$

37. $\boxed{(}\;\boxed{6.4}\;\boxed{+}\;\boxed{3.5}\;\boxed{)}\;\boxed{\div}\;\boxed{(}\;\boxed{3.6}\;\boxed{\times}\;\boxed{3.2}\;\boxed{)}\;\boxed{=}$ 0.859

39. $\boxed{(}\;\boxed{26.2}\;\boxed{-}\;\boxed{9.1}\;\boxed{)}\;\boxed{\div}\;\boxed{(}\;\boxed{8.4}\;\boxed{\div}\;\boxed{7.7}\;\boxed{)}\;\boxed{+}\;\boxed{5.1}\;\boxed{\times}\;\boxed{(}\;\boxed{6.9}\;\boxed{-}\;\boxed{1.6}\;\boxed{)}\;\boxed{=}$ 42.705

41. $\boxed{(}\;\boxed{1728}\;\boxed{\times}\;\boxed{(}\;\boxed{847}\;\boxed{-}\;\boxed{603}\;\boxed{)}\;\boxed{)}\;\boxed{\div}\;\boxed{(}\;\boxed{216}\;\boxed{\times}\;\boxed{(}\;\boxed{98}\;\boxed{-}\;\boxed{38}\;\boxed{)}\;\boxed{)}\;\boxed{+}\;\boxed{6}\;\boxed{\times}$
$\boxed{(}\;\boxed{876}\;\boxed{-}\;\boxed{514}\;\boxed{)}\;\boxed{=}$ 2204.533

43.

x	2	2.5	3	2.7	2.8
y	1.9	3.15	4.4	3.65	3.9

$x = 2.8$

45a. $8 + 2 \cdot 5 = 18$ **b.** $(8+2) \cdot 5 = 50$ **47a.** $\dfrac{24}{2+6} = 3$ **b.** $\dfrac{24}{2} + 6 = 18$

49a. $(9-4)-3 = 2$ **b.** $9-(4-3) = 8$ **51a.** $6 \cdot 8 - 6 = 42$ **b.** $6(8-6) = 12$

53a. $\dfrac{36}{6(3)} = 2$ **b.** $\dfrac{36}{6}(3) = 18$ **55a.** $(2.3 + 5.7)6 - (1.2 + 3.3)2$

b. $= (8)6 - (4.5)2$
$= 48 - 9$
$= 39$ sq cm

Homework 1.9

1a. $5000 - 200(3) = 5000 - 600 = \$4400;$ $5000 - 200(12) = 5000 - 2400 = \$2600$

b. Multiply the number of weeks by 200, then subtract that from 5000.

c.

Number of Weeks	4	5	6	10
Calculation	$5000 - 200(4)$	$5000 - 200(5)$	$5000 - 200(6)$	$5000 - 200(10)$
Saving Left	4200	4000	3800	3000

15	20
$5000 - 200(15)$	$5000 - 200(20)$
2000	1000

d. $5000 - 200w$ **e.** $S = 5000 - 200w$

3a. $0.12(8000 - 2000) = 0.12(6000) = \$720;$ $0.12(10{,}000 - 2000) = 0.12(8000) = \$960$

b. Subtract 2000 from her income, then multiply the result by 0.12.

c.

Income	7000	12,000	15,000	20,000
Calculation	$0.12(7000 - 2000)$	$0.12(12{,}000 - 2000)$	$0.12(15{,}000 - 2000)$	$0.12(20{,}000 - 2000)$
State Tax	600	1200	1560	2160

24,000	30,000
$0.12(24{,}000 - 2000)$	$0.12(30{,}000 - 2000)$
2640	3360

d. $0.12(I - 2000)$ **e.** $T = 0.12(I - 2000)$

5.

z	$5z$	$5z - 3$
2	10	7
4	20	17
5	25	22

7.

Q	$12 + Q$	$2(12 + Q)$
0	12	24
4	16	32
8	20	40

9a. Width: w

$2w - 3$

b. $2(13) - 3$

$= 26 - 3$

$= 23$ in.

11a. Principal: P

$20 + 0.40P$

b. $20 + 0.40(500)$

$= 20 + 200$

$= \$220$

13a. Number of cars; c

Number of trucks: t

$\dfrac{1}{3}(c + t)$

b. $\dfrac{1}{3}(7 + 5)$

$= \dfrac{1}{3}(12)$

$= 4$ vehicles

15a. Number of women: w

Number of men: m

$\dfrac{1}{2}w + \dfrac{2}{3}m$

b. $\dfrac{1}{2}(18) + \dfrac{2}{3}(12)$

$= 9 + 8$

$= 17$

17a. Verbal score: v

Math score: m

$0.80(v - m)$

b. $0.08(680 - 655)$

$= 0.80(25)$

$= 20$

19. $2y + x$

$2(9) + 8$

$= 18 + 8$

$= 26$

21. $4a + 3b$

$4(8) + 3(7)$

$= 32 + 21$

$= 53$

23. $\dfrac{a}{b} - \dfrac{b}{a}$

$\dfrac{8}{6} - \dfrac{6}{8}$

$= \dfrac{4}{3} - \dfrac{3}{4}$

$= \dfrac{16}{12} - \dfrac{9}{12}$

$= \dfrac{7}{12}$

25.
$$\frac{24-2x}{2+y} - \frac{4x+1}{3y}$$
$$\frac{24-2(4)}{2+6} - \frac{4(4)+1}{3(6)}$$
$$= \frac{24-8}{2+6} - \frac{16+1}{18}$$
$$= \frac{16}{8} - \frac{17}{18}$$
$$= 2 - \frac{17}{18}$$
$$= \frac{36}{18} - \frac{17}{18}$$
$$= \frac{19}{18}$$

27.
$$\frac{a}{1-r}$$
$$\frac{6}{1-0.2}$$
$$= \frac{6}{0.8}$$
$$= \frac{15}{2} \text{ or } 7.5$$

29.
$$P = 2l + 2w$$
$$P = 2(8.5) + 2(6.4)$$
$$P = 17 + 12.8$$
$$P = 29.8 \text{ m}$$

31.
$$A = \frac{1}{2}(B+b)h$$
$$A = \frac{1}{2}(9+7)(3)$$
$$A = \frac{1}{2}(16)(3)$$
$$A = 24 \text{ sq cm}$$

33.
$$P = n(p-c)$$
$$P = 300(50-32)$$
$$P = 300(18)$$
$$P = \$5400$$

35.
$$C = \frac{5}{9}(F-32)$$
$$C = \frac{5}{9}(98.6-32)$$
$$C = \frac{5}{9}(66.6)$$
$$C = 37° \text{ C}$$

37a. $P = 100 - 6d$

b.

d	0	2	5	10	15
P	100	88	70	40	10

c.

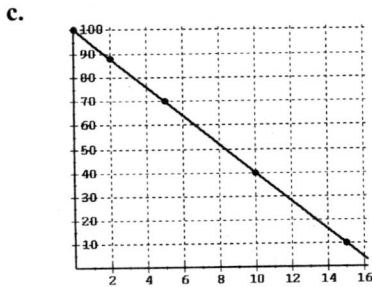

39a. $M = 200 + \dfrac{U}{5}$

b.

U	20	40	80	100	200
M	204	208	216	220	240

c.

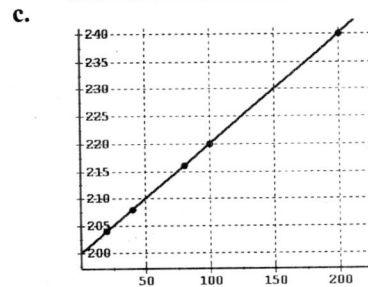

41a. $D = \dfrac{S-40}{6}$

b.

S	70	100	112	130	160
D	5	10	12	15	20

c.

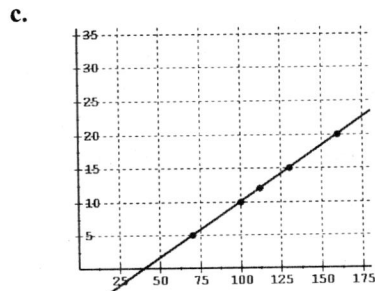

17

43a. $\dfrac{n}{n+4}$ (other columns possible)

b.

n	0	1	2	3	4	5
$\dfrac{n}{n+4}$	0	$\dfrac{1}{5}$	$\dfrac{1}{3}$	$\dfrac{3}{7}$	$\dfrac{1}{2}$	$\dfrac{5}{9}$

45a. $(a+5)(a-5)$ (other columns possible)

b.

a	5	6	7	8	9	10
$(a+5)(a-5)$	0	11	24	39	56	75

47b. $A - 25$

c. $I + 2(A - 25)$

49a. $I - D$

b. $0.08(I - D)$

c. $0.08(I - D) + 300$

51.
$$P = 2l + 2w$$
$$= 2(5) + 2(x)$$
$$= 10 + 2x$$

53.
$$A = \tfrac{1}{2}bh$$
$$= \tfrac{1}{2}(x + 6)(2)$$
$$= x + 6$$

Homework 1.10

1.

x	$2x$	$2x + 4$
3	6	10
6	12	16
5	10	14
8	16	20

3.

q	$q - 3$	$5(q - 3)$
3	0	0
5	2	10
4	1	5
7	4	20

5a. Multiply the value of x by 2 (to get 6), then add 4 to the result (to get 10).

b. Subtract 4 from the value of y (to get 10), then divide the result by 2 (to get 5).

7a. Subtract 3 from the value of q (to get 0), then multiply the result by 5 (to get 0).

b. Divide the value of R by 5 (to get 2), then add 3 to the result (to get 5).

9. Add 6 to 29 (to get 35), then divide 35 by 5 (to get 7). The number is 7.

11.
$$6x - 13 = 5 \quad \text{Add 13 to both sides}$$
$$\underline{+\,13 \qquad +\,13}$$
$$6x = 18 \quad \text{Divide both sides by 6}$$
$$\frac{6x}{6} = \frac{18}{6}$$
$$x = 3$$

13.
$$\frac{2a}{5} = 8 \quad \text{Multiply both sides by 5}$$
$$5\left(\frac{2a}{5}\right) = 5(8)$$
$$2a = 40 \quad \text{Divide both sides by 2}$$
$$\frac{2a}{2} = \frac{40}{2}$$
$$a = 20$$

15.
$$\frac{x}{4} + 2 = 3 \quad \text{Subtract 2 from both sides}$$
$$\underline{-\,2 \qquad -\,2}$$
$$\frac{x}{4} = 1 \quad \text{Multiply both sides by 4}$$
$$4\left(\frac{x}{4}\right) = 4(1)$$
$$x = 4$$

17.
$$6x + 5 = 5 \quad \text{Subtract 5 from both sides}$$
$$\underline{-\,5 \qquad -\,5}$$
$$6x = 0 \quad \text{Divide both sides by 6}$$
$$\frac{6x}{6} = \frac{0}{6}$$
$$x = 0$$

19.

$$24 = 4(p - 7)$$ Divide both sides by 4

$$\frac{24}{4} = \frac{4(p - 7)}{4}$$

$$6 = p - 7$$ Add 7 to both sides

$$\underline{+7 \quad\quad +7}$$

$$13 = p$$

21.

$$0 = \frac{5z}{7}$$ Multiply both sides by 7

$$7(0) = 7\left(\frac{5z}{7}\right)$$

$$0 = 5z$$ Divide both sides by 5

$$\frac{0}{5} = \frac{5z}{5}$$

$$0 = z$$

23.

$$\frac{k + 4}{5} = 9$$ Multiply both sides by 5

$$5\left(\frac{k + 4}{5}\right) = 5(9)$$

$$k + 4 = 45$$ Subtract 4 from both sides

$$\underline{-4 \quad\quad -4}$$

$$k = 41$$

25.

$$\frac{2x}{3} - 5 = 7$$ Add 5 to both sides

$$\underline{+5 \quad\quad +5}$$

$$\frac{2x}{3} = 12$$ Multiply both sides by 3

$$3\left(\frac{2x}{3}\right) = 3(12)$$

$$2x = 36$$ Divide both sides by 2

$$\frac{2x}{2} = \frac{36}{2}$$

$$x = 18$$

27.

$$7 = \frac{4b - 3}{3}$$ Multiply both sides by 3

$$3(7) = 3\left(\frac{4b - 3}{3}\right)$$

$$21 = 4b - 3$$ Add 3 to both sides

$$\underline{+3 \quad\quad +3}$$

$$24 = 4b$$ Divide both sides by 4

$$\frac{24}{4} = \frac{4b}{4}$$

$$6 = b$$

29.

$$11.8w - 37.8 = 120.32$$ Add 37.8 to both sides

$$\underline{+37.8 \quad\quad +37.8}$$

$$11.8w = 158.12$$ Divide both sides by 11.8

$$\frac{11.8w}{11.8} = \frac{158.12}{11.8}$$

$$w = 13.4$$

31a. $540   **b.** $30

c. $360   **d.** 18 payments

e. $B = 540 - 30m$

f. $B = 540 - 30(18)$
$B = 540 - 540$
$B = 0$

33a. About 8.5 lb

b. About 37 wk

c. $W = 3.8 + 0.6t$

d.

$$26 = 3.8 + 0.6t$$

$$\underline{-3.8 \quad\quad -3.8}$$

$$22.2 = 0.6t$$

$$\frac{22.2}{0.6} = \frac{0.6t}{0.6}$$

$$37 = t$$

35a. $s = 20 + 4x$

b.

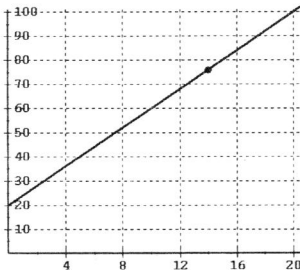

c.
$$s = 20 + 4x$$
$$76 = 20 + 4x$$
$$\underline{-20 \quad -20}$$
$$56 = 4x$$
$$\frac{56}{4} = \frac{4x}{4}$$
$$14 = x$$

37a. $P = 600 + 3r$

b.

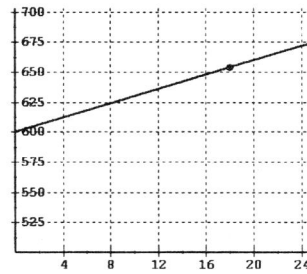

c.
$$P = 600 + 3r$$
$$654 = 600 + 3r$$
$$\underline{-600 \quad -600}$$
$$54 = 3r$$
$$\frac{54}{3} = \frac{3r}{3}$$
$$\$18 = r$$

39a. Amount of each installment: a

b-c.
$$1200 + 36a = 10200$$
$$\underline{-1200 \qquad\qquad -1200}$$
$$36a = 9000$$
$$\frac{36a}{36} = \frac{9000}{36}$$
$$a = \$250$$

41a. Assigned distance: d

b-c.
$$5(d + 0.5) = 22.5$$
$$\frac{5(d + 0.5)}{5} = \frac{22.5}{5}$$
$$d + 0.5 = 4.5$$
$$\underline{-0.5 \quad -0.5}$$
$$d = 4 \text{ mi}$$

43a. Number of seeds planted: p

b-c.
$$0.60p - 38 = 112$$
$$\underline{+38 \quad +38}$$
$$0.60p = 150$$
$$\frac{0.60p}{0.60} = \frac{150}{0.60}$$
$$p = 250 \text{ seeds}$$

45.
$$P = 2l + 2w$$
$$500 = 2l + 2(75)$$
$$500 = 2l + 150$$
$$\underline{-150 \qquad -150}$$
$$350 = 2l$$
$$\frac{350}{2} = \frac{2l}{2}$$
$$175 \text{ yd} = l$$

47.
$$A = \frac{bh}{2}$$
$$12 = \frac{4h}{2}$$
$$12 = 2h$$
$$\frac{12}{2} = \frac{2h}{2}$$
$$6 \text{ m} = h$$

49.
$$A = lw$$
$$36 = (x + 2)4$$
$$\frac{36}{4} = \frac{(x + 2)4}{4}$$
$$9 = x + 2$$
$$\underline{-2 \quad -2}$$
$$7 = x$$

51.
$$A = \frac{bh}{2}$$
$$20 = \frac{(x + 3)2}{2}$$
$$20 = x + 3$$
$$\underline{-3 \quad -3}$$
$$17 = x$$

Chapter 1 Summary and Review

1. The <u>commutative law of addition</u> says that two numbers may be added in either order to give the same answer. If a and b are any numbers, then, $a + b = b + a$.

   The <u>commutative law of multiplication</u> says that two numbers may be multiplied in either order to give the same answer. If a and b are any numbers, then $a \cdot b = b \cdot a$.

2. The <u>associative law of addition</u> says that the sum of three numbers may be grouped either way to give the same answer. If a, b, and c are any numbers, then $(a + b) + c = a + (b + c)$.

   The <u>associative law of multiplication</u> says that the product of three numbers may be grouped either way to give the same answer. If a, b, and c are any numbers, then $(a \cdot b) \cdot c = a \cdot (b \cdot c)$.

20

3. $d = rt \begin{cases} \text{Distance: } d \\ \text{Rate: } r \\ \text{Time: } t \end{cases}$ $P = R - C \begin{cases} \text{Profit: } P \\ \text{Revenue: } R \\ \text{Cost: } C \end{cases}$ $I = Prt \begin{cases} \text{Interest: } I \\ \text{Principal: } P \\ \text{Interest rate: } r \\ \text{Time: } t \end{cases}$

$A = lw \begin{cases} \text{Area: } A \\ \text{Length: } l \\ \text{Width: } w \end{cases}$ $P = rW \begin{cases} \text{Part: } P \\ \text{Percentage rate: } r \\ \text{Whole: } W \end{cases}$ $A = \dfrac{S}{n} \begin{cases} \text{Average: } A \\ \text{Sum of scores: } S \\ \text{Number of scores: } n \end{cases}$

4. Order of operations:
First, perform any operations that appear inside parentheses, or above or below a fraction bar.
Next, perform all multiplications and divisions in order from left to right.
Finally, perform all additions and subtractions in order from left to right.

5. To write an algebraic expression:
Step 1 Identify the unknown quantity and write a short phrase to describe it.
Step 2 Choose a variable to represent the unknown quantity.
Step 3 Use mathematical symbols to represent the relationship described.

6. To graph an equation:
Step 1 Make a table of values.
Step 2 Choose scales for the axes.
Step 3 Plot the points and connect them with a smooth curve.

7. To solve an equation algebraically:
Step 1 Ask yourself what operation has been performed on the variable.
Step 2 Perform the opposite operation on both sides of the equation to isolate the variable.
If two or more operations have been performed on the variable, we perform the opposite operations in reverse order to isolate the variable.

8. To solve an applied problem:
Step 1 Identify the unknown quantity and choose a variable to represent it.
Step 2 Find some quantity that can be described in two different ways, and write an equation using the variable to model the situation.
Step 3 Solve the equation and answer the question in the problem.

9a. The year
c. 47%; 38%
b. Percent of 18- to 21-year-olds that registered and that voted
d. 1980
e. Participation is declining.

10a. The regions
c. 6.1 births; 4.7 births
b. Number of births per woman in 1960-1965 and in 1985-1990
d. East Asia
e. East Asia; Africa

11a.

Year	Degrees per 100 Students
1960	13
1965	18
1970	22
1975	23
1980	22
1985	18
1990	22
1994	26

b. 1975–1985
c. 26 per 100 = 26%
0.26 × 25,000
= 6500 degrees

12a. 12 yr
b-e.

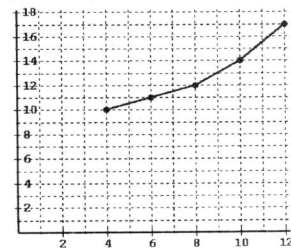

21

13.

Gallons of Gas	3	5	8	10	11	12
Calculation	22×3	22×5	22×8	22×10	22×11	22×12
Miles Driven	66	110	176	220	242	264

a. Multiply 22 by the number of gallons.

b. Miles driven = 22 × gallons

c. $m = 22g$

d.

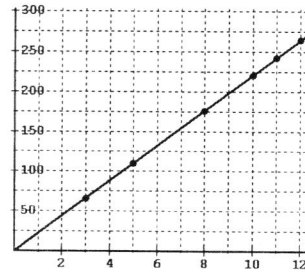

14.

Pages Read	20	50	85	110	135	180
Calculation	$200 - 20$	$200 - 50$	$200 - 85$	$200 - 110$	$200 - 135$	$200 - 180$
Pages Left	180	150	115	90	65	20

a. Subtract the number of pages read from 200.

b. Pages left = 200 − pages read

c. $l = 200 - r$

d.

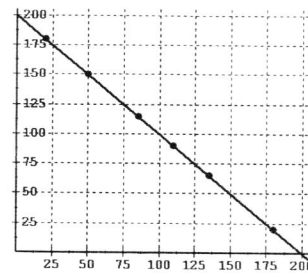

15.

x	4.0	6.0	7.5
y	1.0	1.5	1.875

$y = \dfrac{x}{4}$

16.

x	y
0	12
4	16
8	20
12	24
20	32
40	52

$y = x + 12$

17. $z + 5$ **18.** $0.28t$ **19.** $f - 60$ **20.** $\dfrac{V}{6}$

21a. $d = rt$
$d = 88t$

b. $d = 88(5) = 440$ ft
$d = 88(0.5) = 44$ ft
$d = 88(30) = 2640$ ft

22a. $I = Prt$
$I = (500)(0.07)t$
$I = 35t$

b. $I = 35(1) = \$35$
$I = 35(2) = \$70$
$I = 35(5) = \$175$

22

23a. $n = b - 60$

b.

b	100	120	150	180	200
n	40	60	90	120	140

c.

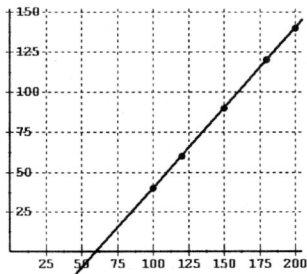

24a.

r	10	20	30	40	60	80	90
t	36	18	12	9	6	4.5	4

b.

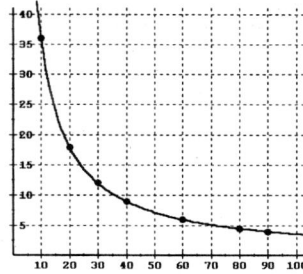

25. $\qquad\qquad y = \frac{5}{2}x$

a. For $(4, 10)$: $\quad y = \frac{5}{2}(4) = 10 \Rightarrow$ yes

b. For $(5, 2)$: $\quad y = \frac{5}{2}(5) = \frac{25}{2} \Rightarrow$ no

c. For $(1.8, 4.5)$: $y = \frac{5}{2}(1.8) = 4.5 \Rightarrow$ yes

d. For $(3, 7.5)$: $\quad y = \frac{5}{2}(3) = \frac{15}{2} = 7.5 \Rightarrow$ yes

26.

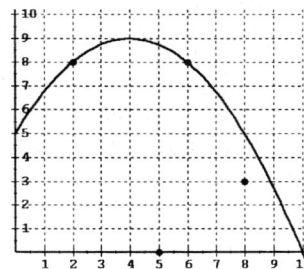

a. $(5, 0) \Rightarrow$ no

b. $(2, 8) \Rightarrow$ yes

c. $(8, 3) \Rightarrow$ no

d. $(6, 8) \Rightarrow$ yes

27a. For $t = 18$:
$$\frac{t}{6} = 3$$
$$\frac{18}{6} \overset{?}{=} 3$$
$$3 = 3 \quad \text{True}$$
Therefore, it is a solution.

b. For $t = 2$:
$$\frac{t}{6} = 3$$
$$\frac{2}{6} \overset{?}{=} 3$$
$$\frac{1}{3} = 3 \quad \text{False}$$
Therefore, it is not a solution.

28a. For $z = 28.8$:
$$8z = 36.8$$
$$8(28.8) \overset{?}{=} 36.8$$
$$230.4 = 36.8 \quad \text{False}$$
Therefore, it is not a solution.

b. For $z = 4.6$:
$$8z = 36.8$$
$$8(4.6) \overset{?}{=} 36.8$$
$$36.8 = 36.8 \quad \text{True}$$
Therefore, it is a solution.

29.
$$
\begin{array}{rl}
12 = & 3 + y \quad \text{Subtract 3 from both sides} \\
\underline{-3} & \underline{-3} \\
9 = & y
\end{array}
$$

30.
$$
\begin{array}{rl}
2.7 = & t - 1.8 \quad \text{Add 1.8 to both sides} \\
\underline{+1.8} & \underline{+1.8} \\
4.5 = & t
\end{array}
$$

23

31.

$$12.2a = 4.88 \qquad \text{Divide both sides by 12.2}$$
$$\frac{12.2a}{12.2} = \frac{4.88}{12.2}$$
$$a = 0.4$$

32.

$$8 = \frac{y}{4} \qquad \text{Multiply both sides by 4}$$
$$4(8) = 4\left(\frac{y}{4}\right)$$
$$32 = y$$

33.

$$\frac{w}{2} = 0 \qquad \text{Multiply both sides by 2}$$
$$2\left(\frac{w}{2}\right) = 2(0)$$
$$w = 0$$

34.

$$9.5 = \frac{b}{0.6} \qquad \text{Multiply both sides by 0.6}$$
$$0.6(9.5) = 0.6\left(\frac{b}{0.6}\right)$$
$$5.7 = b$$

35.

$$I = Prt$$
$$171 = P(0.095)(1)$$
$$171 = 0.095P \qquad \text{Divide both sides by 0.095}$$
$$\frac{171}{0.095} = \frac{0.095P}{0.095}$$
$$1800 = P \qquad \text{She deposited \$1800.}$$

36.

$$d = rt$$
$$2800 = 560 \cdot t \qquad \text{Divide both sides by 560}$$
$$\frac{2800}{560} = \frac{560t}{560}$$
$$5 = t \qquad \text{It takes 5 hr.}$$

37.

$$P = rW$$
$$106{,}000 = 0.53 \cdot W \qquad \text{Divide both sides by 0.53}$$
$$\frac{106{,}000}{0.53} = \frac{0.53W}{0.53}$$
$$200{,}000 = W \qquad \text{200,000 people voted.}$$

38. Grade points for an A: A

$$A - 3.2 = 89.3 \qquad \text{Add 3.2 to both sides}$$
$$\underline{+3.2 \qquad +3.2}$$
$$A \quad = \quad 92.5 \qquad \text{Need 92.5 for an A.}$$

39. Puppy's weight: w

$\dfrac{w}{85} = 0.7$ Multiply both sides by 85

$85\left(\dfrac{w}{85}\right) = 85(0.7)$

$w = 59.5$ Puppy should weigh 59.5 lb.

40. $A = lw$

$180 = l \cdot 12$ Divide both sides by 12

$\dfrac{180}{12} = \dfrac{12l}{12}$

$15 = l$ Garden should be 15 ft long.

41a. $x = 32$ **b.** $x = 24$ **42a.** $x = 4$ **b.** $x = 16$

43. $6 + 18 \div 6 - 3 \cdot 2$ Multiply and divide in order

$= 6 + 3 - 3 \cdot 2$ from left to right

$= 6 + 3 - 6$ Add and subtract in order

$= 9 - 6$ from left to right

$= 3$

44. $\dfrac{2}{3} + \dfrac{1}{3} \cdot 4$ Multiply before add

$= \dfrac{2}{3} + \dfrac{4}{3}$

$= \dfrac{6}{3}$

$= 2$

45. $36 \div (9 - 3 \cdot 2) \cdot 2 - 24 \div 4 \cdot 3$ Multiply inside parentheses

$= 36 \div (9 - 6) \cdot 2 - 24 \div 4 \cdot 3$ Subtract inside parentheses

$= 36 \div 3 \cdot 2 - 24 \div 4 \cdot 3$ Multiply and divide in order from left

$= 12 \cdot 2 - 24 \div 4 \cdot 3$ to right

$= 24 - 24 \div 4 \cdot 3$

$= 24 - 6 \cdot 3$

$= 24 - 18$ Subtract last

$= 6$

46. $\dfrac{1.2(7.7)}{4.3 - 2.3}$ Simplify above and below fraction bar

$= \dfrac{9.24}{2}$ Divide

$= 4.62$

47. $\dfrac{1}{2}d + 4$ **48.** $2p - 100$ **49.** $3(l + 5.6)$ **50.** $0.73(w - 100)$

51. $mx + b$

$\dfrac{1}{2}(3) + \dfrac{5}{2}$

$= \dfrac{3}{2} + \dfrac{5}{2}$

$= \dfrac{8}{2}$

$= 4$

52. $\dfrac{1}{m} - \dfrac{1}{n}$

$\dfrac{1}{4} - \dfrac{1}{6}$

$= \dfrac{3}{12} - \dfrac{2}{12}$

$= \dfrac{1}{12}$

53. $\dfrac{3w + z}{z}$

$\dfrac{3(8) + 6}{6}$

$= \dfrac{24 + 6}{6}$

$= \dfrac{30}{6}$

$= 5$

54. $2(l + w)$

$2\left(\dfrac{1}{3} + \dfrac{1}{6}\right)$

$= 2\left(\dfrac{2}{6} + \dfrac{1}{6}\right)$

$= 2\left(\dfrac{3}{6}\right)$

$= 1$

55. $P(1 + rt)$
$1000[1 + (0.10)(3)]$
$= 1000[1 + 0.30]$
$= 1000[1.30]$
$= \$1300$

56. $t(16t + v)$
$2[16(2) + 10]$
$= 2[32 + 10]$
$= 2[42]$
$= 84$ ft

57. $V = 2000 - 200t$

t	V
0	2000
1	1800
2	1600
5	1000
10	0

(other columns
possible)

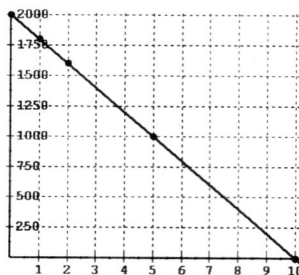

58. $P = 5 + 2s$

s	P
0	5
5	15
10	25
20	45
30	65

(other columns
possible)

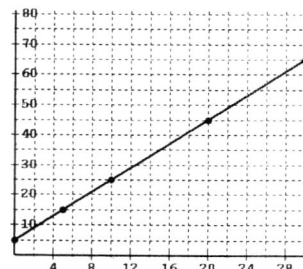

59.
$$\begin{array}{ll} 3x - 4 = \quad 1 & \text{Add 4 to both sides} \\ \underline{+4 \quad +4} & \\ 3x \quad = \quad 5 & \text{Divide both sides by 3} \\ \dfrac{3x}{3} = \dfrac{5}{3} & \\ x = \dfrac{5}{3} & \end{array}$$

60.
$$\begin{array}{ll} 1.2 + 0.4z = \quad 3.2 & \text{Subtract 1.2 from both} \\ \underline{-1.2 \qquad\quad -1.2} & \text{sides} \\ 0.4z = \quad 2.0 & \text{Divide both sides by 0.4} \\ \dfrac{0.4z}{0.4} = \dfrac{2.0}{0.4} & \\ z = 5 & \end{array}$$

61.
$$\begin{array}{ll} \dfrac{7v}{8} - 3 = \quad 4 & \text{Add 3 to both sides} \\ \underline{+3 \quad +3} & \\ \dfrac{7v}{8} = \quad 7 & \text{Multiply both sides by 8} \\ 8\left(\dfrac{7v}{8}\right) = 8(7) & \\ 7v = 56 & \text{Divide both sides by 7} \\ \dfrac{7v}{7} = \dfrac{56}{7} & \\ v = 8 & \end{array}$$

62.
$$\begin{array}{ll} 13 = \dfrac{2}{7}x + 13 & \text{Subtract 13 from both sides} \\ \underline{-13 \qquad -13} & \\ 0 = \dfrac{2}{7}x & \text{Multiply both sides by 7} \\ 7(0) = 7\left(\dfrac{2}{7}x\right) & \\ 0 = 2x & \text{Divide both sides by 2} \\ \dfrac{0}{2} = \dfrac{2x}{2} & \\ 0 = x & \end{array}$$

63. Number of bushels: b
$$\begin{array}{l} 5.75 + 4.50b = \quad 19.25 \\ \underline{-5.75 \qquad\qquad -5.75} \\ 4.50b = \quad 13.50 \\ \dfrac{4.50b}{4.50} = \dfrac{13.50}{4.50} \\ b = 3 \text{ bushels} \end{array}$$

64.
$$\begin{array}{l} P = 2w + 2l \\ 150 = 2w + 2(45) \\ 150 = 2w + 90 \\ \underline{-90 \qquad\quad -90} \\ 60 = 2w \\ \dfrac{60}{2} = \dfrac{2w}{2} \\ 30 \text{ m} = w \end{array}$$

26

CHAPTER 2

Homework 2.1

1.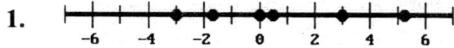

3. $|6| = 6$

5. $-|9| = -9$

7. $|-(-8.5)| = 8.5$

9. $3|-6| = 3 \cdot 6 = 18$

11. $|7 - 3| - 2|-2|$
$= |4| - 2|-2|$
$= 4 - 2 \cdot 2$
$= 4 - 4$
$= 0$

13. $0 > -4$

15. $-5 > -9$

17. $13.6 < 13.66$

19. $|-2| = 2$
Since $-2 < 2$
then $-2 < |-2|$

21. $-|-4.1| = -4.1$
Since $-4.1 > -5.8$
then $-|-4.1| > -5.8$

23.
$-x$
a. For $x = 14$: $-(14) = -14$
b. For $x = -8$: $-(-8) = 8$
c. For $x = -21$: $-(-21) = 21$
d. For $x = 17$: $-(17) = -17$

25.
$-(-x)$
a. For $x = 9$: $-(-9) = 9$
b. For $x = -4$: $-[-(-4)] = -4$
c. For $x = -13$: $-[-(-13)] = -13$
d. For $x = 15$: $-(-15) = 15$

27a. p is positive means $p > 0$.
b. n is negative means $n < 0$.

29. The <u>opposite</u> of a number can be positive or negative. For example, the opposite of 3 is -3, but the opposite of -3 is 3. The <u>absolute value</u> of a number is never negative. For example, $|-3| = 3$ and also $|3| = 3$.

31. $x > -3$
a. $-2.9, -2, -1$ (other answers possible)
b.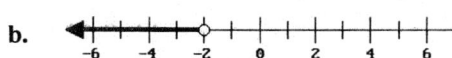

33. $x < 4$
a. $3.9, 3, 2$ (other answers possible)
b.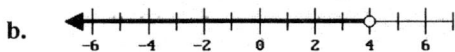

35. Note: If $-2 > x$, then $x < -2$.
a. $-2.1, -3, -4$ (other answers possible)
b.

37. $5 + (-3) = 2$

39. $-5 + (-3) = -8$

41. $-12 + 18 = 6$

43. $-47 + 22 = -25$

45. $6.8 + (-2.7) = 4.1$

47. $-\dfrac{5}{6} + \dfrac{2}{3} = -\dfrac{5}{6} + \dfrac{4}{6} = -\dfrac{1}{6}$

49. $-4 + (-5) + 6$
$= -9 + 6$
$= -3$

51. $-8 + (-8) + (-6)$
$= -16 + (-6)$
$= -22$

53. $9 + (-12) + 7 + (-15)$
$= -3 + 7 + (-15)$
$= 4 + (-15)$
$= -11$

55. $-26 + 13 + (-11) + (-32) + 16 + 20$
$= -13 + (-11) + (-32) + 16 + 20$
$= -24 + (-32) + 16 + 20$
$= -56 + 16 + 20$
$= -40 + 20$
$= -20$

57. $-1000 + 1500 = \$500$

59. $-87 + 127 = 40$ ft

61. $-2 + \left(-1\frac{1}{2}\right)$
$= -\frac{2}{1} + \left(-\frac{3}{2}\right)$
$= -\frac{4}{2} + \left(-\frac{3}{2}\right)$
$= -\frac{7}{2}$
$=$ down $3\frac{1}{2}$

63. $2\frac{1}{2} + \left(-\frac{3}{4}\right)$
$= \frac{5}{2} + \left(-\frac{3}{4}\right)$
$= \frac{10}{4} + \left(-\frac{3}{4}\right)$
$= \frac{7}{4}$
$=$ up $1\frac{3}{4}$

65a. 3%; −3%
   b. Kellogg's Corn Flakes and Crispix
   c. Lucky Charms and Frosted Flakes; −21%
   d. Their original prices were different.

Homework 2.2

1. $4 - 8 = 4 + (-8) = -4$

3. $3 - (-9) = 3 + 9 = 12$

5. $-8 - (-6) = -8 + 6 = -2$

7. $-6 - 5 = -6 + (-5) = -11$

9. $12 + (-6) = 6$

11. $6 - (-4) = 6 + 4 = 10$

13. $-2 - 8 = -2 + (-8) = -10$

15. $-7 + 9 = 2$

17. $-14 - (-3) = -14 + 3 = -11$

19. $-5 + (-4) = -9$

21. $-6 - (-6) = -6 + 6 = 0$

23. $-4 - 4 = -4 + (-4) = -8$

25. $2 - 8 = 2 + (-8) = -6°$

27. $14{,}494 - (-282) = 14{,}494 + 282 = 14{,}776$ ft

29. $-24.20 - 11.20 = -24.20 + (-11.20) = -\$35.40$

31a. $15 - (+5) = 10$ **b.** $15 + (-5) = 10$ **c.** $15 - 5 = 10$

33a. $-6 - (+2) = -8$ **b.** $-6 + (-2) = -8$ **c.** $-6 - 2 = -6 + (-2) = -8$

35.
$$6 - 3 + 4 - 5$$
$$= 6 + (-3) + 4 + (-5)$$
$$= 3 + 4 + (-5)$$
$$= 7 + (-5)$$
$$= 2$$

37.
$$13 - 6 - 12 + 17$$
$$= 13 + (-6) + (-12) + 17$$
$$= 7 + (-12) + 17$$
$$= -5 + 17$$
$$= 12$$

39.
$$120 - 80 + 20 - 40$$
$$= 120 + (-80) + 20 + (-40)$$
$$= 40 + 20 + (-40)$$
$$= 60 + (-40)$$
$$= 20$$

41.
$$3 + 2 - (-4) + (-7)$$
$$= 5 + 4 + (-7)$$
$$= 9 + (-7)$$
$$= 2$$

43.
$$12 - (-7) - (-2) - 4$$
$$= 12 + 7 + 2 - 4$$
$$= 19 + 2 - 4$$
$$= 21 - 4$$
$$= 17$$

45.
$$21 + (-15) - (-2) - 7$$
$$= 6 + 2 - 7$$
$$= 8 - 7$$
$$= 1$$

47.
$$-18 - [8 - 12 - (-4)]$$
$$= -18 - [8 + (-12) + 4]$$
$$= -18 - [-4 + 4]$$
$$= -18 - 0$$
$$= -18$$

49.
$$3 - (-6 + 2) + (-1 - 4)$$
$$= 3 - (-4) + [-1 + (-4)]$$
$$= 3 + 4 + [-5]$$
$$= 7 + [-5]$$
$$= 2$$

51.
$$-7 + [-8 - (-2)] - [6 + (-4)]$$
$$= -7 + [-8 + 2] - [6 + (-4)]$$
$$= -7 + [-6] - 2$$
$$= -13 + (-2)$$
$$= -15$$

53.
$$0 - [5 - (-1)] + [-6 - 3]$$
$$= 0 - [5 + 1] + [-6 + (-3)]$$
$$= 0 - 6 + [-9]$$
$$= 0 + (-6) + (-9)$$
$$= -6 + (-9)$$
$$= -15$$

55. $2 - 7 = -5$ **57.** $-2 - 7 = -9$ **59.** $-2 + 7 = 5$ **61.** $2 - (-7) = 9$

63. $-2 - (-7) = 5$

65.
$$15 - x - y$$
$$15 - (-6) - 8$$
$$= 15 + 6 - 8$$
$$= 21 - 8$$
$$= 13$$

67.
$$p - (4 - m)$$
$$-2 - [4 - (-6)]$$
$$= -2 - [4 + 6]$$
$$= -2 - 10$$
$$= -2 + (-10)$$
$$= -12$$

69. If $x = 3.01$: $x + 2 > 5$ If $x = 4$: $x + 2 > 5$ If $x = 5$: $x + 2 > 5$
                $3.01 + 2 > 5$ $4 + 2 > 5$ $5 + 2 > 5$
                    $5.01 > 5$ **True** $6 > 5$ **True** $7 > 5$ **True**

Therefore, $x = 3.01$ is a solution.   Therefore, $x = 4$ is a solution.   Therefore, $x = 5$ is a solution.
The values 3.01, 4, and 5 satisfy the inequality. (other answers possible)

71. If $x = 1$: $\quad x - 4 < -2$ If $x = 0$: $\quad x - 4 < -2$ If $x = -1$: $\quad x - 4 < -2$

$\qquad\qquad\quad\; 1 - 4 < -2$ $0 - 4 < -2$ $-1 - 4 < -2$

$\qquad\qquad\quad\;\; -3 < -2$ True $-4 < -2$ True $-5 < -2$ True

Therefore, $x = 1$ is a solution.   Therefore, $x = 0$ is a solution.   Therefore, $x = -1$ is a solution.

The values 1, 0, and -1 satisfy the inequality. (other answers possible)

73. If $x = -3.1$: $\quad -12 > x - 9$ If $x = -4$: $\quad -12 > x - 9$

$\qquad\qquad\qquad\;\; -12 > -3.1 - 9$ $-12 > -4 - 9$

$\qquad\qquad\qquad\;\; -12 > -12.1$ True $-12 > -13$ True

Therefore, $x = -3.1$ is a solution.   Therefore, $x = -4$ is a solution.

If $x = -5$: $\quad -12 > x - 9$

$\qquad\qquad\;\; -12 > -5 - 9$

$\qquad\qquad\;\; -12 > -14$ True

Therefore, $x = -5$ is a solution.

The values -3.1, -4, and -5 satisfy the inequality. (other answers possible)

75a. $\bar{x} = \dfrac{4 + 5 + 6 + 7 + 7 + 8 + 8 + 9 + 10 + 10}{10}$

$\qquad = \dfrac{74}{10} = 7.4$

b.

x	$x - \bar{x}$	x	$x - \bar{x}$
4	-3.4	8	0.6
5	-2.4	8	0.6
6	-1.4	9	1.6
7	-0.4	10	2.6
7	-0.4	10	2.6

c. $\dfrac{\text{Average of}}{\text{deviations}} = \dfrac{-3.4 + (-2.4) + (-1.4) + (-0.4) + (-0.4) + 0.6 + 0.6 + 1.6 + 2.6 + 2.6}{10}$

$\qquad\qquad\qquad = \dfrac{-8 + 8}{10} = \dfrac{0}{10} = 0$

The mean is greater than some of the scores and less than other scores. The total amount by which some of the scores exceeds the mean is balanced by the total amount by which the other scores are less than the mean.

77a. -9% **b.** First quarter of 1997   **c.** $-1 - (-7) = -1 + 7 = +6\%$

d. $-9 - (-1) = -9 + 1 = -8\%$ **e.** $6 - (-1) = 6 + 1 = +7\%$

Homework 2.3

1. $(-8)(-4) = 32$ **3.** $\dfrac{12}{-4} = -3$ **5.** $-20 \div (-5) = 4$

7. $-6(-1)(3) = 18$ **9.** $\dfrac{-8}{0}$ is undefined   **11.** $(-5)(0)(6) = 0$

13. $6(-8) = -48$ m   **15.** $\dfrac{-115}{4} = -28.75°$ **17.** $\dfrac{-100{,}000{,}000{,}000}{100{,}000{,}000} = -\$1000$

19a. $(-2)(3)(4) = -24$ **b.** $(-2)(-3)(4) = 24$ **c.** $(-2)(-3)(-4) = -24$

d. $(-2)(-3)(4)(2) = 48$ **e.** $(-2)(-3)(-4)(2) = -48$ **f.** $(-2)(-3)(-4)(-2) = 48$

21a. $-12 - 4 = -16$ **b.** $-12(-4) = 48$

    c. $-12 - (-4) = -12 + 4 = -8$ **d.** $-12 + (-4) = -$

23a. Parts (a) and (d) have the same answer.   **b.** Part (b) is a multiplicat.

25a. $-3 - 3 = -6$ **b.** $-3(-3) = 9$ **c.** $-3 - (-3) = -\mathfrak{d}$

    d. $-3 + (-3) = -6$ **e.** $-3 \div (-3) = 1$

27a. 3 and -3 because $3 + (-3) = 0$.   **b.** 3 and 3 because $3 - 3 = 0$.

    c. 3 and 0 because $3(0) = 0$.   **d.** 0 and 3 because $0 \div 3 = 0$.

    (other answers possible)

29.
$$
\begin{aligned}
-2(-3) - 4 \quad &\text{Multiply} \\
= 6 - 4 \quad &\text{Subtract} \\
= 2
\end{aligned}
$$

31.
$$
\begin{aligned}
5(-4) - 3(-6) \quad &\text{Multiply before subtract} \\
= -20 - (-18) \\
= -20 + 18 \\
= -2
\end{aligned}
$$

33.
$$
\begin{aligned}
(-4 - 3)(-4 + 3) \quad &\text{Simplify inside parentheses} \\
= (-7)(-1) \quad &\text{Multiply} \\
= 7
\end{aligned}
$$

35.
$$
\begin{aligned}
-3(8) - 6(-2) - 5(2) \quad &\text{Multiply first} \\
= -24 - (-12) - 10 \quad &\text{Add and subtract in order} \\
= -24 + 12 - 10 \quad &\quad\text{from left to right} \\
= -12 - 10 \\
= -22
\end{aligned}
$$

37.
$$
\begin{aligned}
\frac{15}{-3} - \frac{4 - 8}{8 - 12} \quad &\text{Simplify above and} \\
&\quad\text{below fraction bar} \\
= \frac{15}{-3} - \frac{-4}{-4} \quad &\text{Divide} \\
= -5 - 1 \\
= -6
\end{aligned}
$$

39.
$$
\begin{aligned}
\frac{2(-3) - 4(-8)}{-4 - (-2)(-3)} \quad &\text{Multiply above and} \\
&\quad\text{below fraction bar} \\
= \frac{-6 - (-32)}{-4 - 6} \quad &\text{Simplify above and} \\
&\quad\text{below fraction bar} \\
= \frac{-6 + 32}{-4 - 6} \\
= \frac{26}{-10} \quad &\text{Reduce} \\
= -\frac{13}{5} \text{ or } \frac{-13}{5}
\end{aligned}
$$

41a.
$$
\begin{aligned}
-3(-4)(-5) \quad &\text{Multiply} \\
= 12(-5) \\
= -60
\end{aligned}
$$

    b.
$$
\begin{aligned}
-3(-4) - 5 \quad &\text{Multiply before subtract} \\
= 12 - 5 \\
= 7
\end{aligned}
$$

    c.
$$
\begin{aligned}
-3(-4 - 5) \quad &\text{Add inside parentheses} \\
= -3(-9) \quad &\text{Multiply} \\
= 27
\end{aligned}
$$

    d.
$$
\begin{aligned}
-3 - (-4 - 5) \quad &\text{Add inside parentheses} \\
= -3 - (-9) \quad &\text{Subtract} \\
= -3 + 9 \\
= 6
\end{aligned}
$$

$$-3 - (-4)(-5) \quad \text{Multiply before subtract}$$
$$= -3 - 20$$
$$= -23$$

f. $\quad (-3 - 4)(-5) \quad \text{Add inside parentheses}$
$$= -7(-5) \quad \text{Multiply}$$
$$= 35$$

43a. $\quad -\dfrac{2}{5} = \dfrac{-2}{5} = \dfrac{2}{-5} = -0.4$

45a. $\quad \dfrac{-3}{4}x = \dfrac{-3}{4}(5) = \dfrac{-3}{4} \cdot \dfrac{5}{1} = \dfrac{-15}{4} \left.\right\}$ All

b. No. $-\dfrac{2}{5} = -0.4$ a negative number.

$\qquad \dfrac{-2}{-5} = \dfrac{2}{5} = 0.4$ a positive number.

b. $\quad \dfrac{-3x}{4} = \dfrac{-3(5)}{4} = \dfrac{-15}{4} \left.\right\}$ the

c. $\quad -0.75x = -0.75(5) = -3.75 \left.\right\}$ same

47. If $x = 5$:

$$\dfrac{-8}{5}x = \dfrac{-8}{5}(5) = \dfrac{-8}{5} \cdot \dfrac{5}{1} = -8 \left.\right\}$$ Not

$$\dfrac{-8}{5x} = \dfrac{-8}{5(5)} = \dfrac{-8}{25} = -\dfrac{8}{25} \left.\right\}$$ the same

49. $2x + 5$
$2(-3) + 5$
$= -6 + 5$
$= -1$

51. $12x - 3xy$
$12(-3) - 3(-3)(2)$
$= -36 - (-18)$
$= -36 + 18$
$= -18$

53. $\dfrac{y - 3}{x - 4}$
$\dfrac{2 - 3}{-9 - 4}$
$= \dfrac{\cancel{-1}}{\cancel{-13}}$
$= \dfrac{1}{13}$

55. $\dfrac{5}{9}(F - 32)$
$\dfrac{5}{9}(-13 - 32)$
$= \dfrac{5}{9}(-45)$
$= -25$

57. $\dfrac{1}{2}t(t - 1)$
$\dfrac{1}{2}\left(\dfrac{2}{3}\right)\left(\dfrac{2}{3} - 1\right)$
$= \dfrac{1}{2}\left(\dfrac{2}{3}\right)\left(\dfrac{2}{3} - \dfrac{3}{3}\right)$
$= \dfrac{1}{2}\left(\dfrac{2}{3}\right)\left(\dfrac{-1}{3}\right)$
$= -\dfrac{1}{9}$ or $\dfrac{-1}{9}$

59. $\dfrac{2(-5) + 4(-3) + 1(-2) + 1(3) + 1(6) + 3(12)}{12}$
$= \dfrac{-10 + (-12) + (-2) + 3 + 6 + 36}{12}$
$= \dfrac{-24 + 45}{12} = \dfrac{21}{12} = 1.75$

61. If $x = 0$: $\quad 4x < 7 \qquad$ If $x = 1$: $\quad 4x < 7 \qquad$ If $x = 1.5$: $\quad 4x < 7$
$\qquad\qquad\qquad 4(0) < 7 \qquad\qquad\qquad\qquad 4(1) < 7 \qquad\qquad\qquad\qquad 4(1.5) < 7$
$\qquad\qquad\qquad 0 < 7 \quad \text{True} \qquad\qquad\qquad 4 < 7 \quad \text{True} \qquad\qquad\qquad 6 < 7 \quad \text{True}$

Therefore, $x = 0$ is a solution. $\quad$ Therefore, $x = 1$ is a solution. $\quad$ Therefore, $x = 1.5$ is a solution.
The values 0, 1, and 1.5 satisfy the inequality. (other answers possible)

63. If $x = -4.1$: $\quad -2x > 8 \qquad$ If $x = -5$: $\quad -2x > 8 \qquad$ If $x = -6$: $\quad -2x > 8$
$\qquad\qquad\qquad\quad -2(-4.1) > 8 \qquad\qquad\qquad -2(-5) > 8 \qquad\qquad\qquad -2(-6) > 8$
$\qquad\qquad\qquad\quad 8.2 > 8 \quad \text{True} \qquad\qquad\qquad 10 > 8 \quad \text{True} \qquad\qquad\qquad 12 > 8 \quad \text{True}$

Therefore, $x = -4.1$ is a solution. Therefore, $x = -5$ is a solution. Therefore, $x = -6$ is a solution.
The values -4.1, -5, and -6 satisfy the inequality. (other answers possible)

65. If $x = 2.1$: $\quad -3x < -6$ $\qquad$ If $x = 3$: $\quad -3x < -6$ $\qquad$ If $x = 4$: $\quad -3x < -6$
$\qquad\qquad\qquad -3(2.1) < -6$ $\qquad\qquad\qquad\qquad -3(3) < -6$ $\qquad\qquad\qquad\qquad -3(4) < -6$
$\qquad\qquad\qquad\quad -6.3 < -6$ True $\qquad\qquad\qquad -9 < -6$ True $\qquad\qquad\qquad -12 < -6$ True
Therefore, $x = 2.1$ is a solution. $\quad$ Therefore, $x = 3$ is a solution. $\quad$ Therefore, $x = 4$ is a solution.
The values 2.1, 3, and 4 satisfy the inequality. (other answers possible)

Homework 2.4

1a.

d	B
-15	3500
-5	2500
0	2000
$.5$	1500
15	500
20	0
25	-500

b. $\quad B = 2000 - 100d$

c.

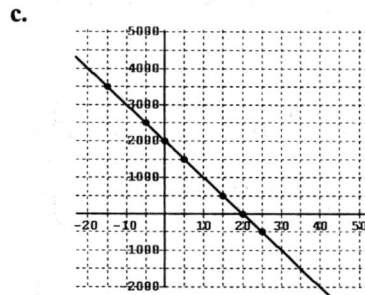

d. $3000

e. 10 days from now

f. $2500 - (-2000) = 2500 + 2000 = \$4500$

3a.

F	E
-5	-45
-2	-18
0	0
1	9
2	18
4	36
6	54

b. $\quad E = 9F$

c.

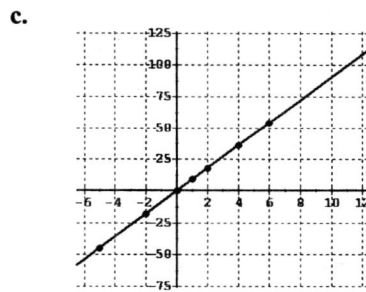

d. -27 ft

e. Floor 8

f. $90 - (-45) = 90 + 45 = 135$ ft

5. $y = x + 3$ (other columns possible)

x	-4	-2	0	2	4
y	-1	1	3	5	7

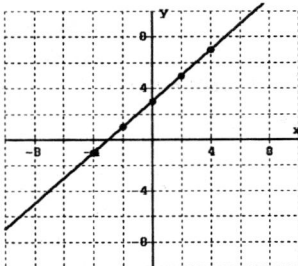

7. $y = 2x + 1$ (other columns possible)

x	-2	-1	0	1	2
y	-3	-1	1	3	5

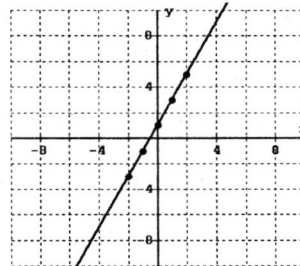

9. $y = -\frac{1}{2}x - 5$ (other columns possible)

x	-10	-2	0	2	10
y	0	-4	-5	-6	-10

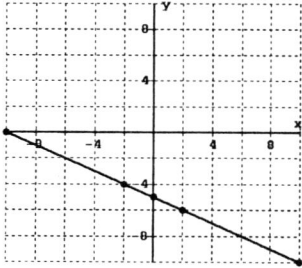

11. $y = \frac{5}{4}x - 4$ (other columns possible)

x	-4	-2	0	4	8
y	-9	-6.5	-4	1	6

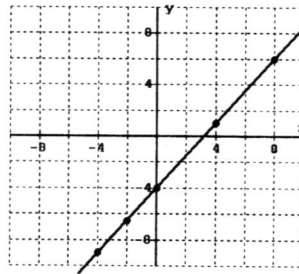

13.

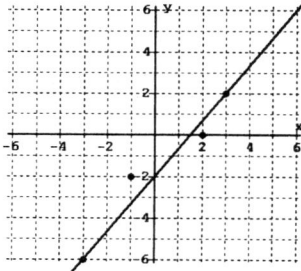

a. $(2, 0) \Rightarrow$ No **b.** $(3, 2) \Rightarrow$ Yes
c. $(-1, -2) \Rightarrow$ No **d.** $(-3, -6) \Rightarrow$ Yes

15.

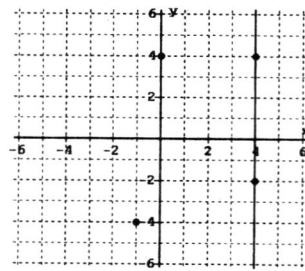

a. $(4, -2) \Rightarrow$ Yes **b.** $(4, 4) \Rightarrow$ Yes
c. $(0, 4) \Rightarrow$ No **d.** $(-1, -4) \Rightarrow$ No

17. Graph D

19. Graph B

21.

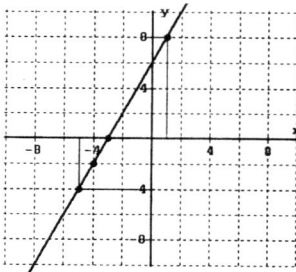

a. When $x = -5$, $y = -4$.
b. When $y = -4$, $x = -5$.
c.
$$y = 2x + 6$$
For $(-5, -4)$: $-4 \overset{?}{=} 2(-5) + 6$
$$-4 \overset{?}{=} -10 + 6$$
$$-4 = -4 \qquad \text{True}$$
d. When $y = 8$, $x = 1$.
e. Any point on the graph above $(-5, -4)$ will have $y > -4$.
Therefore, $x = -4.9$ or -4. (other answers possible)

23.

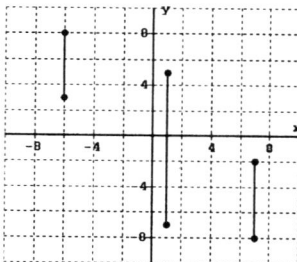

a. Distance $= 5$
b. Distance $= 12$
c. Distance $= 6$

25a. $|8 - 3| = |5| = 5$
b. $|5 - (-7)| = |5 + 7| = |12| = 12$
c. $|-2 - (-8)| = |-2 + 8| = |6| = 6$

27a. On a vertical line the x-coordinates are the same.
b. To find the distance between two points on a vertical line, subtract the y-coordinates and take the absolute value.

Homework 2.5

1.
$$x - 9 = -4 \quad \text{Add 9 to both sides}$$
$$\underline{+9 \qquad +9}$$
$$x \quad = \quad 5$$

3.
$$-9z = 12 \quad \text{Divide both sides by } -9$$
$$\frac{-9z}{-9} = \frac{12}{-9} \quad \text{Reduce}$$
$$z = \frac{-4}{3}$$

5.
$$\frac{-a}{4} = 8 \quad \text{Multiply both sides by 4}$$
$$4\left(\frac{-a}{4}\right) = 4(8)$$
$$-a = 32 \quad \text{Note: } -a = -1 \cdot a$$
$$-1 \cdot a = 32 \quad \text{Divide both sides by } -1$$
$$\frac{-1 \cdot a}{-1} = \frac{32}{-1}$$
$$a = -32$$

7.
$$9 - x = \quad 3 \quad \text{Subtract 9 from both sides}$$
$$\underline{-9 \qquad\quad -9}$$
$$-x = -6 \quad \text{Divide both sides by } -1$$
$$\frac{-x}{-1} = \frac{-6}{-1}$$
$$x = 6$$

9.
$$3c - 7 = -13 \quad \text{Add 7 to both sides}$$
$$\underline{+7 \qquad +7}$$
$$3c \quad = -6 \quad \text{Divide both sides by 3}$$
$$\frac{3c}{3} = \frac{-6}{3}$$
$$c = -2$$

11.
$$-5 = -2 - 3t \quad \text{Add 2 to both sides}$$
$$\underline{+2 \qquad +2}$$
$$-3 = \quad -3t \quad \text{Divide both sides by } -3$$
$$\frac{-3}{-3} = \frac{-3t}{-3}$$
$$1 = t$$

13.
$$1 - \frac{b}{3} = -5 \quad \text{Subtract 1 from both sides}$$
$$\underline{-1 \qquad\quad -1}$$
$$-\frac{b}{3} = -6 \quad \text{Rewrite fraction in standard form}$$
$$\frac{-b}{3} = -6 \quad \text{Multiply both sides by 3}$$
$$3\left(\frac{-b}{3}\right) = 3(-6)$$
$$-b = -18 \quad \text{Divide both sides by } -1$$
$$\frac{-b}{-1} = \frac{-18}{-1}$$
$$b = 18$$

15.
$$\frac{3y}{5} + 2 = -4 \quad \text{Subtract 2 from both sides}$$
$$\underline{-2 \qquad -2}$$
$$\frac{3y}{5} = -6 \quad \text{Multiply both sides by 5}$$
$$5\left(\frac{3y}{5}\right) = 5(-6)$$
$$3y = -30 \quad \text{Divide both sides by 3}$$
$$\frac{3y}{3} = \frac{-30}{3}$$
$$y = -10$$

35

17.

$$\frac{5x}{2} + 10 = 0 \qquad \text{Subtract 10 from both sides}$$

$$\underline{\qquad -10 \qquad -10}$$

$$\frac{5x}{2} = -10 \qquad \text{Multiply both sides by 2}$$

$$2\left(\frac{5x}{2}\right) = 2(-10)$$

$$5x = -20 \qquad \text{Divide both sides by 5}$$

$$\frac{5x}{5} = \frac{-20}{5}$$

$$x = -4$$

19. If $-x = -3$, then $x = 3$.

21. Subtracting 6 from the right side of the first equation should give -18, not -6.

$$6 - 3x = -12 \qquad \text{Subtract 6 from both sides}$$

$$\underline{-6 \qquad\qquad -6}$$

$$-3x = -18 \qquad \text{Divide both sides by } -3$$

$$\frac{-3x}{-3} = \frac{-18}{-3}$$

$$x = 6$$

23. The first step should be to add 2 to both sides of the first equation, not multiply by 3.

$$-2 + \frac{2}{3}x = -4 \qquad \text{Add 2 to both sides}$$

$$\underline{+2 \qquad\qquad +2}$$

$$\frac{2}{3}x = -2 \qquad \text{Multiply both sides by 3}$$

$$3\left(\frac{2}{3}x\right) = 3(-2)$$

$$2x = -6 \qquad \text{Divide both sides by 2}$$

$$\frac{2x}{2} = \frac{-6}{2}$$

$$x = -3$$

25a.

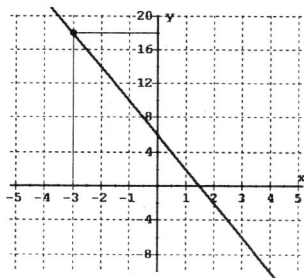

$$x = -3$$

$$-4x + 6 = 18 \qquad \text{Subtract 6 from both sides}$$

$$\underline{\qquad -6 \qquad -6}$$

$$-4x = 12 \qquad \text{Divide both sides by } -4$$

$$\frac{-4x}{-4} = \frac{12}{-4}$$

$$x = -3$$

b.

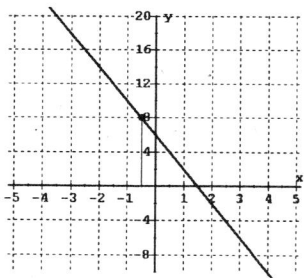

$$x = \frac{-1}{2}$$

$$-4x + 6 = 8 \qquad \text{Subtract 6 from both sides}$$

$$\underline{\qquad -6 \qquad -6}$$

$$-4x = 2 \qquad \text{Divide both sides by } -4$$

$$\frac{-4x}{-4} = \frac{2}{-4}$$

$$x = \frac{-1}{2}$$

c.

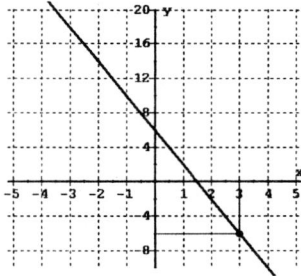

$x = 3$

$$-4x + 6 = -6 \quad \text{Subtract 6 from both sides}$$
$$\underline{\;-6 \quad\quad -6}$$
$$-4x \quad\;\; = -12 \quad \text{Divide both sides by } -4$$
$$\frac{-4x}{-4} = \frac{-12}{-4}$$
$$x = 3$$

27a.

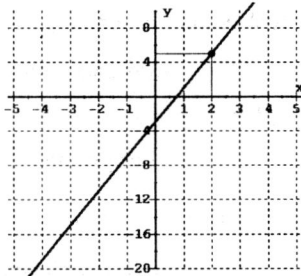

$x = 2$

$$-3 + 4x = \quad 5 \quad \text{Add 3 to both sides}$$
$$\underline{+3 \quad\quad\quad +3}$$
$$4x = \quad 8 \quad \text{Divide both sides by 4}$$
$$\frac{4x}{4} = \frac{8}{4}$$
$$x = 2$$

b.

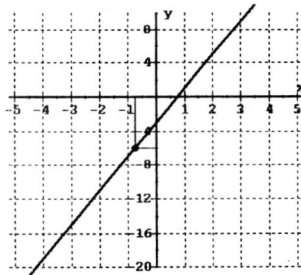

$x = \dfrac{-3}{4}$

$$-3 + 4x = -6 \quad \text{Add 3 to both sides}$$
$$\underline{+3 \quad\quad\quad +3}$$
$$4x = -3 \quad \text{Divide both sides by 4}$$
$$\frac{4x}{4} = \frac{-3}{4}$$
$$x = \frac{-3}{4}$$

c.

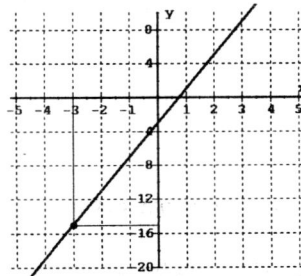

$x = -3$

$$-3 + 4x = -15 \quad \text{Add 3 to both sides}$$
$$\underline{+3 \quad\quad\quad\; +\,3}$$
$$4x = -12 \quad \text{Divide both sides by 4}$$
$$\frac{4x}{4} = \frac{-12}{4}$$
$$x = -3$$

29.

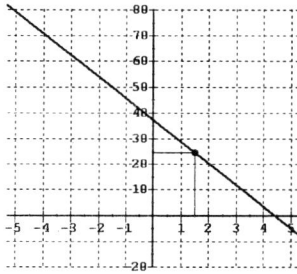

$t = 1.5$

$$
\begin{array}{ll}
\cdot\ 37.21 - 8.4t = \quad 24.61 & \text{Subtract 37.21 from both sides} \\
\underline{-\ 37.21 \qquad\qquad -\ 37.21} & \\
-8.4t = -12.60 & \text{Divide both sides by } -8.4 \\
\dfrac{-8.4t}{-8.4} = \dfrac{-12.6}{-8.4} & \\
t = 1.5 &
\end{array}
$$

31.

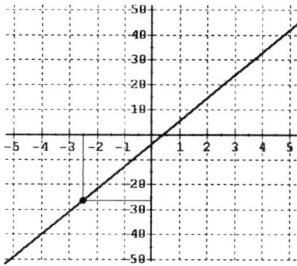

$x = -2.5$

$$
\begin{array}{ll}
-26.4 = -3.65 + 9.1x & \text{Add 3.65 to both sides} \\
\underline{+\ 3.65 \quad +\ 3.65} & \\
-22.75 = \qquad\quad 9.1x & \text{Divide both sides by 9.1} \\
\dfrac{-22.75}{9.1} = \dfrac{9.1x}{9.1} & \\
-2.5 = x &
\end{array}
$$

33a. $s = 100 - 5x$

b.

x	2	5	6	12
s	90	75	70	40

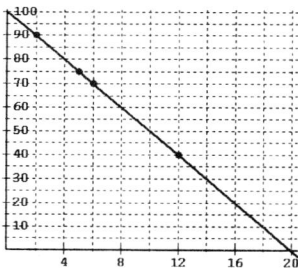

c.
$$
\begin{array}{l}
65 = \quad 100 - 5x \\
\underline{-100 \quad -100} \\
-35 = \qquad\ -5x \\
\dfrac{-35}{-5} = \dfrac{-5x}{-5} \\
7 = x
\end{array}
$$

35a. $D = 10 - \dfrac{1}{4}w$

b.

w	2	8	10	28
D	9.5	8	7.5	3

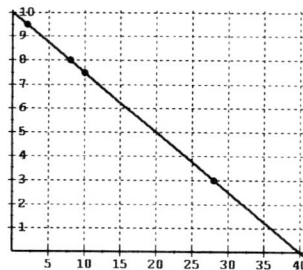

c.
$$
\begin{array}{l}
3.5 = \quad 10 - \dfrac{1}{4}w \\
\underline{-10 \qquad -10} \\
-6.5 = \qquad\ -\dfrac{1}{4}w \\
4(-6.5) = 4\left(\dfrac{-1}{4}w\right) \\
-26 = -w \\
\dfrac{-26}{-1} = \dfrac{-w}{-1} \\
26 \text{ wk} = w
\end{array}
$$

38

37. Average temperature change per day: t

$$
\begin{array}{rcl}
-6 + 4t &=& 26 \\
+6 && +6 \\
\hline
4t &=& 32 \\
\dfrac{4t}{4} &=& \dfrac{32}{4} \\
t &=& 8^\circ
\end{array}
$$

39. Number of weeks: w

$$
\begin{array}{rcl}
196 - 4w &=& 162 \\
-196 && -196 \\
\hline
-4w &=& -34 \\
\dfrac{-4w}{-4} &=& \dfrac{-34}{-4} \\
w &=& 8.5 \text{ wk}
\end{array}
$$

41.

$$
\begin{array}{rcl}
P &=& R - C \\
-45,000 &=& 600,000 - C \\
-600,000 && -600,000 \\
\hline
-645,000 &=& -C \\
\dfrac{-645,000}{-1} &=& \dfrac{-C}{-1} \\
\$645,000 &=& C
\end{array}
$$

43. Cost of ticket: t

$$
\begin{array}{rcl}
900 &=& 20t - 1800 \\
+1800 && +1800 \\
\hline
2700 &=& 20t \\
\dfrac{2700}{20} &=& \dfrac{20t}{20} \\
\$135 &=& t
\end{array}
$$

45.

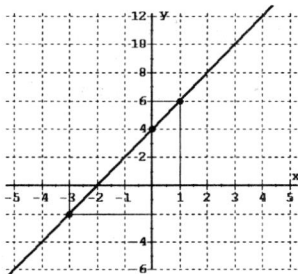

a. When $y = 6$, $x = 1$.

b. When $y = 4$, $x = 0$. Since $mx + k < 4$, any x-values on the graph below $(0, 4)$ will satisfy the inequality. Therefore, $x = -0.1, -1$, or -2. (other answers possible)

c. When $y = -2$, $x = -3$. Since $mx + k > -2$, any x-values on the graph above $(-3, -2)$ will satisfy the inequality. Therefore, $x = -2.9, -2$, or -1. (other answers possible)

Midchapter 2 Review

1. The <u>natural numbers</u> are the counting numbers: $1, 2, 3, 4, \ldots$.
The <u>whole numbers</u> are the natural numbers and zero: $0, 1, 2, 3, 4, \ldots$.
The <u>integers</u> are the natural numbers, zero, and the negatives of the natural numbers: $\ldots, -3, -2, -1, 0, 1, 2, 3, \ldots$.

2. No. To find the <u>opposite</u> of a number, we change its sign.
For example, the opposite of 3 is -3.
The <u>absolute value</u> of a number is its distance from zero, and distances are positive.
For example, $|3| = 3$.

3a. Multiplication and division: $(-6)(-2) = 12$ and $\dfrac{-6}{-2} = 3$

b. Addition: $\quad -6 + (-2) = -8$

c. Subtraction: $\quad -6 - (-2) = -6 + 2 = -4$ and $-2 - (-6) = -2 + 6 = 4$

4. quadrants (or regions); origin

5a.
$$
\begin{aligned}
&6 - 2(-4) \\
&= 6 - (-8) \\
&= 6 + 8 \\
&= 14
\end{aligned}
$$

b.
$$
\begin{aligned}
&6 - 2|-4| \\
&= 6 - 2 \cdot 4 \\
&= 6 - 8 \\
&= -2
\end{aligned}
$$

c.
$$
\begin{aligned}
&6 - (2 - 4) \\
&= 6 - (-2) \\
&= 6 + 2 \\
&= 8
\end{aligned}
$$

d.
$$
\begin{aligned}
&6 - |2 - 4| \\
&= 6 - |-2| \\
&= 6 - 2 \\
&= 4
\end{aligned}
$$

6a. $-3|5-9|$
$= -3|-4|$
$= -3 \cdot 4$
$= -12$

b. $-3|5|-|9|$
$= -3 \cdot 5 - 9$
$= -15 - 9$
$= -24$

c. $|-3||-5|-9$
$= 3 \cdot 5 - 9$
$= 15 - 9$
$= 6$

d. $|-3-5||5-9|$
$= |-8||-4|$
$= 8 \cdot 4$
$= 32$

7. $-48 + 37 - 25 - 54$
$= -11 - 25 - 54$
$= -36 - 54$
$= -90$

8. $-7.9 + (-2) - (-5) - 2.7$
$= -9.9 - (-5) - 2.7$
$= -9.9 + 5 - 2.7$
$= -4.9 - 2.7$
$= -7.6$

9. $-5[4 - 2(3)] + 6$
$= -5[4 - 6] + 6$
$= -5[-2] + 6$
$= 10 + 6$
$= 16$

10. $3(-2) - 2[(6-8) + 5(-3)]$
$= 3(-2) - 2[-2 + 5(-3)]$
$= 3(-2) - 2[-2 + (-15)]$
$= 3(-2) - 2[-17]$
$= -6 - (-34)$
$= -6 + 34$
$= 28$

11. $\dfrac{2 - (-9)(-4)}{1 - 2(-8)}$
$= \dfrac{2 - 36}{1 - (-16)}$
$= \dfrac{2 - 36}{1 + 16}$
$= \dfrac{-34}{17}$
$= -2$

12. $\dfrac{-8 - (-2)(-4)}{4 - 3(-2)}$
$= \dfrac{-8 - 8}{4 - (-6)}$
$= \dfrac{-8 - 8}{4 + 6}$
$= \dfrac{-16}{10}$
$= \dfrac{-8}{5}$ or -1.6

13. $\dfrac{-3}{5} + \dfrac{1}{2}\left[-\dfrac{1}{3} - \left(\dfrac{-1}{3}\right)\right]$
$= \dfrac{-3}{5} + \dfrac{1}{2}\left[-\dfrac{1}{3} + \left(\dfrac{1}{3}\right)\right]$
$= \dfrac{-3}{5} + \dfrac{1}{2}[0]$
$= \dfrac{-3}{5} + 0$
$= \dfrac{-3}{5}$

14. $\left(\dfrac{-5}{4} + \dfrac{1}{4}\right)\left[\dfrac{7}{3} - \left(\dfrac{-2}{3}\right)\right]$
$= \left(\dfrac{-5}{4} + \dfrac{1}{4}\right)\left[\dfrac{7}{3} + \left(\dfrac{2}{3}\right)\right]$
$= \left(\dfrac{-4}{4}\right)\left[\dfrac{9}{3}\right]$
$= (-1)[3]$
$= -3$

15. $(m + n)(m - n)$
$[-8 + (-2)][-8 - (-2)]$
$= [-8 + (-2)][-8 + 2]$
$= [-10][-6]$
$= 60$

16. $-\dfrac{3k}{2 - l}$
$-\dfrac{3(-7)}{2 - (-4)}$
$= -\dfrac{-21}{2 + 4}$
$= -\dfrac{-21}{6}$
$= \dfrac{7}{2}$ or 3.5

17. $1.8C + 32$
$1.8(-23.6) + 32$
$= -42.48 + 32$
$= -10.48$

18. $\dfrac{5(F - 32)}{9}$
$\dfrac{5(-13.27 - 32)}{9}$
$= \dfrac{5(-45.27)}{9}$
$= \dfrac{-226.35}{9}$
$= -25.15$

19. $-13.26 - 15.00 = -\$28.26$

20. $3.30 - 3.45 = -0.15$ kg

21. If $x = -9.1$: $x - 3 < -12$
$$-9.1 - 3 < -12$$
$$-12.1 < -12 \quad \text{True}$$
Therefore, $x = -9.1$ is a solution.
If $x = -10$: $x - 3 < -12$
$$-10 - 3 < -12$$
$$-13 < -12 \quad \text{True}$$
Therefore, $x = -10$ is a solution.
If $x = -11$: $x - 3 < -12$
$$-11 - 3 < -12$$
$$-14 < -12 \quad \text{True}$$
Therefore, $x = -11$ is a solution.
The values -9.1, -10, and -11 satisfy the inequality.   (other answers possible)

22. If $x = 4.1$: $-3x < -12$
$$-3(4.1) < -12$$
$$-12.3 < -12 \quad \text{True}$$
Therefore, $x = 4.1$ is a solution.
If $x = 5$: $-3x < -12$
$$-3(5) < -12$$
$$-15 < -12 \quad \text{True}$$
Therefore, $x = 5$ is a solution.
If $x = 6$: $-3x < -12$
$$-3(6) < -12$$
$$-18 < -12 \quad \text{True}$$
Therefore, $x = 4.1$ is a solution.
The values 4.1, 5, and 6 satisfy the inequality.   (other answers possible)

23a.

d	-3	0	4	8	12
W	-28	-22	-14	-6	2

b. $W = 2d - 22$

c.

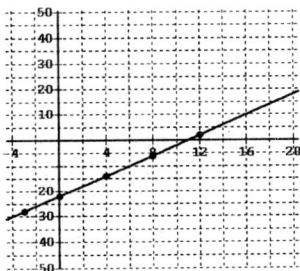

d. -8 ft

e.   In 11 days

f.   On the 6th day, the water level is -10 ft, and on the 14th day, the water level is 6 ft.
The change $= 6 - (-10) = 6 + 10 = 16$ ft

24.

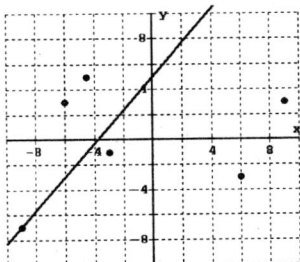

a. $(-3, -1) \Rightarrow$ No
b. $(-9, -7) \Rightarrow$ Yes
c. $(-4.5, 5) \Rightarrow$ No
d. $(6, -3) \Rightarrow$ No
e. $(-6, 3) \Rightarrow$ No
f. $(9, 3) \Rightarrow$ No

25.   Distance $= |-6 - 7| + |-5 - 8|$
$$= |-13| + |-13|$$
$$= 13 + 13$$
$$= 26$$

26a.

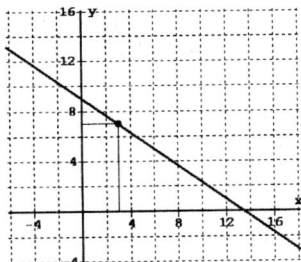

When $x = 3$, $y = 7$.

41

b.

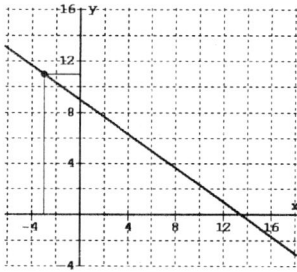

When $y = 11$, $x = -3$

$$\frac{-2}{3}x + 9 = 11 \quad \text{Subtract 9 from both sides}$$

$$\underline{-9 \quad -9}$$

$$\frac{-2}{3}x = 2 \quad \text{Multiply both sides by 3}$$

$$3\left(\frac{-2}{3}x\right) = 3(2)$$

$$-2x = 6 \quad \text{Divide both sides by } -2$$

$$\frac{-2x}{-2} = \frac{6}{-2}$$

$$x = -3$$

c.

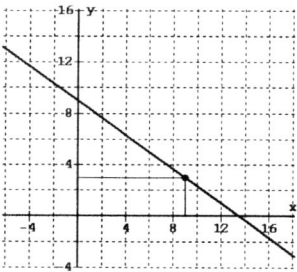

When $y = 3$, $x = 9$

$$\frac{-2}{3}x + 9 = 3 \quad \text{Subtract 9 from both sides}$$

$$\underline{-9 \quad -9}$$

$$\frac{-2}{3}x = -6 \quad \text{Multiply both sides by 3}$$

$$3\left(\frac{-2}{3}x\right) = 3(-6)$$

$$-2x = -18 \quad \text{Divide both sides by } -2$$

$$\frac{-2x}{-2} = \frac{-18}{-2}$$

$$x = 9$$

27.

x	0	1	2	3	4
y	8	6	4	2	0

(other columns possible)

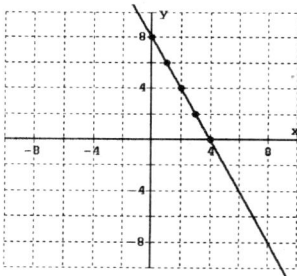

28.

x	3	0	-3	-6	-9
y	-8	-6	-4	-2	0

(other columns possible)

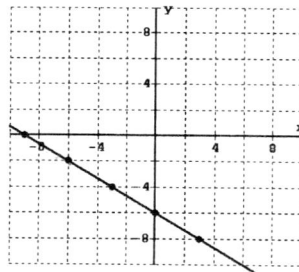

29.

$$-4 = -3c + 2 \quad \text{Subtract 2 from both sides}$$

$$\underline{-2 \qquad\qquad -2}$$

$$-6 = -3c \quad \text{Divide both sides by } -3$$

$$\frac{-6}{-3} = \frac{-3c}{-3}$$

$$2 = c$$

30.

$$30 - 7x = 2 \quad \text{Subtract 30 from both sides}$$

$$\underline{-30 \qquad\qquad -30}$$

$$-7x = -28 \quad \text{Divide both sides by } -7$$

$$\frac{-7x}{-7} = \frac{-28}{-7}$$

$$x = 4$$

31.

$$7 - \frac{2y}{3} = -5 \qquad \text{Subtract 7 from both sides}$$

$$\underline{-7 \qquad\qquad -7}$$

$$-\frac{2y}{3} = -12 \qquad \begin{array}{l}\text{Rewrite fraction in}\\ \text{standard form}\end{array}$$

$$\frac{-2y}{3} = -12 \qquad \text{Multiply both sides by 3}$$

$$3\left(\frac{-2y}{3}\right) = 3(-12)$$

$$-2y = -36 \qquad \text{Divide both sides by } -2$$

$$\frac{-2y}{-2} = \frac{-36}{-2}$$

$$y = 18$$

32.

$$\frac{3x}{2} + 3 = 2 \qquad \text{Subtract 3 from both sides}$$

$$\underline{\quad -3 \quad -3}$$

$$\frac{3x}{2} = -1 \qquad \text{Multiply both sides by 2}$$

$$2\left(\frac{3x}{2}\right) = 2(-1)$$

$$3x = -2 \qquad \text{Divide both sides by 3}$$

$$\frac{3x}{3} = \frac{-2}{3}$$

$$x = \frac{-2}{3}$$

33a.

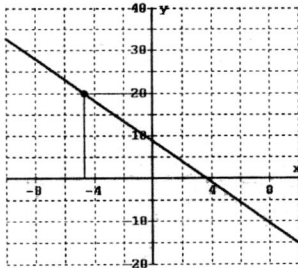

$$x \approx -4.7$$

b.

$$8.8 - 2.4x = 20$$

$$\underline{-8.8 \qquad\qquad -8.8}$$

$$-2.4x = 11.2$$

$$\frac{-2.4x}{-2.4} = \frac{11.2}{-2.4}$$

$$x \approx -4.7$$

34a. $h = 500 - 15m$

b.

m	0	10	20	30
h	500	350	200	50

(other columns possible)

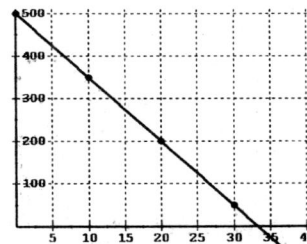

c.

$$230 = 500 - 15m$$

$$\underline{-500 \qquad -500}$$

$$-270 = -15m$$

$$\frac{-270}{-15} = \frac{-15m}{-15}$$

$$18\,\text{min} = m$$

43

35. Average temperature change per day: t

$$-5 + 7t = 16$$
$$\underline{+5 \qquad\quad +5}$$
$$7t = 21$$
$$\frac{7t}{7} = \frac{21}{7}$$
$$t = 3°$$

36.
$$P = rW$$
$$-1.5 = 0.60W$$
$$\frac{-1.5}{0.60} = \frac{0.60W}{0.60}$$
$$-2.5 = W$$

Homework 2.6

1a. $x + 10 \leq -5$ Subtract 10 from both sides
$$\underline{-10 \qquad -10}$$
$$x \qquad \leq -15$$

b.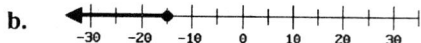

c. $x = -15.1$ is a solution, and $x = -14.9$ is not a solution. (other answers possible)

3a. $-3y < 15$ Divide both sides by -3, and
$$\frac{-3y}{-3} > \frac{-15}{-3}$$ reverse direction of inequality
$$y > -5$$

b.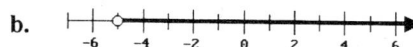

c. $y = -4.9$ is a solution, and $y = -5$ is not a solution. (other answers possible)

5a. $\dfrac{x}{3} \leq 4$ Multiply both sides by 3
$$3\left(\frac{x}{3}\right) \leq 3(4)$$
$$x \leq 12$$

b.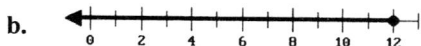

c. $x = 12$ is a solution, and $x = 12.1$ is not a solution. (other answers possible)

7a. $-8t \geq -60$ Divide both sides by -8, and
$$\frac{-8t}{-8} \leq \frac{-60}{-8}$$ reverse direction of inequality
$$t \leq 7.5$$

b.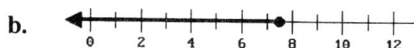

c. $t = 7$ is a solution, and $t = 8$ is not a solution. (other answers possible)

9a.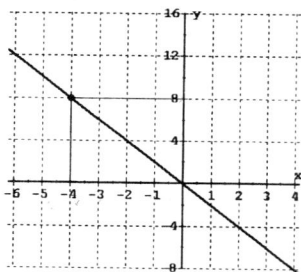

When $y = 8$, $x = -4$. For $-2x \geq 8$, any x-values on the graph above $(-4, 8)$ will satisfy the inequality. Therefore, $x \leq -4$.

c. $-2x \geq 8$ Divide both sides by -2, and
$$\frac{-2x}{-2} \leq \frac{8}{-2}$$ reverse direction of inequality
$$x \leq -4$$

11a.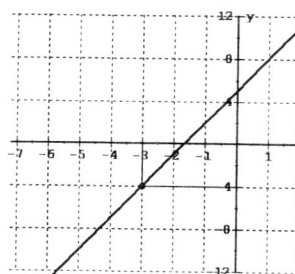

When $y = -4$, $x = -3$. For $3x + 5 \leq -4$, any x-values on the graph below $(-3, -4)$ will satisfy the inequality. Therefore, $x \leq -3$.

c. $3x + 5 \leq -4$ Subtract 5 from both sides
$$\underline{-5 \qquad -5}$$
$$3x \qquad \leq -9$$ Divide both sides by 3
$$\frac{3x}{3} \leq \frac{-9}{3}$$
$$x \leq -3$$

13a.

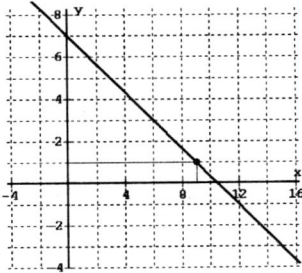

When $y = 1$, $x = 9$. For $7 - \frac{2x}{3} > 1$, any x-values on the graph above $(9, 1)$ will satisfy the inequality. Therefore, $x < 9$.

c.

$$7 - \frac{2x}{3} > 1 \quad \text{Subtract 7 from both sides}$$

$$\underline{-7 \qquad\qquad -7}$$

$$-\frac{2x}{3} > -6 \quad \text{Multiply both sides by 3}$$

$$3\left(\frac{-2x}{3}\right) > 3(-6)$$

$$-2x > -18 \quad \text{Divide both sides by } -2\text{, and}$$

$$\frac{-2x}{-2} < \frac{-18}{-2} \quad \text{reverse direction of inequality}$$

$$x < 9$$

15.

$$2x + 3 > \quad 7 \quad \text{Subtract 3 from both sides}$$

$$\underline{-3 \qquad -3}$$

$$2x \quad > \quad 4 \quad \text{Divide both sides by 2}$$

$$\frac{2x}{2} > \frac{4}{2}$$

$$x > 2$$

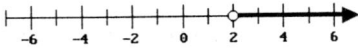

17.

$$-3x + 2 \leq \quad 11 \quad \text{Subtract 2 from both sides}$$

$$\underline{-2 \qquad -2}$$

$$-3x \quad \leq \quad 9 \quad \text{Divide both sides by } -3\text{, and}$$

$$\frac{-3x}{-3} \geq \frac{9}{-3} \quad \text{reverse direction of inequality}$$

$$x \geq -3$$

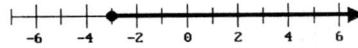

19.

$$-3 > \frac{2x}{3} + 1 \quad \text{Subtract 1 from both sides}$$

$$\underline{-1 \qquad\qquad -1}$$

$$-4 > \frac{2x}{3} \quad \text{Multiply both sides by 3}$$

$$3(-4) > 3\left(\frac{2x}{3}\right)$$

$$-12 > 2x \quad \text{Divide both sides by 2}$$

$$\frac{-12}{2} > \frac{2x}{2}$$

$$-6 > x \quad \text{Note: If } -6 > x, \text{ then } x < -6$$

$$x < -6 \quad \begin{array}{l}\text{Switching sides of inequality,} \\ \text{reverse direction of inequality}\end{array}$$

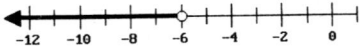

21. $T = 56 - 4d$

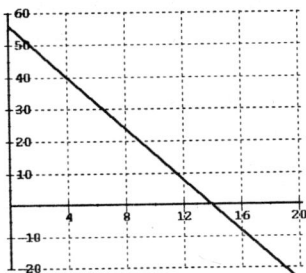

a.

$$32 = \quad 56 - 4d$$

$$\underline{-56 \quad -56}$$

$$-24 = \quad\quad -4d$$

$$\frac{-24}{-4} = \frac{-4d}{-4}$$

$$6 = d$$

b.

$$-12 > \quad 56 - 4d$$

$$\underline{-56 \quad -56}$$

$$-68 > \quad\quad -4d$$

$$\frac{-68}{-4} < \frac{-4d}{-4}$$

$$17 < d$$

$$d > 17 \text{ days}$$

23. $h = -200 + 15m$

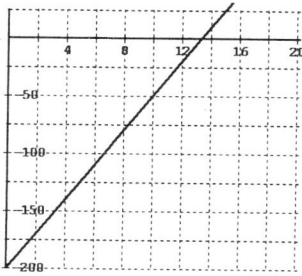

a.
$$-20 > -200 + 15m$$
$$\underline{+\,200 \qquad +\,200}$$
$$180 > \qquad\quad 15m$$
$$\frac{180}{15} > \frac{15m}{15}$$
$$12 > m$$
$$m < 12 \text{ min}$$

b.
$$0 = -200 + 15m$$
$$\underline{+\,200 \qquad +\,200}$$
$$200 = \qquad\quad 15m$$
$$\frac{200}{15} = \frac{15m}{15}$$
$$13\tfrac{1}{3} \text{ min} = m$$

25. $\quad -3 \le 3x \le 12 \qquad$ Divide each side by 3
$$\frac{-3}{3} \le \frac{3x}{3} \le \frac{12}{3}$$
$$-1 \le x \le 4$$

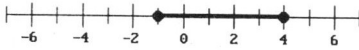

27. $\quad 23 > \quad 9 - 2b \ge \quad 13 \qquad$ Subtract 9 from each side
$$\underline{-\,9 \quad -\,9 \qquad\quad -\,9}$$
$$14 > \qquad -2b \ge \quad 4 \qquad \text{Divide each side by } -2, \text{ and}$$
$$\frac{14}{-2} < \frac{-2b}{-2} \le \frac{4}{-2} \qquad \text{reverse direction of inequality}$$
$$-7 < \quad b \quad \le -2$$

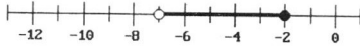

29. $\quad -8 \le \dfrac{5w + 3}{4} < -3 \qquad$ Multiply each side by 4
$$4(-8) \le 4\left(\frac{5w + 3}{4}\right) < 4(-3)$$
$$-32 \le \quad 5w + 3 \quad < -12 \qquad \text{Subtract 3 from each side}$$
$$\underline{-\,3 \qquad\quad -\,3 \qquad\quad -\,3}$$
$$-35 \le \qquad 5w \qquad < -15 \qquad \text{Divide each side by 5}$$
$$\frac{-35}{5} \le \frac{5w}{5} < \frac{-15}{5}$$
$$-7 \le \quad w \quad < -3$$

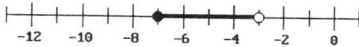

31. x

33. $y, n,$ and p

35. $n = 53° \qquad$ Corresponding angles are equal.

37. A straight angle measures $180°$.
$$z + 53 = 180$$
$$\underline{-\,53 \quad -\,53}$$
$$z = 127°$$

39. A straight angle measures $180°$.
$$m + n = 180$$
$$m + 53 = 180$$
$$\underline{-\,53 \quad -\,53}$$
$$m = 127°$$

46

Homework 2.7

1a.
$$2(0) + 4y = 8 \qquad 2x + 4(0) = 8$$
$$4y = 8 \qquad\qquad 2x = 8$$
$$y = 2 \qquad\qquad x = 4$$
y-intercept: $(0, 2)$ x-intercept: $(4, 0)$

b.

x	y
0	2
4	0

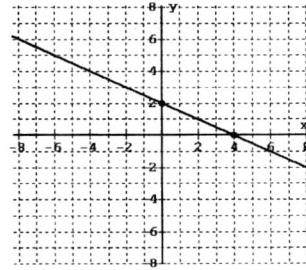

3a.
$$(0) + 2y + 10 = 0 \qquad x + 2(0) + 10 = 0$$
$$2y + 10 = 0 \qquad\qquad x + 10 = 0$$
$$2y = -10 \qquad\qquad x = -10$$
$$y = -5 \qquad x\text{-intercept: } (-10, 0)$$
y-intercept: $(0, -5)$

b.

x	y
0	-5
-10	0

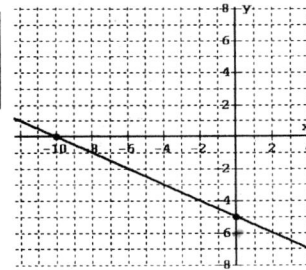

5a.
$$2(0) = 14 + 7y \qquad 2x = 14 + 7(0)$$
$$0 = 14 + 7y \qquad\quad 2x = 14$$
$$-14 = 7y \qquad\qquad x = 7$$
$$-2 = y \qquad\quad x\text{-intercept: } (7, 0)$$
y-intercept: $(0, -2)$

b.

x	y
0	-2
7	0

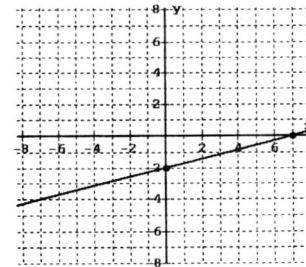

7a.
$$y = -4(0) + 8 \qquad 0 = -4x + 8$$
$$y = 8 \qquad\qquad -8 = -4x$$
y-intercept: $(0, 8)$ $2 = x$
$$x\text{-intercept: } (2, 0)$$

b.

x	y
0	8
2	0

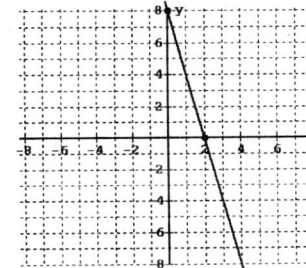

9a.

$$\frac{0}{2} + \frac{y}{3} = 1 \qquad \frac{x}{2} + \frac{0}{3} = 1$$

$$\frac{y}{3} = 1 \qquad\qquad \frac{x}{2} = 1$$

$$y = 3 \qquad\qquad x = 2$$

y-intercept: $(0, 3)$ x-intercept: $(2, 0)$

b.

x	y
0	3
2	0

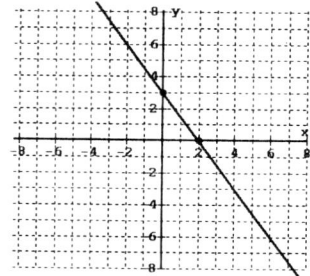

11a.

$$3(0) - 2y = 120 \qquad 3x - 2(0) = 120$$

$$-2y = 120 \qquad\qquad 3x = 120$$

$$y = -60 \qquad\qquad x = 40$$

y-intercept: $(0, -60)$ x-intercept: $(40, 0)$

b.

x	y
0	-60
40	0

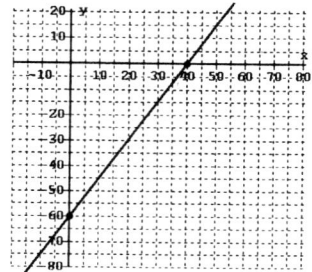

13a. 360 mi **b.** 24 hr

15a.

w	G
0	200
$13\frac{1}{3}$	0

b.

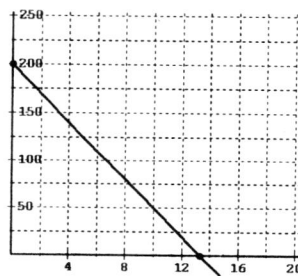

c. The G-intercept at $G = 200$ shows that there were 200 gallons in the tank when they turned on the furnace.

The w-intercept at $w = 13\frac{1}{3}$ shows that the fuel will run out after $13\frac{1}{3}$ weeks.

17a.

w	B
0	225
-9	0

b.

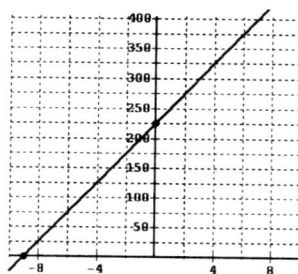

c. The B-intercept at $B = 225$ shows that she has $225 this week.

The w-intercept at $w = -9$ shows that she had a zero balance 9 weeks ago.

48

19a.

d	P
0	-600
15	0

b.

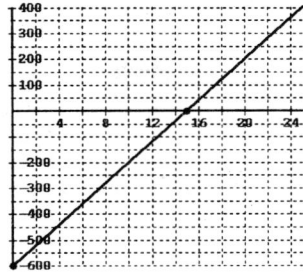

c. The P-intercept at $P = -600$ shows that he spent \$600 on equipment. The d-intercept at $d = 15$ shows that he must groom 15 dogs to break even.

21. Graph (d)　　　**23.** Graph (c)　　　**25.** Graph (a)

27a. A graph crosses the x-axis when its y-coordinate is zero.
$$2x - 3(0) = 25$$
$$2x = 25$$
$$x = 12.5$$
x-intercept: $(12.5, 0)$

b. A graph crosses the y-axis when its x-coordinate is zero.
$$1.4(0) + 3.6y = -18$$
$$3.6y = -18$$
$$y = -5$$
y-intercept: $(0, -5)$

29. Increasing

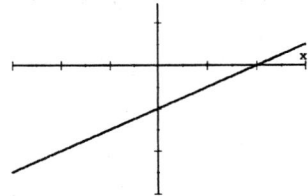

31a.
$$-2x = 6 \qquad \text{Divide both sides by } -2$$
$$\frac{-2x}{-2} = \frac{6}{-2}$$
$$x = -3$$

b.
$$-2x = k \qquad \text{Divide both sides by } -2$$
$$\frac{-2x}{-2} = \frac{k}{-2}$$
$$x = \frac{-k}{2}$$

33a.
$$15 - 4x = 3 \qquad \text{Subtract 15 from both sides}$$
$$15 - 4x - 15 = 3 - 15$$
$$-4x = -12 \qquad \text{Divide both sides by } -4$$
$$\frac{-4x}{-4} = \frac{-12}{-4}$$
$$x = 3$$

b.
$$15 - 4x = k \qquad \text{Subtract 15 from both sides}$$
$$15 - 4x - 15 = k - 15$$
$$-4x = k - 15 \qquad \text{Divide both sides by } -4$$
$$\frac{-4x}{-4} = \frac{k - 15}{-4}$$
$$x = \frac{k - 15}{-4}$$

35a.
$$9 + 3x = -1 \qquad \text{Subtract 9 from both sides}$$
$$9 + 3x - 9 = -1 - 9$$
$$3x = -10 \qquad \text{Divide both sides by 3}$$
$$\frac{3x}{3} = \frac{-10}{3}$$
$$x = \frac{-10}{3}$$

49

b.
$$9 + kx = -1 \qquad \text{Subtract 9 from both sides}$$
$$9 + kx - 9 = -1 - 9$$
$$kx = -10 \qquad \text{Divide both sides by } k$$
$$\frac{kx}{k} = \frac{-10}{k}$$
$$x = \frac{-10}{k}$$

37a. A straight angle measures $180°$.
$$A + 56 + 63 = \quad 180$$
$$A + 119 = \quad 180$$
$$\underline{\quad -119 \quad -119}$$
$$A \quad = \quad 61°$$

b. $B = 56°$ Vertical angles are equal.
c. $C = 63°$ Vertical angles are equal.
d. $D = 61°$ Vertical angles are equal.

39a. A straight angle measures $180°$.
$$v + 130 = \quad 180$$
$$\underline{\quad -130 \quad -130}$$
$$v = \quad 50°$$

b. Sum of angles of a triangle is $180°$.
$$w + 95 + 50 = \quad 180$$
$$w + 145 = \quad 180$$
$$\underline{\quad -145 \quad -145}$$
$$w = \quad 35°$$

c. $x = 35°$ **d.** $y = 50°$

e. A straight angle measures $180°$.
$$z + 35 = \quad 180$$
$$\underline{\quad -35 \quad -35}$$
$$z = \quad 145°$$

41a. $x = 41°$ **b.** $y = 43°$

Homework 2.8
For Problems **1–5**, other answers possible.

1. For $x = 0$:
$$2 + 7x = 2 + 7(0) = 2 + 0 = 2$$
$$9x = 9(0) = 0$$

3. For $a = 0$:
$$-(a - 3) = -(0 - 3) = -(-3) = 3$$
$$-a - 3 = -0 - 3 = 0 - 3 = -3$$

5. For $x = 0$:
$$5(x + 3) = 5(0 + 3) = 5(3) = 15$$
$$5x + 3 = 5(0) + 3 = 0 + 3 = 3$$

7. $-6x + 2x = (-6 + 2)x = -4x$

9. $-7.6a - 5.2a = (-7.6 - 5.2)a = -12.8a$

11. $3t - 4t + 2t = (3 - 4 + 2)t = 1 \cdot t = t$

13. $-ab + 5ab - (-3ab)$
$$= -ab + 5ab + 3ab$$
$$= (-1 + 5 + 3)ab$$
$$= 7ab$$

15. $3 + 4y - (-8y) - 7$
$$= 3 + 4y + 8y - 7$$
$$= 12y - 4$$

17. $-2st - 2 + 5s - 6st - (-4s)$
$$= -2st - 2 + 5s - 6st + 4s$$
$$= -8st + 9s - 2$$

19a. $-3y + 2 + 7y - 6y - 4y - 8$
$$= -3(2.5) + 2 + 7(2.5) - 6(2.5) - 4(2.5) - 8$$
$$= -7.5 + 2 + 17.5 - 15 - 10 - 8$$
$$= -21$$

b. $-3y + 2 + 7y - 6y - 4y - 8 = -6y - 6$

c. $-6y - 6$
$$= -6(2.5) - 6$$
$$= -15 - 6$$
$$= -21$$

50

21. $4x + (5x - 2)$
$= 4x + 5x - 2$
$= 9x - 2$

23. $22y - 34 - (16y - 24)$
$= 22y - 34 - 16y + 24$
$= 6y - 10$

25. $6a - 5 - 2a - (2a - 5)$
$= 6a - 5 - 2a - 2a + 5$
$= 2a$

27a. $3x$

b. $x + 3x = 4x$

29a. $10t - 0.2t = 9.8t$

b. $9.8t = 9.8(10) = 98$ ft

c. $9.8t = 147$
$$\frac{9.8t}{9.8} = \frac{147}{9.8}$$
$t = 15$ sec

31a. $847m - (251m + 1355)$
$= 847m - 251m - 1355$
$= 596m - 1355$

b.
$$\begin{array}{r} 596m - 1{,}355 = \;\; 16{,}525 \\ + 1{,}355 \quad + 1{,}355 \\ \hline 596m \quad\;\; = \;\; 17{,}880 \end{array}$$
$$\frac{596m}{596} = \frac{17{,}880}{596}$$
$m = 30$ stereos

33a. $0.08x$

b. $x + 0.08x = 1 \cdot x + 0.08x = 1.08x$

c. $1.08x = 928.80$
$$\frac{1.08x}{1.08} = \frac{928.80}{1.08}$$
$x = \$860$

35.
$$\begin{array}{r} 4m - 3 = \;\; 2m + 5 \\ - 2m \quad\;\; - 2m \\ \hline 2m - 3 = \qquad 5 \\ + 3 \qquad\;\; + 3 \\ \hline 2m \quad = \qquad 8 \end{array}$$
Subtract $2m$ from both sides
Add 3 to both sides
Divide both sides by 2
$$\frac{2m}{2} = \frac{8}{2}$$
$m = 4$

37.
$$\begin{array}{r} 15 - 9t = \;\; 33 - 5t \\ + 9t \qquad\;\; + 9t \\ \hline 15 \quad = \;\; 33 + 4t \\ - 33 \qquad - 33 \\ \hline - 18 \quad = \qquad 4t \end{array}$$
Add $9t$ to both sides
Subtract 33 from both sides
Divide both sides by 4
$$\frac{-18}{4} = \frac{4t}{4}$$
$$\frac{-9}{2} = t$$

39.
$$\begin{array}{r} - 6s = \;\; 3s \\ + 6s \;\; + 6s \\ \hline 0 = \;\; 9s \end{array}$$
Add $6s$ to both sides
Divide both sides by 9
$$\frac{0}{9} = \frac{9s}{9}$$
$0 = s$

41.
$$\begin{array}{r} 3x + 5 > \;\; 2x + 3 \\ - 2x \qquad - 2x \\ \hline x + 5 > \qquad 3 \\ - 5 \qquad\;\; - 5 \\ \hline x \qquad > \;\; - 2 \end{array}$$
Subtract $2x$ from both sides
Subtract 5 from both sides

missing

51

43.

$$-8g + 35 = 9g - 13 + g \quad \text{Combine like terms}$$

$$
\begin{array}{ll}
-8g + 35 = 10g - 13 & \text{Add } 8g \text{ to both sides} \\
\underline{+8g +8g } & \\
 35 = 18g - 13 & \text{Add 13 to both sides} \\
\underline{+13 +13 } & \\
 48 = 18g & \text{Divide both sides by 18}
\end{array}
$$

$$\frac{48}{18} = \frac{18g}{18}$$

$$\frac{8}{3} = g$$

45.

$$-15y + 5 - 2y - 4 \geq -12y + 21 \quad \text{Combine like terms}$$

$$
\begin{array}{ll}
-17y + 1 \geq -12y + 21 & \text{Add } 17y \text{ to both sides} \\
\underline{+17y +17y } & \\
 1 \geq 5y + 21 & \text{Subtract 21 from both sides} \\
\underline{-21 -21 } & \\
-20 \geq 5y & \text{Divide both sides by 5}
\end{array}
$$

$$\frac{-20}{5} \geq \frac{5y}{5}$$

$$-4 \geq y$$

$$y \leq -4 \qquad \begin{array}{l} \text{Switching sides of inequality,} \\ \text{reverse direction of inequality} \end{array}$$

47. Perimeter $= 3x + 4y + 3x + 4y = 6x + 8y$ **49.** Perimeter $= 5a + 3a + 5a + 4 + 3 = 13a + 7$

51a. Corresponding angles are equal.

$$
\begin{array}{rl}
4x - 4 = & 3x + 28 \\
\underline{-3x } & \underline{-3x } \\
x - 4 = & 28 \\
\underline{+4 } & \underline{+4 } \\
x = & 32°
\end{array}
$$

b. Vertical angles are equal.

$$y = 3x + 28 \qquad \text{Substitute } x = 32$$
$$y = 3(32) + 28$$
$$y = 96 + 28$$
$$y = 124°$$

Homework 2.9

1a. $8(4c) = (8 \cdot 4)c = 32c$

b. $8(4 + c) = 8(4) + 8(c) = 32 + 8c$
The distributive law is used in part (b).

3a. $2(-8 - t) = 2(-8) + 2(-t) = -16 - 2t$

b. $2(-8t) = [2 \cdot (-8)]t = -16t$
The distributive law is used in part (a).

5. $5(2y - 3) = 5(2y) + 5(-3) = 10y - 15$

7. $-2(4x + 8) = -2(4x) - 2(8) = -8x - 16$

9. $-(5b - 3) = -1(5b) - 1(-3) = -5b + 3$

11. $(-6 + 2t)(-6) = -6(-6) - 6(2t) = 36 - 12t$

13. $-6(x + 1) + 2x$
$= -6x - 6 + 2x$
$= -4x - 6$

15. $5 - 2(4x - 9) + 9x$
$= 5 - 8x + 18 + 9x$
$= 23 + x$

17. $-4(3 + 2z) + 2z - 3(2z + 1)$
$= -12 - 8z + 2z - 6z - 3$
$= -15 - 12z$

19a. $(2x + 7) - 4(4x - 2) - (-2x + 3)$
$= [2(-3) + 7] - 4[4(-3) - 2] - [-2(-3) + 3]$
$= [-6 + 7] - 4[-12 - 2] - [6 + 3]$
$= 1 - 4[-14] - 9$
$= 1 + 56 - 9$
$= 48$

b. $(2x + 7) - 4(4x - 2) - (-2x + 3)$
$= 2x + 7 - 16x + 8 + 2x - 3$
$= -12x + 12$

c. $-12x + 12$
$= -12(-3) + 12$
$= 36 + 12$
$= 48$

21a. $2(xy)$
$= 2xy$
$= 2(3)(9)$
$= 54$

b. $2(x + y)$
$= 2x + 2y$
$= 2(3) + 2(9)$
$= 6 + 18$
$= 24$

c. $2 - xy$
$= 2 - (3)(9)$
$= 2 - 27$
$= -25$

d. $-2xy$
$= -2(3)(9)$
$= -54$

23. (d) $\quad 5(3 + a) = 5(3) + 5(a) = 15 + 3a$

25a. $260 - a$

b. $2a + 3(260 - a)$
$= 2a + 780 - 3a$
$= -a + 780$

c.
$$-a + 780 = 660$$
$$\underline{ - 780 - 780}$$
$$-a = -120$$
$$\frac{-a}{-1} = \frac{-120}{-1}$$
$$a = 120 \text{ calories}$$

27. Width: w
Length: $2w + 6$

a. $2(2w + 6) + 2(w)$
$= 4w + 12 + 2w$
$= 6w + 12$

b.
$$6w + 12 = 42 \quad \text{Length} = 2w + 6$$
$$\underline{ - 12 - 12} \qquad\quad = 2(5) + 6$$
$$6w = 30 \qquad\quad = 10 + 6$$
$$\frac{6w}{6} = \frac{30}{6} \qquad\qquad\quad = 16 \text{ yd}$$
$$w = 5 \text{ yd}$$

29a. $47 - x$
b. $10x$
c. $6(47 - x) = 282 - 6x$
d. $10x + 282 - 6x = 4x + 282$

31.
$$4x + 282 = 330 \quad \text{Subtract 282 from both sides}$$
$$\underline{ - 282 - 282}$$
$$4x = 48 \quad \text{Divide both sides by 4}$$
$$\frac{4x}{4} = \frac{48}{4}$$
$$w = 12 \text{ seats}$$

33.

	Rate	Time	Distance
Downstream	$b + 8$	5	$5(b + 8)$
Upstream	$b - 8$	9	$9(b - 8)$

$$5(b + 8) = 9(b - 8) \quad \text{Distribute}$$
$$5b + 40 = 9b - 72 \quad \text{Subtract } 5b \text{ from both sides}$$
$$\underline{- 5b - 5b}$$
$$40 = 4b - 72 \quad \text{Add 72 to both sides}$$
$$\underline{+ 72 + 72}$$
$$112 = 4b \quad\quad \text{Divide both sides by 4}$$
$$\frac{112}{4} = \frac{4b}{4}$$
$$28 \text{ mph} = b$$

35a.

	Rate	Time	Distance
Upstream	$9-c$	5	$5(9-c)$
Downstream	$9+c$	4	$4(9+c)$

$$5(9-c) = 4(9+c) \quad \text{Distribute}$$

$$45 - 5c = \quad 36 + 4c \quad \text{Add } 5c \text{ to both sides}$$

$$\underline{ + 5c \qquad\qquad + 5c }$$

$$45 \quad = \quad 36 + 9c \quad \text{Subtract 36 from both sides}$$

$$\underline{- 36 \qquad\quad - 36 }$$

$$9 \quad = \quad\quad 9c \quad \text{Divide both sides by 9}$$

$$\frac{9}{9} = \frac{9c}{9}$$

$$1 \text{ mph} = c$$

b. Distance $= 5(9-c) = 5(9-1) = 5(8) = 40$ mi

37.
$$6(3y-4) = -60 \quad \text{Distribute}$$
$$18y - 24 = -60 \quad \text{Add 24 to both sides}$$
$$\underline{ +24 \qquad +24}$$
$$18y \quad = -36 \quad \text{Divide both sides by 18}$$
$$\frac{18y}{18} = \frac{-36}{18}$$
$$y = -2$$

39.
$$5w - 64 = -2(3w-1) \quad \text{Distribute}$$
$$5w - 64 = -6w + 2 \quad \text{Add } 6w \text{ to both sides}$$
$$\underline{+6w \qquad\quad +6w }$$
$$11w - 64 = \quad 2 \quad \text{Add 64 to both sides}$$
$$\underline{ +64 \qquad +64}$$
$$11w \quad = \quad 66 \quad \text{Divide both sides by 11}$$
$$\frac{11w}{11} = \frac{66}{11}$$
$$w = 6$$

41.
$$-22c + 5(3c+4) = 20 + 8c \quad \text{Distribute}$$
$$-22c + 15c + 20 = 20 + 8c \quad \text{Combine like terms}$$
$$-7c + 20 = \quad 20 + 8c \quad \text{Add } 7c \text{ to both sides}$$
$$\underline{+7c \qquad\qquad +7c }$$
$$20 = \quad 20 + 15c \quad \text{Subtract 20 from both sides}$$
$$\underline{-20 \quad -20 }$$
$$0 = \quad\quad 15c \quad \text{Divide both sides by 15}$$
$$\frac{0}{15} = \frac{15c}{15}$$
$$0 = c$$

43.
$$4 - 3(2t-4) > -2(4-3t) \quad \text{Distribute}$$
$$4 - 6t + 12 > -8 + 6t \quad \text{Combine like terms}$$
$$-6t + 16 > \quad -8 + 6t \quad \text{Add } 6t \text{ to both sides}$$
$$\underline{+6t \qquad\qquad +6t }$$
$$16 > \quad -8 + 12t \quad \text{Add 8 to both sides}$$
$$\underline{+8 \quad +8 }$$
$$24 > \quad\quad 12t \quad \text{Divide both sides by 12}$$
$$\frac{24}{12} > \frac{12t}{12}$$
$$2 > t \quad \text{Note: If } 2 > t, \text{ then } t < 2$$
$$t < 2 \quad \text{Switching sides of inequality,}$$
$$\text{reverse direction of inequality}$$

45.
$$0.25(x+3) - 0.45(x-3) = 0.30 \qquad \text{Distribute}$$
$$0.25x + 0.75 - 0.45x + 1.35 = 0.30 \qquad \text{Combine like terms}$$
$$-0.20x + 2.10 = 0.30 \qquad \text{Subtract 2.10 from both sides}$$
$$\underline{\; -2.10 \qquad -2.10}$$
$$-0.20x = -1.80 \qquad \text{Divide both sides by } -0.20$$
$$\frac{-0.20x}{-0.20} = \frac{-1.80}{-0.20}$$
$$x = 9$$

47. Area by adding small rectangles $= 5(2x) + 5(3) = 10x + 15$
Area using distributive law $= 5(2x + 3) = 10x + 15$

49. Area by adding small rectangles $= 8x(3y) + 8x(2) = 24xy + 16x$
Area using distributive law $= 8x(3y + 2) = 24xy + 16x$

Homework 2.10

1a. M **b.** A **c.** D

3.

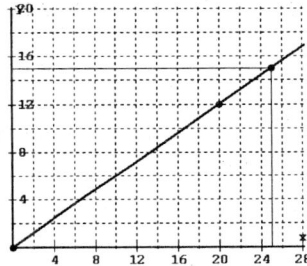

15 oz

5.

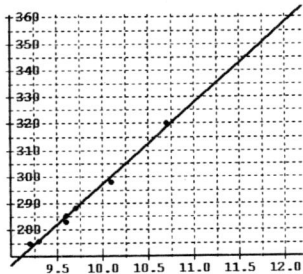

a. ≈ 358 mi
b. ≈ 10.4 gal

7.

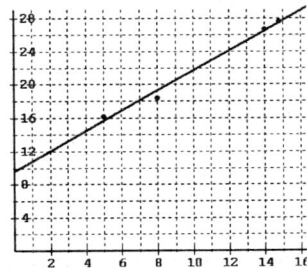

a. ≈ 24 oz
b. ≈ 8.5 cm
c. The y-intercept represents the length of the rubber band with no weight stretching it.

9a. 3 yr; 10 yr
b. About \$24,000; About \$9000
c. About $28,000 - 8000 = \$20,000$; About $39,000 - 13,000 = \$26,000$
d. $\dfrac{8000}{13,000} \approx 0.6$; $\dfrac{28,000}{39,000} \approx 0.7$

Chapter 2 Summary and Review

1. If we add or subtract the same quantity on both sides of an inequality, the direction of the inequality is unchanged.

If $a < b$

then $a + c < b + c$

and $a - c < b - c$

If we multiply or divide both sides of an inequality by the same positive quantity, the direction of the inequality is unchanged.

If $a < b$ and $c > 0$

then $ac < bc$

and $\dfrac{a}{c} < \dfrac{b}{c}$

If we multiply or divide both sides of an inequality by the same negative quantity, the direction of the inequality must be reversed.

If $a < b$ and $c < 0$

then $ac > bc$

and $\dfrac{a}{c} > \dfrac{b}{c}$

2. $a(b + c) = ab + ac$

3. To add signed numbers:

If the numbers have the same signs, add their absolute values. The sum has the same sign as the numbers.
$$2 + 5 = 7 \quad \text{and} \quad -2 + (-5) = -7$$

If the numbers have different signs, subtract their absolute values. The sum has the same sign as the number with the larger absolute value.
$$-2 + 5 = 3 \quad \text{and} \quad 2 + (-5) = -3$$

4. To subtract signed numbers:

To subtract b from a, change the subtraction to addition and change the sign of b.
$$-2 - (-5) = -2 + 5 = 3$$

5. To multiply or divide signed numbers:

If the numbers have the same signs, their product or quotient is positive.
$$6(2) = 12 \quad \text{and} \quad -6(-2) = 12$$
$$\frac{6}{2} = 3 \quad \text{and} \quad \frac{-6}{-2} = 3$$

If the numbers have different signs, their product or quotient is negative.
$$6(-2) = -12 \quad \text{and} \quad -6(2) = -12$$
$$\frac{6}{-2} = -3 \quad \text{and} \quad \frac{-6}{2} = -3$$

6. Division by zero is undefined. $\frac{2}{0}$ is undefined.

7. Quadrant I: Both coordinates are positive.

Quadrant II: The x-coordinate is negative, and the y-coordinate is positive.

Quadrant III: Both coordinates are negative.

Quadrant IV: The x-coordinate is positive, and the y-coordinate is negative.

8. To solve an equation by using a graph, locate the y-value on the vertical or y-axis. Move horizontally from that location to find the corresponding point on the graph. From there, move vertically to the horizontal or x-axis to find the x-coordinate of that point. The x-coordinate is the solution to the equation.

9. To graph a line using the intercept method:

Step 1 Find the x- and y-intercepts.

To find the y-intercept, substitute 0 for x and solve for y.

To find the x-intercept, substitute 0 for y and solve for x.

Step 2 Draw a line through the two intercepts.

Step 3 Find a third point as a check. Choose any convenient value for x and solve for y.

10. To add or subtract like terms, add or subtract their numerical coefficients. Do not change the variable factors.

11. A regression line is a "line of best fit" obtained from a set of data. It is used to estimate or to predict other values of the variables.

12. Interpolation gives estimates for values that are between two data points.
Extrapolation gives predictions for values that are outside of the data points.

13a. $-x = -(-3) = 3$

b. $|x| = |-3| = 3$

c. $-|-x| = -|-(-3)| = -|3| = -3$

d. $-(-x) = -[-(-3)] = -[3] = -3$

14a.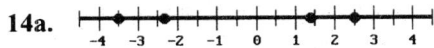

b.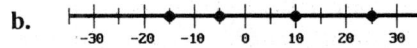

15. $-|-2| = -2$ and $-|-3| = -3$
Since $\quad\quad\quad -2 > -3$
then $\quad\quad -|-2| > -|-3|$

16. $-2.02 > -2.1$

17. $2 - (-5) = 2 + 5 = 7$
Since $\quad\quad\quad\quad 7 > -7$
then $\quad 2 - (-5) > -7$

18. $-6\left(\frac{-1}{3}\right) = 2$
Since $\quad\quad\quad\quad 2 > -2$
then $\quad -6\left(\frac{-1}{3}\right) > -2$

19. $28 - 14 - 9 + 15$ Add and subtract in order
$= 14 - 9 + 15 \quad\quad$ from left to right
$= 5 + 15$
$= 20$

20. $11 - 14 + (-24) - (-18)$ Add and subtract in
$= -3 + (-24) + 18 \quad\quad$ order left to right
$= -27 + 18$
$= -9$

21. $12 - [6 - (-2) - 5]$ Simplify inside brackets
$= 12 - [6 + 2 - 5]$
$= 12 - 3 \quad\quad\quad\quad$ Subtract
$= 9$

22. $-2 + [-3 - (-14) + 6]$ Simplify inside brackets
$= -2 + [-3 + 14 + 6]$
$= -2 + 17 \quad\quad\quad\quad$ Add
$= 15$

23. $5 - (-4)3 - 7(-2) \quad$ Multiply first
$= 5 - (-12) - (-14) \quad$ Subtract
$= 5 + 12 + 14$
$= 31$

24. $5 - (-4)(3 - 7)(-2) \quad$ Subtract inside parentheses
$= 5 - (-4)(-4)(-2) \quad$ Multiply
$= 5 - (-32) \quad\quad\quad\quad$ Subtract
$= 5 + 32$
$= 37$

25. $\dfrac{6(-3) - 8}{-4(-3 - 5)}$ Multiply above fraction bar and subtract inside parentheses

$= \dfrac{-18 - 8}{-4(-8)}$ Subtract above fraction bar and multiply below bar

$= \dfrac{-26}{32}$ Reduce

$= \dfrac{-13}{16}$

26. $\dfrac{6(-2) - 8(-9)}{5(2 - 7)}$ Multiply above fraction bar and subtract inside parentheses

$= \dfrac{-12 + 72}{5(-5)}$ Add above fraction bar and multiply below bar

$= \dfrac{60}{-25}$ Reduce

$= \dfrac{-12}{5}$

27. $2 - ab - 3a$

$2 - (-5)(-4) - 3(-5)$ Multiply first

$= 2 - 20 - (-15)$ Subtract left to right

$= -18 + 15$

$= -3$

28. $(8 - 6xy)xy$

$[8 - 6(-2)(2)](-2)(2)$ Multiply inside brackets

$= [8 - (-24)](-2)(2)$ Simplify inside brackets

$= [8 + 24](-2)(2)$

$= 32(-2)(2)$ Multiply

$= -128$

29. $\dfrac{-3 - y}{4 - x}$

$\dfrac{-3 - 2}{4 - (-1)}$ Simplify above and below fraction bar

$= \dfrac{-5}{5}$ Divide

$= -1$

30. $\dfrac{5}{9}(F - 32) + 273$

$\dfrac{5}{9}(-22 - 32) + 273$ Subtract inside parentheses

$= \dfrac{5}{9}(-54) + 273$ Multiply

$= -30 + 273$ Add

$= 243$

31. $-4 + 10 = 6°$

32. $-7 + 3 = -4°$

33. $280{,}000 - (-180{,}000) = \$460{,}000$

34. $-8.5 - 1.1 = -9.6°$

35. $3(-4) = -12$ yd

36. $-150(6 - 1) = -150(5) = -750$ ft

37. If $x = 6.9$: $x - 2 < 5$
$6.9 - 2 < 5$
$4.9 < 5$ True
Therefore, $x = 6.9$ is a solution.
If $x = 6$: $x - 2 < 5$
$6 - 2 < 5$
$4 < 5$ True
Therefore, $x = 6$ is a solution.
If $x = 5$: $x - 2 < 5$
$5 - 2 < 5$
$3 < 5$ True
Therefore, $x = 5$ is a solution.
The values 6.9, 6, and 5 satisfy the
inequality.   (other answers possible)

38. If $x = -3$: $4x \geq -12$
$4(-3) \geq -12$
$-12 \geq -12$ True
Therefore, $x = -3$ is a solution.
If $x = -2$: $4x \geq -12$
$4(-2) \geq -12$
$-8 \geq -12$ True
Therefore, $x = -2$ is a solution.
If $x = -1$: $4x \geq -12$
$4(-1) \geq -12$
$-4 \geq -12$ True
Therefore, $x = -1$ is a solution.
The values -3, -2, and -1 satisfy the
inequality.   (other answers possible)

39. If $x = 3$: $-9 \leq -3x$
$-9 \leq -3(3)$
$-9 \leq -9$ True
Therefore, $x = 3$ is a solution.
If $x = 2$: $-9 \leq -3x$
$-9 \leq -3(2)$
$-9 \leq -6$ True
Therefore, $x = 2$ is a solution.
If $x = 1$: $-9 \leq -3x$
$-9 \leq -3(1)$
$-9 \leq -3$ True
Therefore, $x = 1$ is a solution.
The values 3, 2, and 1 satisfy the
inequality.   (other answers possible)

40. If $x = -20.1$: $-15 > 5 + x$
$-15 > 5 + (-20.1)$
$-15 > -15.1$ True
Therefore, $x = -20.1$ is a solution.
If $x = -21$: $-15 > 5 + x$
$-15 > 5 + (-21)$
$-15 > -16$ True
Therefore, $x = -21$ is a solution.
If $x = -22$: $-15 > 5 + x$
$-15 > 5 + (-22)$
$-15 > -17$ True
Therefore, $x = -22$ is a solution.
The values -20.1, -21, and -22 satisfy the
inequality.   (other answers possible)

41a.

h	-4	-2	0	1	3	5	8
T	30	24	18	15	9	3	-6

b. $T = 18 - 3h$

c.

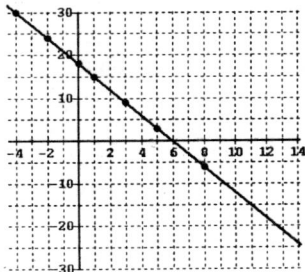

d. $24°$ **e.**   11 pm   **f.** $-9 - 9 = -18°$

59

42.

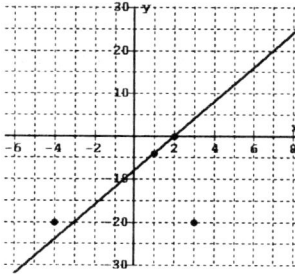

    a. For $(3, -20)$:   Point is not on line. ($\neq$) $\Rightarrow$ False
    b. For $(2, 0)$:   Point is on line. ($=$) $\Rightarrow$ True
    c. For $(-4, -20)$: Line is below point. ($<$) $\Rightarrow$ False
    d. For $(1, -4)$:   Line contains point. ($=$) $\Rightarrow$ False

43a. Distance $= |-12 - (-7)|$
                $= |-12 + 7|$
                $= |-5|$
                $= 5$

b. Distance $= |-2 - 8|$
              $= |-10|$
              $= 10$

44. $Ax + By = C$

45.

x	y
0	7
1	5
2	3

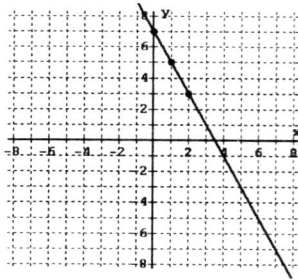

46.

x	y
0	-2
3	2
6	6

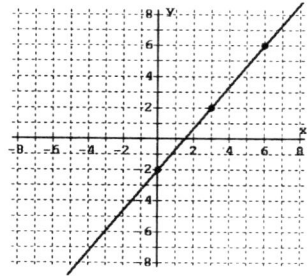

47.

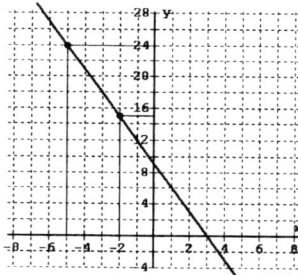

  a. $x = -5$ **b.** $x \leq -2$

48.

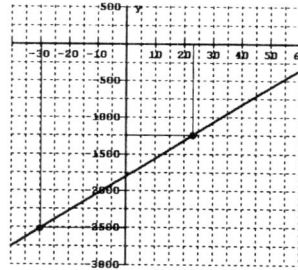

  a. About $x = 23$ **b.** About $x < -30$

49a. $-5(-6m) = [-5 \cdot (-6)]m = 30m$
   b. $-5(6 - m) = -5(6) - 5(-m) = -30 + 5m$
   The distributive law is used in part (b).

50a. $9(-3 - w) = 9(-3) + 9(-w) = -27 - 9w$
   b. $9(-3w) = [9 \cdot (-3)]w = -27w$
   The distributive law is used in part (a).

51. $(4m + 2n) - (2m - 5n)$
      $= 4m + 2n - 2m + 5n$
      $= 2m + 7n$

52. $(-5c - 6) + (-11c + 15)$
      $= -5c - 6 - 11c + 15$
      $= -16c + 9$

53. $-7w - 2(4w - 13)$
      $= -7w - 8w + 26$
      $= -15w + 26$

54. $4(3z - 10) + 5(-z - 6)$
      $= 12z - 40 - 5z - 30$
      $= 7z - 70$

55. $4z - 6 = -10$ Add 6 to both sides

$$\begin{array}{rr} +6 & +6 \\ \hline 4z \;=& -4 \end{array}$$ Divide both sides by 4

$$\frac{4z}{4} = \frac{-4}{4}$$

$$z = -1$$

56. $3 - 5x = -17$ Subtract 3 from both sides

$$\begin{array}{rr} -3 & -3 \\ \hline -5x =& -20 \end{array}$$ Divide both sides by -5

$$\frac{-5x}{-5} = \frac{-20}{-5}$$

$$x = 4$$

57. $-1 = \dfrac{5w}{3} + 4$ Subtract 4 from both sides

$$\begin{array}{rr} -4 & -4 \\ \hline -5 =& \dfrac{5w}{3} \end{array}$$ Multiply both sides by 3

$$3(-5) = 3\left(\frac{5w}{3}\right)$$

$$-15 = 5w$$ Divide both sides by 5

$$\frac{-15}{5} = \frac{5w}{5}$$

$$-3 = w$$

58. $4 - \dfrac{2z}{5} = 8$ Subtract 4 from both sides

$$\begin{array}{rr} -4 & -4 \\ \hline -\dfrac{2z}{5} =& 4 \end{array}$$ Rewrite fraction in standard form

$$\frac{-2z}{5} = 4$$ Multiply both sides by 5

$$5\left(\frac{-2z}{5}\right) = 5(4)$$

$$-2z = 20$$ Divide both sides by -2

$$\frac{-2z}{-2} = \frac{20}{-2}$$

$$z = -10$$

59. $3h - 2 = 5h + 10$ Subtract $3h$ from both sides

$$\begin{array}{rr} -3h & -3h \\ \hline -2 =& 2h + 10 \end{array}$$ Subtract 10 from both sides

$$\begin{array}{rr} -10 & -10 \\ \hline -12 =& 2h \end{array}$$ Divide both sides by 2

$$\frac{-12}{2} = \frac{2h}{2}$$

$$-6 = h$$

60. $7 - 9w = w + 7$ Add $9w$ to both sides

$$\begin{array}{rr} +9w & +9w \\ \hline 7 \;=& 10w + 7 \end{array}$$ Subtract 7 from both sides

$$\begin{array}{rr} -7 & -7 \\ \hline 0 \;=& 10w \end{array}$$ Divide both sides by 10

$$\frac{0}{10} = \frac{10w}{10}$$

$$0 = w$$

61. $5p + 10(17 - p) = 2p - 5$ Distribute

$5p + 170 - 10p = 2p - 5$ Collect like terms

$$\begin{array}{rr} -5p + 170 =& 2p - 5 \end{array}$$ Add $5p$ to both sides

$$\begin{array}{rr} +5p & +5p \\ \hline 170 =& 7p - 5 \end{array}$$ Add 5 to both sides

$$\begin{array}{rr} +5 & +5 \\ \hline 175 =& 7p \end{array}$$ Divide both sides by 7

$$\frac{175}{7} = \frac{7p}{7}$$

$$25 = p$$

62.

$$-3(k-2)-4(2k+5)=10+3k \qquad \text{Distribute}$$
$$-3k+6-8k-20=10+3k \qquad \text{Collect like terms}$$
$$-11k-14= \qquad 10+3k \qquad \text{Add } 11k \text{ to both sides}$$
$$\underline{+11k \qquad\qquad\qquad +11k}$$
$$-14= \qquad 10+14k \qquad \text{Subtract 10 from both sides}$$
$$\underline{-10 \qquad -10}$$
$$-24= \qquad\qquad 14k \qquad \text{Divide both sides by 14}$$
$$\frac{-24}{14}=\frac{14k}{14} \qquad \text{Reduce fraction}$$
$$\frac{-12}{7}=k$$

63a. $S = 7800 - 600m$

b.

m	S
0	7800
13	0

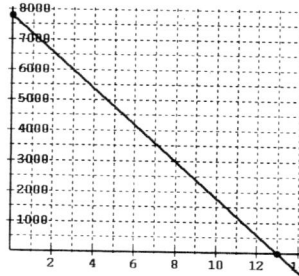

c. The S-intercept at $S = 7800$ shows
that she started with \$7800.
The m-intercept at $m = 13$ shows
that she will use all the money in 13 mo.

d.
$$2400 = \qquad 7800 - 600m$$
$$\underline{-7800 \qquad -7800}$$
$$-5400 = \qquad\qquad -600m$$
$$\frac{-5400}{-600}=\frac{-600m}{-600}$$
$$9 \text{ mo} = m$$

64a. $h = -156 + 4m$

b.

m	h
0	-156
39	0

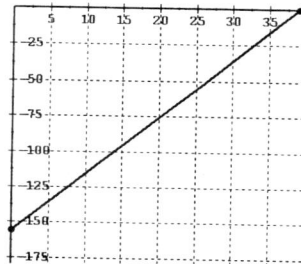

c. The h-intercept at $h = -156$ shows
that he started 156 ft down.
The m-intercept at $m = 39$ shows
that it will take him 39 min to climb out.

d.
$$-20 = -156 + 4m$$
$$\underline{+156 \qquad +156}$$
$$136 = \qquad\qquad 4m$$
$$\frac{136}{4}=\frac{4m}{4}$$
$$34 \text{ min} = m$$

65.

x	y
0	-3
2	0

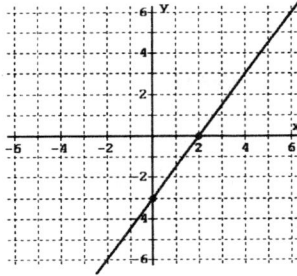

66.

x	y
0	-8
$\dfrac{8}{3}$	0

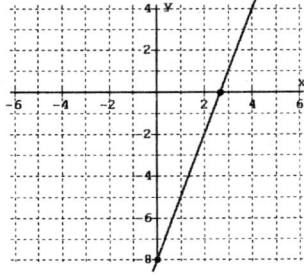

67a. $3x$

b. $2(3x) + 2(x) = 6x + 2x = 8x$

c. $8x = 48$

$$\frac{8x}{8} = \frac{48}{8}$$

$x = 6$ cm and

$3x = 18$ cm

68a. $y - 132$

b. $y + y - 132 = 2y - 132$

c.

$$
\begin{aligned}
2y - 132 &= 12{,}822\\
+132 \quad &\quad +132\\
\hline
2y \quad &= 12{,}954
\end{aligned}
$$

$$\frac{2y}{2} = \frac{12{,}954}{2}$$

$y = 6477$ votes for winner and

$y - 132 = 6345$ votes for opponent

69a. $30 - x$ **b.** $1200x$ **c.** $800(30 - x)$

d. $1200x + 800(30 - x)$
$= 1200x + 24{,}000 - 800x$
$= 400x + 24{,}000$

e.

$$
\begin{aligned}
400x + 24{,}000 &= \quad 28{,}800\\
-24{,}000 \quad &\quad -24{,}000\\
\hline
400x \qquad &= \quad 4{,}800
\end{aligned}
$$

$$\frac{400x}{400} = \frac{4800}{400}$$

$x = 12$ computers with speakers and

$30 - x = 18$ computers without

70.

	Rate	Time	Distance
Upstream	$b - 4$	$\frac{3}{2}$	$\frac{3}{2}(b - 4)$
Downstream	$b + 4$	$\frac{1}{2}$	$\frac{1}{2}(b + 4)$

$\dfrac{3}{2}(b - 4) = \dfrac{1}{2}(b + 4)$ Distribute

$\dfrac{3}{2}b - \dfrac{3}{2} \cdot 4 = \dfrac{1}{2}b + \dfrac{1}{2} \cdot 4$

$\dfrac{3}{2}b - 6 = \dfrac{1}{2}b + 2$ Add 6 to both sides

$$
\begin{aligned}
+6 \qquad\quad +6 \\
\hline
\dfrac{3}{2}b \quad = \quad \dfrac{1}{2}b + 8
\end{aligned}
$$

Subtract $\frac{1}{2}b$ from both sides

$$
\begin{aligned}
-\dfrac{1}{2}b \qquad -\dfrac{1}{2}b \\
\hline
b \quad = \qquad 8 \text{ mph}
\end{aligned}
$$

63

71.

$$2 - 3z \leq -7 \qquad \text{Subtract 2 from both sides}$$

$$\underline{-2 \qquad\qquad -2}$$

$$-3z \leq -9 \qquad \text{Divide both sides by } -3\text{, and}$$

$$\frac{-3z}{-3} \geq \frac{-9}{-3} \qquad \text{reverse direction of inequality}$$

$$z \geq 3$$

72.

$$\frac{t}{-3} - 1.7 > 2.8 \qquad \text{Add 1.7 to both sides}$$

$$\underline{+1.7 \qquad +1.7}$$

$$\frac{t}{-3} > 4.5 \qquad \text{Multiply both sides by } -3\text{, and}$$

$$-3\left(\frac{t}{-3}\right) < -3(4.5) \qquad \text{reverse direction of inequality}$$

$$t < -13.5$$

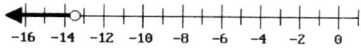

73.

$$3k - 13 < 5 + 6k \qquad \text{Subtract } 3k \text{ from both sides}$$

$$\underline{-3k \qquad\qquad -3k}$$

$$-13 < 5 + 3k \qquad \text{Subtract 5 from both sides}$$

$$\underline{-5 \qquad -5}$$

$$-18 < 3k \qquad \text{Divide both sides by 3}$$

$$\frac{-18}{3} < \frac{3k}{3}$$

$$-6 < k \qquad \text{Note: If } -6 < k \text{, then } k > -6$$

$$k > -6 \qquad \text{Switching sides of inequality,}$$
$$\text{reverse direction of inequality}$$

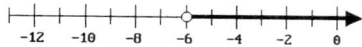

74.

$$4(3a - 7) < -18 + 2a \qquad \text{Distribute}$$

$$12a - 28 < -18 + 2a \qquad \text{Subtract } 2a \text{ from both sides}$$

$$\underline{-2a \qquad\qquad -2a}$$

$$10a - 28 < -18 \qquad \text{Add 28 to both sides}$$

$$\underline{+28 \qquad +28}$$

$$10a < 10 \qquad \text{Divide both sides by 10}$$

$$\frac{10a}{10} < \frac{10}{10}$$

$$a < 1$$

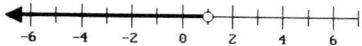

75.
$$-9 < \quad 5 - 2n \le -1 \qquad \text{Subtract 5 from each side}$$
$$\underline{-5} \quad \underline{-5} \qquad\qquad \underline{-5}$$
$$-14 < \qquad -2n \le -6 \qquad \text{Divide each side by } -2, \text{and}$$
$$\frac{-14}{-2} > \quad \frac{-2n}{-2} \quad \ge \frac{-6}{-2} \qquad \text{reverse direction of inequality}$$
$$7 > \qquad n \quad \ge 3$$

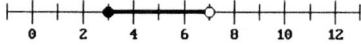

76.
$$15 \ge -6 + 3m \ge -6 \qquad \text{Add 6 to both sides}$$
$$\underline{+6} \quad \underline{+6} \qquad\qquad \underline{+6}$$
$$21 \ge \qquad 3m \ge \quad 0 \qquad \text{Divide both sides by 3}$$
$$\frac{21}{3} \ge \quad \frac{3m}{3} \quad \ge \frac{0}{3}$$
$$7 \ge \qquad m \quad \ge 0$$

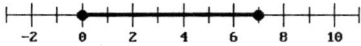

77.

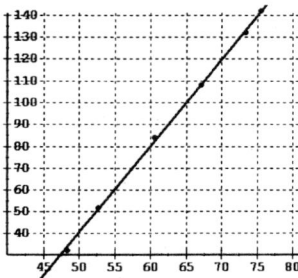

a. 120 chirps/min **b.** 65° F

78.

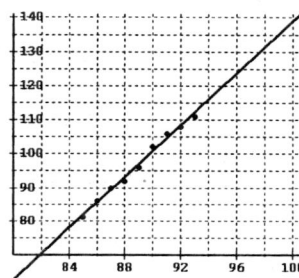

a. 1983 **b.** 1999

65

CHAPTER 3

Homework 3.1

1. A <u>ratio</u> is a comparison of two quantities. Written as a fraction, the ratio of a to b is $\frac{a}{b}$. A <u>proportion</u> is a statement that two ratios are equal. $\frac{a}{b} = \frac{c}{d}$

3. 125 employees used public transportation. $300 - 125 = 175$ employees did not.

$$\frac{125}{175} = \frac{5}{7}$$

reduce

5. $\dfrac{\$56.68}{6.5 \text{ hr}} = \$8.72/\text{hr}$

7. $\dfrac{1\frac{3}{8} \text{ cups nitrogen}}{\frac{3}{4} \text{ cups potash}} = \dfrac{\frac{11}{8}}{\frac{3}{4}} = \dfrac{11}{8} \div \dfrac{3}{4} = \dfrac{11}{8} \cdot \dfrac{4}{3} = \dfrac{11}{6}$

9. $\dfrac{0.6 \text{ mg niacin}}{0.14 \text{ mg thiamin}} = \dfrac{0.6(100)}{0.14(100)} = \dfrac{60}{14} = \dfrac{30}{7}$ or ≈ 4.3

11. $\dfrac{x}{16} = \dfrac{9}{24}$
$24x = 144$
$x = 6$

13. $\dfrac{182}{65} = \dfrac{21}{w}$
$182w = 1365$
$w = \dfrac{1365}{182}$
$w = \dfrac{15}{2}$
or $w = 7.5$

15. $\dfrac{a}{a+2} = \dfrac{2}{3}$
$3a = 2(a+2)$
$3a = 2a + 4$
$a = 4$

17. $\dfrac{0.3}{0.5} = \dfrac{b+2}{12-b}$
$0.3(12-b) = 0.5(b+2)$
$3.6 - 0.3b = 0.5b + 1$
$3.6 = 0.8b + 1$
$2.6 = 0.8b$
$3.25 = b$

19.

Time	Distance	$\dfrac{\text{Distance}}{\text{Time}}$
1	45	$\frac{45}{1} = 45$
2	90	$\frac{90}{2} = 45$
4	180	$\frac{180}{4} = 45$
5	225	$\frac{225}{5} = 45$

Yes

21.

Length	Area	$\dfrac{\text{Area}}{\text{Length}}$
3	9	$\frac{9}{3} = 3$
4	16	$\frac{16}{4} = 4$
8	64	$\frac{64}{8} = 8$
10	100	$\frac{100}{10} = 10$

No

23.

Rate	Time	$\dfrac{\text{Time}}{\text{Rate}}$
20	40	$\frac{40}{20} = 2$
40	20	$\frac{20}{40} = 0.5$
50	16	$\frac{16}{50} = 0.32$
80	10	$\frac{10}{80} = 0.125$

No

25. $\dfrac{3 \text{ lb}}{225 \text{ cups}} = \dfrac{x \text{ lb}}{3000 \text{ cups}}$
$9000 = 225x$
$40 \text{ lb} = x$

27. $\dfrac{32 \ \ell}{184 \text{ km}} = \dfrac{x \ \ell}{575 \text{ km}}$
$18400 = 184x$
$100 \ \ell = x$

29. $\dfrac{\frac{3}{4} \text{ cm}}{10 \text{ km}} = \dfrac{6 \text{ cm}}{x \text{ km}}$
$\dfrac{3}{4}x = 60$
$3x = 240$
$x = 80 \text{ km}$

31. $\dfrac{1200 \text{ voters}}{863 \text{ in favor}} = \dfrac{8000 \text{ voters}}{x \text{ in favor}}$
$1200x = 6,904,000$
$x \approx 5753 \text{ in favor}$

35.
$$V = \frac{1}{3}\pi r^2 h$$
$$500 = \frac{1}{3}\pi r^2 (16)$$
$$1500 = 16\pi r^2$$
$$\frac{1500}{16\pi} = r^2$$
$$\sqrt{\frac{1500}{16\pi}} = r$$
$$\sqrt{29.84155...} = r$$
$$5.463 \text{ cm} \approx r$$

37.
$$V = \frac{1}{3}s^2 h$$
$$116,500,000 = \frac{1}{3}s^2(177)$$
$$349,500,000 = 177s^2$$
$$\frac{349,500,000}{177} = s^2$$
$$\sqrt{\frac{349,500,000}{177}} = s$$
$$\sqrt{1,974,576.271...} = s$$
$$1405 \text{ ft} \approx r$$

39.

x	$x+x$	$2x$	x^2
3	6	6	9
5	10	10	25
-4	-8	-8	16
-1	-2	-2	1

(a) $x + x = 2x$

41.

x	x^2+x^2	x^4	$2x^2$
2	8	16	8
3	18	81	18
-2	8	16	8
-1	2	1	2

(b) $x^2 + x^2 = 2x^2$

43.

x	$x+x^2$	$x \cdot x^2$	x^3
1	2	1	1
4	20	64	64
-3	6	-27	-27
-1	0	-1	-1

(b) $x \cdot x^2 = x^3$

45. $5a^2 - 7a^2 = -2a^2$

47. Cannot be simplified

49. $-m^2 - m^2 = -2m^2$

51. $3k(4k) = 12k^2$

53. Cannot be simplified

55. $3k^2 + 4k^2 = 7k^2$

57. $\sqrt{5}(\sqrt{5}) = (\sqrt{5})^2 = 5$ **59.** $\sqrt{x}(\sqrt{x}) = (\sqrt{x})^2 = x$ **61.** $\dfrac{6}{\sqrt{6}} = \dfrac{\sqrt{6}\sqrt{6}}{\sqrt{6}} = \sqrt{6}$

63.
$$A = 2a(2a) + 2a(3a) + 2(2a)(3a)$$
$$= 4a^2 + 6a^2 + 12a^2$$
$$= 22a^2$$

65.
$$A = 2(2m)(\sqrt{3}m) + 2(2m)(\sqrt{3})$$
$$+ 2(\sqrt{3}m)(\sqrt{3})$$
$$= 4\sqrt{3}m^2 + 4\sqrt{3}m + 6m$$

67.

$3g^2 - 54 = 0$	Add 54 to both sides
$3g^2 = 54$	Divide both sides by 3
$g^2 = 18$	Take square roots
$g = \pm\sqrt{18}$	Simplify the square root
$g \approx \pm 4.243$	

69.

$2.4m^2 = 126$	Divide both sides by 2.4
$m^2 = 52.5$	Take square roots
$m = \pm\sqrt{52.5}$	Simplify the square root
$m \approx \pm 7.246$	

71.

$2x^2 - 200 = x^2 + 25$	Subtract x^2 from both sides
$x^2 - 200 = 25$	Add 200 to both sides
$x^2 = 225$	Take square roots
$x = \pm\sqrt{225}$	Simplify the square root
$x = \pm 15$	

7a. $\sqrt{9 - 4(-18)}$ Multiply under radical first

$= \sqrt{9 + 72}$ Add under radical

$= \sqrt{81}$ Simplify root

$= 9$

b. $\sqrt{\dfrac{4(50) - 56}{16}}$ Multiply under radical

$= \sqrt{\dfrac{200 - 56}{16}}$ Subtract under radical

$= \sqrt{\dfrac{144}{16}}$ Divide under radical

$= \sqrt{9}$ Simplify root

$= 3$

9a. $5\sqrt[3]{8} - \dfrac{\sqrt[3]{64}}{8}$ Simplify cube roots

$= 5 \cdot 2 - \dfrac{4}{8}$ Multiply and reduce fraction

$= 10 - \dfrac{1}{2}$ LCD $= 2$

$= \dfrac{20}{2} - \dfrac{1}{2}$ Subtract

$= \dfrac{19}{2}$

b. $\dfrac{3 + \sqrt[3]{-729}}{6 - \sqrt[3]{-27}}$ Simplify cube roots

$= \dfrac{3 + (-9)}{6 - (-3)}$ Simplify in numerator and denominator

$= \dfrac{-6}{9}$ Reduce fraction

$= \dfrac{-2}{3}$

11. In a <u>linear equation</u>, such as $2x + 3 = 0$, the variable cannot have an exponent other than 1 In a <u>quadratic equation</u>, such as $x^2 + 2x + 3 = 0$ the variable must have an exponent of 2.

13. $x^2 = 121$ Take square root of both sides

$x = \pm\sqrt{121}$ Simplify the square root

$x = \pm 11$

15. $98 = 2a^2$ Divide both sides by 2

$49 = a^2$ Take square root of both sides

$\pm\sqrt{49} = a$ Simplify the square root

$\pm 7 = a$

17. $0 = 3n^2 - 15$ Add 15 to both sides

$15 = 3n^2$ Divide both sides by 3

$5 = n^2$ Take square roots

$\pm\sqrt{5} = n$ Simplify the square root

$\pm 2.236 \approx n$

19. $400 + \dfrac{k^2}{6} = 625$ Subtract 400 from both sides

$\dfrac{k^2}{6} = 225$ Multiply both sides by 6

$k^2 = 1350$ Take square roots

$k = \pm\sqrt{1350}$ Simplify the square root

$k \approx \pm 36.742$

21. $55 - 3z^2 = 7$ Subtract 55 from both sides

$-3z^2 = -48$ Divide both sides by -3

$z^2 = 16$ Take square roots

$z = \pm\sqrt{16}$ Simplify the square root

$z = \pm 4$

23. 8.660   **25.** -3.055 **27.** 1.899   **29.** -15.544 **31.** 1.293   **33.** -1.512

33.
$$\frac{1 \text{ in.}}{2.54 \text{ cm}} = \frac{x \text{ in.}}{35 \text{ cm}}$$
$$35 = 2.54x$$
$$13.78 \text{ in.} \approx x$$

35. Money to education: x
Money to administration: $24,000,000 - x$
$$\frac{x}{24,000,000 - x} = \frac{4}{3}$$
$$3x = 4(24,000,000 - x)$$
$$3x = 96,000,000 - 4x$$
$$7x = 96,000,000$$
$$x \approx \$13,714,286$$

37a.
$$\frac{18 \text{ cans puree}}{12 \text{ cans catsup}} = \frac{3}{2};$$
$$\frac{12 \text{ cans catsup}}{1000 \text{ servings}} = \frac{5 \text{ cans catsup}}{x \text{ servings}};$$
$$12x = 5000$$
$$x = 416\frac{2}{3} \text{ servings}$$

$$\frac{x \text{ cans puree}}{5 \text{ cans catsup}} = \frac{3}{2}$$
$$2x = 15$$
$$x = 7\frac{1}{2} \text{ cans}$$

b. $3 \text{ lb} \times 16\frac{\text{oz}}{\text{lb}} = 48 \text{ oz of chili powder}$
$$\frac{48 \text{ oz}}{1000 \text{ servings}} = \frac{x \text{ oz}}{50 \text{ servings}};$$
$$2400 = 1000x$$
$$2.4 \text{ oz} = x$$

$$\frac{310 \text{ lb}}{1000 \text{ servings}} = \frac{x \text{ lb}}{50 \text{ servings}}$$
$$15,500 = 1000x$$
$$15.5 \text{ lb} = x$$

c.
$$\frac{5.5 \text{ cans}}{1000 \text{ servings}} = \frac{x \text{ cans}}{4500 \text{ servings}};$$
$$24,750 = 1000x$$
$$24.75 \text{ cans} = x$$

$$\frac{20 \text{ lb}}{1000 \text{ servings}} = \frac{x \text{ lb}}{4500 \text{ servings}}$$
$$90,000 = 1000x$$
$$90 \text{ lb} = x$$

Homework 3.2

1. Similar; Ratio $= \dfrac{2.5}{4} = \dfrac{2.5(10)}{4(10)} = \dfrac{25}{40} = \dfrac{5}{8}$

3. Not similar because corresponding ratios are not equal: $\dfrac{7}{12} \neq \dfrac{7}{8}$

5. Similar; Ratio $= \dfrac{5}{4}$

7. Not similar because corresponding ratios are not equal: $\dfrac{9}{6} \neq \dfrac{10}{7}$

9. (b), (c), and (d)

11. $A = B = 37°$

13. $A = B = C = 45°$; $d = 5$

15. $A = 60°$; $\dfrac{w}{3} = \dfrac{7}{6}$
$$6w = 21$$
$$w = 3.5$$

17. Yes, their corresponding sides are proportional: $\dfrac{3}{6} = \dfrac{4}{8} = \dfrac{6}{12}$

19. Yes, their corresponding angles are equal.

21.
$$\frac{h}{8} = \frac{18}{12}$$
$$12h = 144$$
$$h = 12$$

23.
$$\frac{q + 12}{q} = \frac{11}{5}$$
$$5(q + 12) = 11q$$
$$5q + 60 = 11q$$
$$60 = 6q$$
$$10 = q$$

25.
$$\frac{h}{h - 10} = \frac{12}{8}$$
$$8h = 12(h - 10)$$
$$8h = 12h - 120$$
$$-4h = -120$$
$$h = 30$$

67

27a. $\dfrac{775}{882.75} \approx 0.88 = 88\%$

b. $\dfrac{m}{175} = \dfrac{88}{100}$
$100m = 15400$
$m = 154 \text{ ft}$

29. $\dfrac{h}{6} = \dfrac{1080}{35}$
$35h = 6480$
$h \approx 185 \text{ ft}$

31. $\dfrac{x}{\frac{4}{9}} = \dfrac{4.5}{2}$
$2x = 2$
$x = 1 \text{ mi}$

33. $\dfrac{y}{6} = \dfrac{x}{3}$
$3y = 6x$
$\dfrac{3y}{3} = \dfrac{6x}{3}$
$y = 2x$

35. $\dfrac{y}{x} = \dfrac{12}{17}$
$17y = 12x$
$\dfrac{17y}{17} = \dfrac{12x}{17}$
$y = \dfrac{12x}{17}$

37. $\dfrac{h}{12} = \dfrac{10}{16}$
$16h = 120$
$h = 7.5$

39. $\dfrac{c}{c+25} = \dfrac{12}{32}$
$32c = 12(c+25)$
$32c = 12c + 300$
$20c = 300$
$c = 15$

41. $\dfrac{12}{s+12} = \dfrac{6}{9}$
$108 = 6(s+12)$
$108 = 6s + 72$
$36 = 6s$
$6 = s$

43. $\dfrac{y}{3} = \dfrac{x}{5}$
$5y = 3x$
$\dfrac{5y}{5} = \dfrac{3x}{5}$
$y = \dfrac{3x}{5}$

45. $\dfrac{y-5}{3} = \dfrac{x}{4}$
$4(y-5) = 3x$
$4y - 20 = 3x$
$4y = 3x + 20$
$\dfrac{4y}{4} = \dfrac{3x+20}{4}$
$y = \dfrac{3x+20}{4}$

Homework 3.3

1. If two variables are proportional, their ratios are always the same.

3. Figures 3.77 and 3.80 illustrate direct variation.

5. If S varies directly with w, the graph of S verses w is a straight line through the origin.

7.

Month	HCF	Amount	$\dfrac{\text{Amount}}{\text{HCF}}$
March	43	76.90	$\frac{76.90}{43} \approx 1.79$
May	77	156.51	$\frac{156.51}{77} \approx 2.03$
June	101	220.17	$\frac{220.17}{101} \approx 2.18$

No, the ratios are not constant.

9a. $\dfrac{16}{2} = \dfrac{24}{3} = 8 = k$

b. $d = 8t$

c.

t	d
2	16
3	24

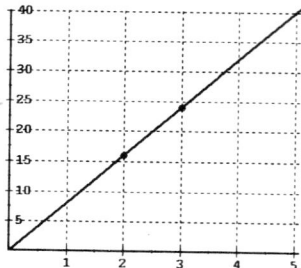

11a. $\dfrac{0.72}{12} = \dfrac{0.90}{15} = 0.06 = k$

b. $t = 0.06p$

c.

p	t
12	0.72
15	0.90

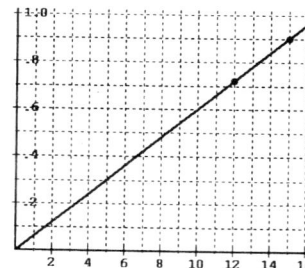

68

13a.

Time	Distance	Time	Distance
2	16	4	32
3	24	6	48
5	40	10	80

b. The distance is doubled when the time is doubled.

15a.

Price	Sales Tax	Price	Sales Tax
12	0.72	6	0.36
20	1.20	10	0.60
30	1.80	15	0.90

b. The sales tax is halved when the price is halved.

17. If y varies directly with x, and we multiply the value of x by a constant n, then the value of y is also multiplied by the same constant n.

19a. No

b.

Units	Tuition	Units	Tuition
3	620	6	740
5	700	10	900
8	820	16	1140

c. No

21. No $\left(\dfrac{9}{6} \neq \dfrac{12}{10}\right)$

23. Yes $\left(\dfrac{10}{20} = \dfrac{15}{30}\right)$

25. Yes $\left(\dfrac{4.5}{2} = \dfrac{18}{8}\right)$

27. No $\left(\dfrac{18}{4} \neq \dfrac{32}{6}\right)$

29. No $\left(\dfrac{10}{1} \neq \dfrac{40}{2}\right)$

31. No $\left(\dfrac{20}{2} \neq \dfrac{30}{4}\right)$

33.

x	y	z	w
-4	-8	-12	-6
-2	-4	-6	-3
0	0	0	0
2	4	6	3
4	8	12	6

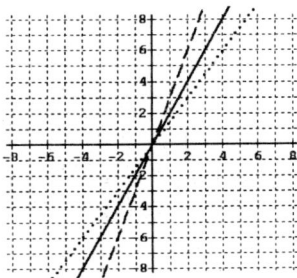

35.

x	y	z	w
-6	2	4	8
-3	1	2	4
0	0	0	0
3	-1	-2	-4
6	-2	-4	-8

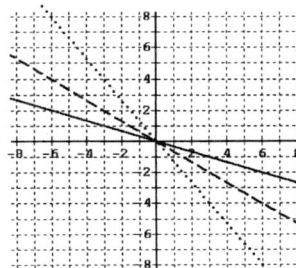

37. The origin $(0, 0)$

Homework 3.4

1a. $\dfrac{\Delta y}{\Delta x} = \dfrac{6}{4} = \dfrac{3}{2}$

b. $\dfrac{\Delta y}{\Delta x} = \dfrac{-8}{4} = -2$

3a. $\dfrac{\Delta y}{\Delta x} = \dfrac{-2}{8} = \dfrac{-1}{4}$

b. $\dfrac{\Delta y}{\Delta x} = \dfrac{9}{3} = 3$

5.

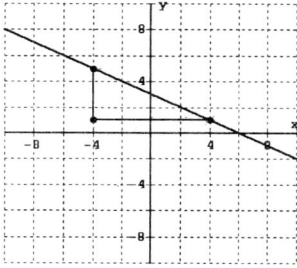

$$\frac{\Delta y}{\Delta x} = \frac{-4}{8} = \frac{-1}{2}$$

7.

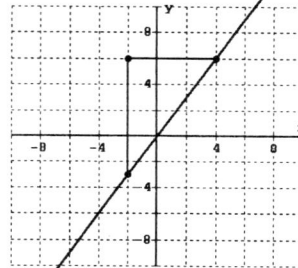

$$\frac{\Delta y}{\Delta x} = \frac{9}{6} = \frac{3}{2}$$

9a.

x	y
0	4
6	0

b.

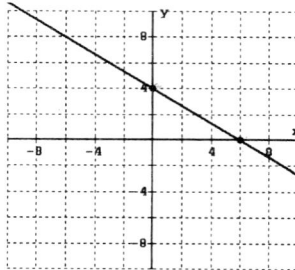

c. $\dfrac{\Delta y}{\Delta x} = \dfrac{-4}{6} = \dfrac{-2}{3}$

d. $\dfrac{\Delta y}{\Delta x} = \dfrac{-4}{6} = \dfrac{-2}{3}$

11a.

x	y
0	-5
2	0

b.

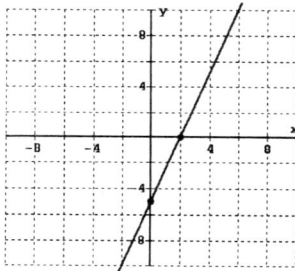

c. $\dfrac{\Delta y}{\Delta x} = \dfrac{5}{2}$

d. $\dfrac{\Delta y}{\Delta x} = \dfrac{15}{6} = \dfrac{5}{2}$

13a.

x	y
0	5
5	0

b.

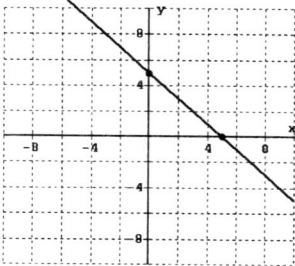

c. $\dfrac{\Delta y}{\Delta x} = \dfrac{-5}{5} = -1$

d. $\dfrac{\Delta y}{\Delta x} = \dfrac{-11}{11} = -1$

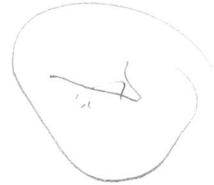

15a.

x	y
0	-2
4	0

b.

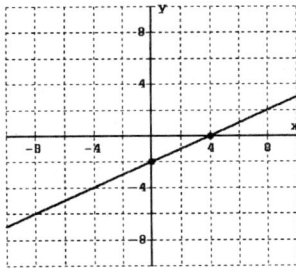

c. $\dfrac{\Delta y}{\Delta x} = \dfrac{2}{4} = \dfrac{1}{2}$

d. $\dfrac{\Delta y}{\Delta x} = \dfrac{-5}{-10} = \dfrac{1}{2}$

17. $\dfrac{5}{3}$

19. -2

21.
$$\dfrac{\Delta y}{7} = \dfrac{5}{2}$$
$$2 \cdot \Delta y = 35$$
$$\dfrac{2 \cdot \Delta y}{2} = \dfrac{35}{2}$$
$$\Delta y = \dfrac{35}{2}$$

23.
$$\dfrac{-6}{\Delta x} = \dfrac{-4}{1}$$
$$-6 = -4 \cdot \Delta x$$
$$\dfrac{-6}{-4} = \dfrac{-4 \cdot \Delta x}{-4}$$
$$\dfrac{3}{2} = \Delta x$$

25. $3 \text{ mi} \times 5280\frac{\text{ft}}{\text{mi}} = 15{,}840 \text{ ft}$
$$\dfrac{\Delta y}{15{,}840} = \dfrac{6}{100}$$
$$100 \cdot \Delta y = 95{,}040$$
$$\dfrac{100 \cdot \Delta y}{100} = \dfrac{95{,}040}{100}$$
$$\Delta y = 950.4 \text{ ft}$$

27. $m = 0.125 = \dfrac{125}{1000}$
$$\dfrac{2}{\Delta x} = \dfrac{125}{1000}$$
$$2000 = 125 \cdot \Delta x$$
$$\dfrac{2000}{125} = \dfrac{125 \cdot \Delta x}{125}$$
$$16 \text{ ft} = \Delta x$$

29. $m = \dfrac{\Delta y}{\Delta x} = \dfrac{30}{2} = 15$

31.

x	-2	0	3	4
y	56	32	-4	-16

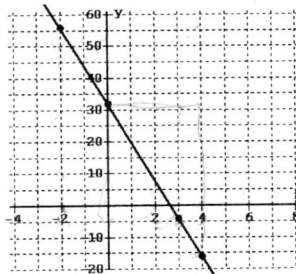

$$m = \dfrac{\Delta y}{\Delta x} = \dfrac{-24}{2} = -12$$

$\dfrac{-60}{5} = -12$

$\dfrac{-36}{3} = -12$

33. $m = \dfrac{\Delta y}{\Delta x} = \dfrac{2}{3}$

35. $m = \dfrac{\Delta y}{\Delta x} = \dfrac{50}{-30} = \dfrac{-5}{3}$

37a. $d = 25g$

b. $m = \dfrac{\Delta d}{\Delta g} = \dfrac{75 \text{ mi}}{3 \text{ gal}} = \dfrac{25 \text{ mi}}{1 \text{ gal}}$

c. The car gets 25 miles per gallon.

39a. $T = 4p$

b. $m = \dfrac{\Delta T}{\Delta p} = \dfrac{20 \text{ cents}}{5 \text{ dollars}} = \dfrac{4 \text{ cents}}{1 \text{ dollar}}$

c. The sales tax is 4 cents per dollar.

41. $m = \dfrac{\Delta y}{\Delta x} = \dfrac{0}{1} = 0$

43. $m = \dfrac{\Delta y}{\Delta x} = \dfrac{1}{0}$ undefined

45a. All the y-coordinates on a horizontal line are the same, so the change in y's is zero.

b. $m = \dfrac{\Delta y}{\Delta x} = \dfrac{0}{\Delta x} = 0$

Homework 3.5

1a.   I.

x	-1	0	1	2	3
y	-8	-6	-4	-2	0

II.

x	-1	0	1	2	3
y	-1	1	3	5	7

III.

x	-1	0	1	2	3
y	1	3	5	7	9

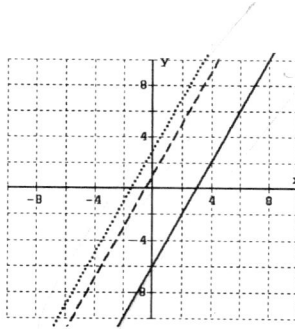

b.   I. $m = 2$

II. $m = 2$

III. $m = 2$

3a.   I.

x	-6	-4	-2	0	2
y	5	2	-1	-4	-7

II.

x	-6	-4	-2	0	2
y	11	8	5	2	-1

III.

x	-6	-4	-2	0	2
y	15	12	9	6	3

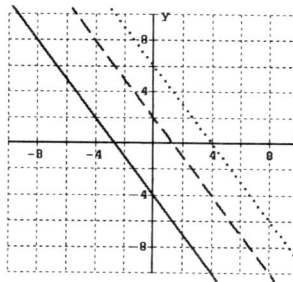

b.   I. $m = \dfrac{-3}{2}$

II. $m = \dfrac{-3}{2}$

III. $m = \dfrac{-3}{2}$

5a.   I.

x	-4	-2	0	2	4
y	1	$\frac{3}{2}$	2	$\frac{5}{2}$	3

II.

x	-4	-2	0	2	4
y	0	1	2	3	4

III.

x	-4	-2	0	2	4
y	-2	0	2	4	6

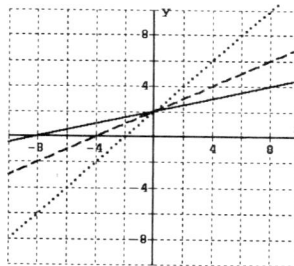

b.   I. $m = \dfrac{1}{4}$

II. $m = \dfrac{1}{2}$

III. $m = 1$

7a.   I.

x	-6	-3	0	3	6
y	16	7	-2	-11	-20

II.

x	-6	-3	0	3	6
y	10	4	-2	-8	-14

III.

x	-6	-3	0	3	6
y	8	3	-2	-7	-12

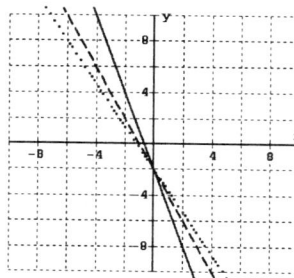

b.   I. $m = -3$

II. $m = -2$

III. $m = \dfrac{-5}{3}$

72

9. $y = 3x + 4;$ The slope is 3 and the y-intercept is $(0, 4)$.

11.

$6x + 3y = 5$	Subtract $6x$ from both sides

$$3y = -6x + 5 \quad \text{Divide both sides by 3}$$

$$\frac{3y}{3} = \frac{-6x + 5}{3} \quad \begin{array}{l}\text{Divide 3 into each} \\ \text{term on the right}\end{array}$$

$$y = \frac{-6x}{3} + \frac{5}{3} \quad \text{Simplify each quotient}$$

$$y = -2x + \frac{5}{3} \quad \text{The slope is } -2 \text{ and the } y\text{-intercept is } \left(0, \frac{5}{3}\right).$$

13.

$$2x - 3y = 6 \quad \text{Subtract } 2x \text{ from both sides}$$

$$-3y = -2x + 6 \quad \text{Divide both sides by } -3$$

$$\frac{-3y}{-3} = \frac{-2x + 6}{-3} \quad \begin{array}{l}\text{Divide } -3 \text{ into each} \\ \text{term on the right}\end{array}$$

$$y = \frac{-2x}{-3} + \frac{6}{-3} \quad \text{Simplify each quotient}$$

$$y = \frac{2}{3}x - 2 \quad \text{The slope is } \frac{2}{3} \text{ and the } y\text{-intercept is } (0, -2)$$

15.

$$5x = 4y \quad \text{Divide both sides by 4}$$

$$\frac{5x}{4} = \frac{4y}{4}$$

$$\frac{5}{4}x = y$$

$$y = \frac{5}{4}x \quad \text{The slope is } \frac{5}{4} \text{ and the } y\text{-intercept is } (0, 0)$$

17a. slope $= \frac{6}{2} = 3;$ y-intercept $= (0, -4)$ **b.** $y = 3x - 4$

19a. slope $= \frac{4}{6} = \frac{2}{3};$ y-intercept $= (0, -3)$ **b.** $y = \frac{2}{3}x - 3$

21a. slope $= \frac{-8}{5};$ y-intercept $= (0, 1)$ **b.** $y = \frac{-8}{5}x + 1$

23a.

x	y
0	3
4	0

b.

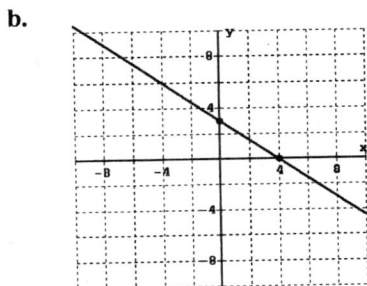

c. $\frac{\Delta y}{\Delta x} = \frac{-3}{4}$

d. $3x + 4y = 12$

$$4y = -3x + 12$$

$$\frac{4y}{4} = \frac{-3x + 12}{4}$$

$$y = \frac{-3x}{4} + \frac{12}{4}$$

$$y = \frac{-3}{4}x + 3$$

25a.

x	y
0	8
$\frac{8}{3}$	0

b.

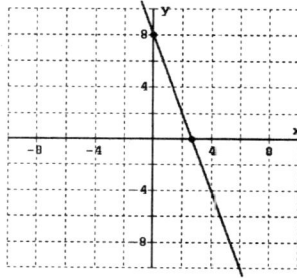

c. $\dfrac{\Delta y}{\Delta x} = \dfrac{-8}{\frac{8}{3}} = -8 \div \dfrac{8}{3} = \dfrac{-8}{1} \cdot \dfrac{3}{8} = -3$

d. $y + 3x - 8 = 0$

$\qquad y = -3x + 8$

27a. $3x - 5y = 0$

$\qquad -5y = -3x$

$\qquad \dfrac{-5y}{-5} = \dfrac{-3x}{-5}$

$\qquad\quad y = \dfrac{3}{5}x$

b. y-intercept $= (0, 0)$

$\quad$ slope $= \dfrac{3}{5}$

c. $(5, 3)$

$\quad (10, 6)$

$\quad (-5, -3)$

$\quad (-10, -6)$

d.

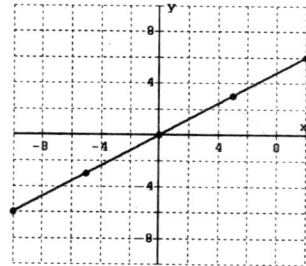

29a. $5x + 4y = 0$

$\qquad 4y = -5x$

$\qquad \dfrac{4y}{4} = \dfrac{-5x}{4}$

$\qquad\quad y = \dfrac{-5}{4}x$

b. y-intercept $= (0, 0)$

$\quad$ slope $= \dfrac{-5}{4}$

c. $(4, -5)$

$\quad (8, -10)$

$\quad (-4, 5)$

$\quad (-8, 10)$

d.

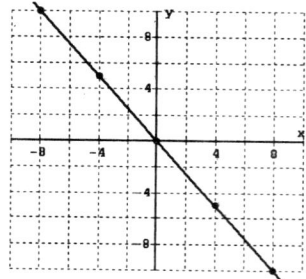

31. $y = 3 - x = -x + 3$

y-intercept $= (0, 3)$

slope $= -1 = \dfrac{-1}{1}$ or $\dfrac{1}{-1}$ $\dfrac{\text{Y}}{\text{X}}$

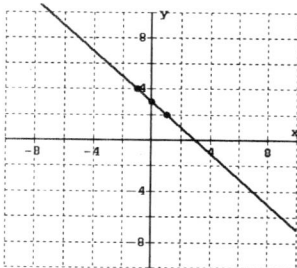

33. $y = 3x - 1$

y-intercept $= (0, -1)$

slope $= 3 = \dfrac{3}{1}$ or $\dfrac{-3}{-1}$

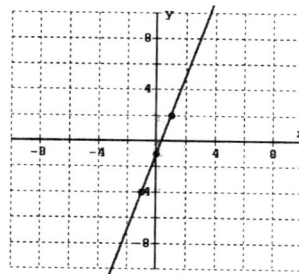

74

35. $y = \dfrac{3}{4}x + 2$

y-intercept $= (0, 2)$

slope $= \dfrac{3}{4}$ or $\dfrac{-3}{-4}$

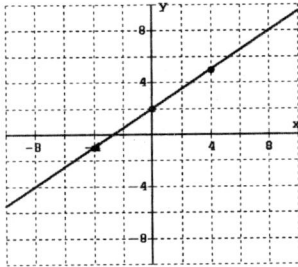

37. point $= (-1, -2)$

slope $= \dfrac{3}{2}$ or $\dfrac{-3}{-2}$

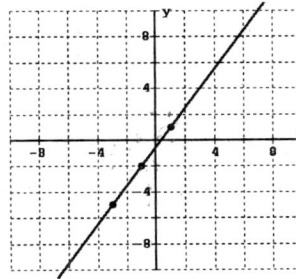

39. point $= (-2, 6)$

slope $= -2 = \dfrac{-2}{1}$ or $\dfrac{2}{-1}$

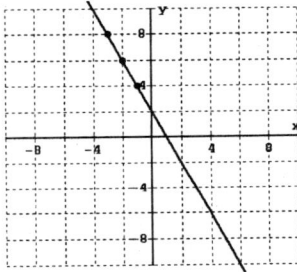

41. point $= (3, 4)$

slope $= \dfrac{-4}{3}$ or $\dfrac{4}{-3}$

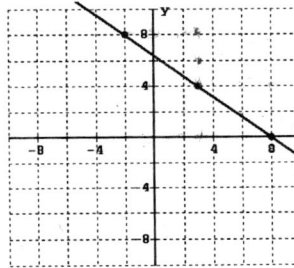

Midchapter 3 Review

1. A <u>ratio</u> is a comparison of two quantities. Written as a fraction, the ratio of a to b is $\frac{a}{b}$.
A <u>proportion</u> is a statement that two ratios are equal. $\frac{a}{b} = \frac{c}{d}$

2. Two variables are <u>proportional</u> if their ratios are always the same.

3. equal; proportional

4. Yes. If two pairs of corresponding angles in the triangles are equal, then the third pair also must be equal because the sum of the angles in each triangle is 180°. Thus, the triangles are similar.

5. $y = kx$

6. Plot the point $(-4, 1)$. Write the slope as a fraction $\frac{\Delta y}{\Delta x}$. Starting at $(-4, 1)$, move Δy units in the y-direction, then Δx units in the x-direction.

7. $y = mx + b$ For example, if $y = \frac{1}{2}x + 3$, the slope is $\frac{1}{2}$ and the y-intercept is $(0, 3)$.

8. A line whose <u>slope is undefined</u> is a <u>vertical</u> line. A line whose <u>slope is zero</u> is a <u>horizontal</u> line.

9. slope; zero

10. Yes. In the equation $y = a + bx$, the coefficient of x is the slope. Thus, the slope is b.

For example, $y = 5 + 2x$
$$y = 2x + 5 \quad \text{Slope-intercept form}$$
The slope is 2 (the coefficient of x).

11. 14 men
$35 - 14 = 21$ women
$$\frac{14}{21} = \frac{2}{3}$$

12. Zach's rate $= \dfrac{8.4 \text{ gal}}{210 \text{ mi}} = 0.04$ gal per mi

Tasha's rate $= \dfrac{8 \text{ gal}}{204 \text{ mi}} \approx 0.039$ gal per mi

Zach's rate of fuel consumption is higher.

13.
$$\frac{x}{25} = \frac{36}{5}$$
$$5x = 900$$
$$x = 180$$

14.
$$\frac{105}{y} = \frac{15}{17}$$
$$1785 = 15y$$
$$119 = y$$

15.
$$\frac{4}{7} = \frac{z-6}{z}$$
$$4z = 7z - 42$$
$$-3z = -42$$
$$z = 14$$

16.
$$\frac{a-3}{a+4} = \frac{34.8}{95.7}$$
$$95.7a - 287.1 = 34.8a + 139.2$$
$$60.9a - 287.1 = 139.2$$
$$60.9a = 426.3$$
$$a = 7$$

17.

Time	Cost	$\dfrac{\text{Cost}}{\text{Time}}$
5	1.50	$\frac{1.50}{5} = 0.30$
10	2.50	$\frac{2.50}{10} = 0.25$
15	3.50	$\frac{3.50}{15} \approx 0.233$
20	4.50	$\frac{4.50}{20} = 0.225$

Variables are not proportional.

18.

Speed	Distance	$\dfrac{\text{Distance}}{\text{Speed}}$
30	42	$\frac{42}{30} = 1.4$
40	56	$\frac{56}{40} = 1.4$
50	70	$\frac{70}{50} = 1.4$
60	84	$\frac{84}{60} = 1.4$

Variables are proportional.

19.
$$\frac{7.4 \text{ m}}{660 \text{ bricks}} = \frac{9.25 \text{ m}}{x \text{ bricks}}$$
$$7.4x = 6105$$
$$x = 825 \text{ bricks}$$

20. Longer piece: x
Shorter piece: $90 - x$
$$\frac{x}{90-x} = \frac{6}{2}$$
$$2x = 540 - 6x$$
$$8x = 540$$
$$x = 67.5 \text{ in.}$$
$$90 - x = 22.5 \text{ in.}$$

21a.
$$\frac{3 \text{ cm}}{5 \text{ mi}} = \frac{6 \text{ cm}}{l \text{ mi}} \qquad \frac{3 \text{ cm}}{5 \text{ mi}} = \frac{9 \text{ cm}}{w \text{ mi}}$$
$$3l = 30 \qquad\qquad 3w = 45$$
$$l = 10 \text{ mi} \qquad\quad w = 15 \text{ mi}$$

b.
Perimeter of town
$P = 2l + 2w$
$= 2(10) + 2(15)$
$= 20 + 30$
$= 50 \text{ mi}$

Perimeter on map
$P = 2l + 2w$
$= 2(6) + 2(9)$
$= 12 + 18$
$= 30 \text{ cm}$

d.
Area of town
$A = lw$
$= 10(15)$
$= 150 \text{ sq mi}$

Area on map
$A = lw$
$= 6(9)$
$= 54 \text{ sq cm}$

c.
$$\frac{\text{Perimeter of town}}{\text{Perimeter on map}} = \frac{50 \text{ mi}}{30 \text{ cm}} = \frac{5 \text{ mi}}{3 \text{ cm}}$$

e.
$$\frac{\text{Area of town}}{\text{Area on map}} = \frac{150 \text{ sq mi}}{54 \text{ sq cm}} = \frac{25 \text{ sq mi}}{9 \text{ sq cm}}$$

22a.

$$\frac{\frac{1}{3}\text{ in.}}{2\text{ mi}} = \frac{1\text{ in.}}{a\text{ mi}}$$
$$\frac{a}{3} = 2$$
$$a = 6\text{ mi}$$

$$\frac{\frac{1}{3}\text{ in.}}{2\text{ mi}} = \frac{\frac{4}{3}\text{ in.}}{b\text{ mi}}$$
$$\frac{b}{3} = \frac{8}{3}$$
$$b = 8\text{ mi}$$

$$\frac{\frac{1}{3}\text{ in.}}{2\text{ mi}} = \frac{\frac{5}{3}\text{ in.}}{c\text{ mi}}$$
$$\frac{c}{3} = \frac{10}{3}$$
$$c = 10\text{ mi}$$

b.

Perimeter of lake
$$P = a + b + c$$
$$= 6 + 8 + 10$$
$$= 24\text{ mi}$$

Perimeter on map
$$P = a + b + c$$
$$= 1 + \frac{4}{3} + \frac{5}{3}$$
$$= \frac{3}{3} + \frac{4}{3} + \frac{5}{3}$$
$$= \frac{12}{3} = 4\text{ in.}$$

d.

Area of lake
$$A = \frac{1}{2}bh$$
$$= \frac{1}{2}(6)(8)$$
$$= 24\text{ sq mi}$$

Area on map
$$A = \frac{1}{2}bh$$
$$= \frac{1}{2}(1)(\frac{4}{3})$$
$$= \frac{2}{3}\text{ sq in.}$$

c.

$$\frac{\text{Perimeter of lake}}{\text{Perimeter on map}} = \frac{24\text{ mi}}{4\text{ in.}} = \frac{6\text{ mi}}{1\text{ in.}}$$

e.

$$\frac{\text{Area of lake}}{\text{Area on map}} = \frac{24\text{ sq mi}}{\frac{2}{3}\text{ sq in.}} = 24 \div \frac{2}{3}$$
$$= \frac{24}{1} \cdot \frac{3}{2} = \frac{36\text{ sq mi}}{1\text{ sq in.}}$$

23.
$$\frac{x}{x+6} = \frac{2}{6}$$
$$6x = 2x + 12$$
$$4x = 12$$
$$x = 3$$

24.
$$\frac{x}{x-6} = \frac{9}{6}$$
$$6x = 9x - 54$$
$$-3x = -54$$
$$x = 18$$

25.
$$m = 0.516 = \frac{516}{1000}$$
$$\frac{\Delta y}{11,380} = \frac{516}{1000}$$
$$1000 \cdot \Delta y = 5,872,080$$
$$\frac{1000 \cdot \Delta y}{1000} = \frac{5,872,080}{1000}$$
$$\Delta y \approx 5872$$

Elevation at top of tram $= \Delta y + 2643$
$$= 5872 + 2643$$
$$= 8515\text{ ft}$$

26.
$$m = 0.32 = \frac{32}{100}, \quad \Delta x = \frac{1}{2}(1132) = 566$$
$$\frac{\Delta y}{566} = \frac{32}{100}$$
$$100 \cdot \Delta y = 18,112$$
$$\frac{100 \cdot \Delta y}{100} = \frac{18,112}{100}$$
$$\Delta y = 181.12\text{ ft}$$

27a.

x	y
0	-5
4	0

b. $m = \dfrac{5}{4}$

c. $y = \dfrac{5}{4}x - 5$

d.

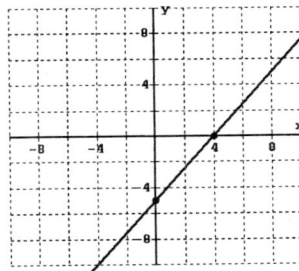

28a.

x	y
0	5
−3	0

b. $m = \dfrac{5}{3}$

c. $y = \dfrac{5}{3}x + 5$

d.

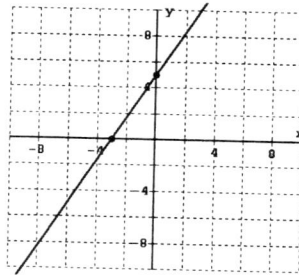

29a. $m = \dfrac{\Delta y}{\Delta x} = \dfrac{-3}{5}$

y-intercept $= (0,3)$

b. $y = \dfrac{-3}{5}x + 3$

30a. $m = \dfrac{\Delta y}{\Delta x} = \dfrac{-5}{3}$

y-intercept $= (0,0)$

b. $y = \dfrac{-5}{3}x$

31. The slope of -5000 shows that the water is draining at 5000 gallons per hour.
The y-intercept of $(0, 500,000)$ shows that the pool started with 500,000 gallons of water.

32. Using 6 in. $= \frac{1}{2}$ ft, $\quad h = \frac{1}{2}t + 3$

33.

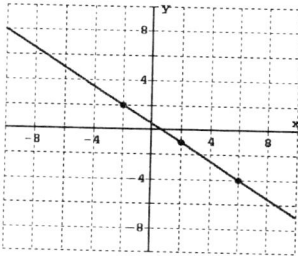

34.

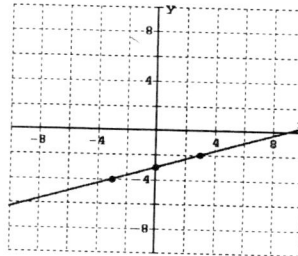

Homework 3.6

1. False

3. True

5. False

7. True

9. Check $(4, -2)$:
$$x + 2y = -8$$
$$4 + 2(-2) \overset{?}{=} -8$$
$$4 + (-4) \overset{?}{=} -8$$
$$0 = -8 \quad \text{False}$$

Check $(4, -2)$:
$$2x - y = 4$$
$$2(4) - (-2) \overset{?}{=} 4$$
$$8 + 2 \overset{?}{=} 4$$
$$10 = 4 \quad \text{False}$$

Therefore, $(4, -2)$ is not a solution.

11. Check $(3, -2)$:
$$x = 5y + 13$$
$$3 \overset{?}{=} 5(-2) + 13$$
$$3 \overset{?}{=} -10 + 13$$
$$3 = 3 \quad \text{True}$$

Check $(3, -2)$:
$$2x + 3y = 0$$
$$2(3) + 3(-2) \overset{?}{=} 0$$
$$6 + (-6) \overset{?}{=} 0$$
$$0 = 0 \quad \text{True}$$

Therefore, $(3, -2)$ is a solution.

13. Intercept method because the equation is in general form, $Ax + By = C$.

15. Slope-intercept method because the equation is in slope-intercept form, $y = mx + b$

78

17. Slope-intercept method because both intercepts are the same point $(0, 0)$.

19. 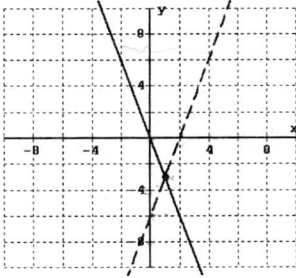 $(1, -3)$

21. 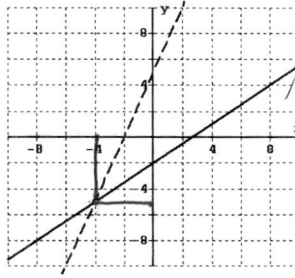 $(-4, -5)$

23. 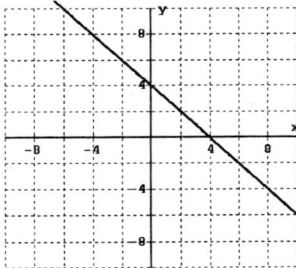 Dependent or infinitely many solutions

25. 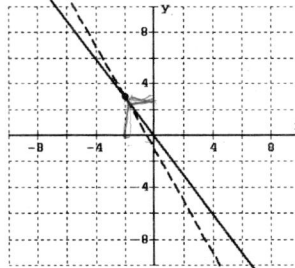 $(-2, 3)$

27. From their slope-intercept forms, both lines have slope 3, but the y-intercepts are different, so their graphs are parallel lines.

29a. Plan A: $y = 0.03x + 20,000$
Plan B: $y = 0.05x + 15,000$

b.

x	Earnings Under Plan A	Earnings Under Plan B
0	20,000	15,000
50	21,500	17,500
100	23,000	20,000
150	24,500	22,500
200	26,000	25,000
250	27,500	27,500
300	29,000	30,000
350	30,500	32,500
400	32,000	35,000

c.

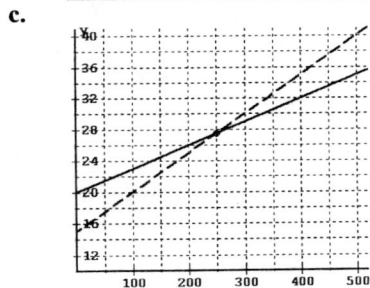

d. $250,000

e. He should prefer Plan A if he expects less than $250,000 in sales.

f. More than $300,000;
More than $333,333

g.
$$0.05x + 15,000 > 30,000$$
$$\underline{-15,000 \qquad -15,000}$$
$$0.05x \qquad\qquad > \qquad 15,000$$
$$\frac{0.05x}{0.05} > \frac{15,000}{0.05}$$
$$x > 300,000$$

$$0.03x + 20,000 > 30,000$$
$$\underline{-20,000 \qquad -20,000}$$
$$0.03x \qquad\qquad > \qquad 10,000$$
$$\frac{0.03x}{0.03} > \frac{10,000}{0.03}$$
$$x > 333,333$$

31a. Cost: $y = 60x + 6000$
Revenue: $y = 80x$

b.

x	Cost	Revenue
0	6,000	0
50	9,000	4,000
100	12,000	8,000
150	15,000	12,000
200	18,000	16,000
250	21,000	20,000
300	24,000	24,000
350	27,000	28,000
400	30,000	32,000

c.

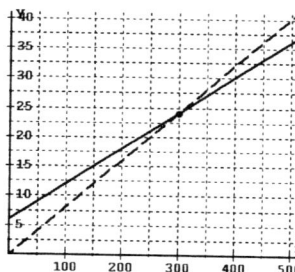

d. More than 250 clarinets

e. 300 clarinets

f. Profit = Revenue − Cost

$$\underset{\text{selling } 500}{\text{Profit}} = 40,000 - 36,000$$
$$= \$4000$$
$$\underset{\text{selling } 200}{\text{Profit}} = 16,000 - 18,000$$
$$= -\$2000 \text{ or a loss of } \$2000$$

33. Cost of one rose: r
Cost of one carnation: c
First bouquet: $4r + 8c = 14$
Second bouquet: $6r + 9c = 18$

$(1.5, 1)$

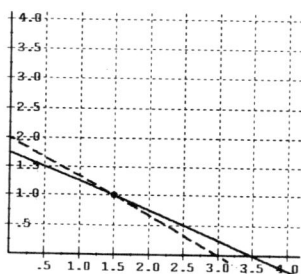

A rose costs $1.50, and a carnation costs $1.00.

Check $(1.5, 1)$:
$$4r + 8c = 14$$
$$4(1.5) + 8(1) \overset{?}{=} 14$$
$$6 + 8 \overset{?}{=} 14$$
$$14 = 14 \quad \text{True}$$

Check $(1.5, 1)$:
$$6r + 9c = 18$$
$$6(1.5) + 9(1) \overset{?}{=} 18$$
$$9 + 9 \overset{?}{=} 18$$
$$18 = 18 \quad \text{True}$$

35. Measure of base angle: b
Measure of vertex angle: v
First equation: $v = b - 15$
Second equation: $2b + v = 180$

$(65, 50)$

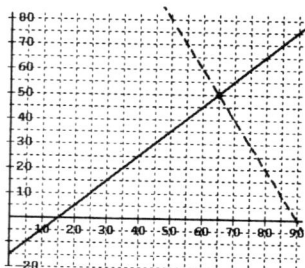

The base angles are 65°, and the vertex angle is 50°.

Check $(65, 50)$:
$$v = b - 15$$
$$50 \overset{?}{=} 65 - 15$$
$$50 = 50 \quad \text{True}$$

Check $(65, 50)$:
$$2b + v = 180$$
$$2(65) + 50 \overset{?}{=} 180$$
$$130 + 50 \overset{?}{=} 180$$
$$180 = 180 \quad \text{True}$$

Homework 3.7

1. Substitute $2x$ for y into Equation 2:

$$3x + y = 10$$
$$3x + 2x = 10$$
$$5x = 10$$
$$x = 2$$

Substitute 2 for x into:

$$y = 2x$$
$$y = 2(2)$$
$$y = 4$$

The solution is $(2, 4)$.

3. Substitute $2a + 4$ for b into Equation 1:

$$2a + 3b = 4$$
$$2a + 3(2a + 4) = 4$$
$$2a + 6a + 12 = 4$$
$$8a + 12 = 4$$
$$8a = -8$$
$$a = -1$$

Substitute -1 for a into:

$$b = 2a + 4$$
$$b = 2(-1) + 4$$
$$b = -2 + 4$$
$$b = 2$$

The solution is $(-1, 2)$.

5. Solve Equation 2 for x:

$$x + 1 = 3y$$
$$x = 3y - 1$$

Substitute $3y - 1$ for x into Equation 1:

$$2x - 3y = 4$$
$$2(3y - 1) - 3y = 4$$
$$6y - 2 - 3y = 4$$
$$3y - 2 = 4$$
$$3y = 6$$
$$y = 2$$

Substitute 2 for y into:

$$x = 3y - 1$$
$$x = 3(2) - 1$$
$$x = 6 - 1$$
$$x = 5$$

The solution is $(5, 2)$.

7. Solve Equation 1 for b:

$$2a + b = 16$$
$$b = 16 - 2a$$

Substitute $16 - 2a$ for b into Equation 2:

$$4a - 4 = -3b$$
$$4a - 4 = -3(16 - 2a)$$
$$4a - 4 = -48 + 6a$$
$$-4 = -48 + 2a$$
$$44 = 2a$$
$$22 = a$$

Substitute 22 for a into:

$$b = 16 - 2a$$
$$b = 16 - 2(22)$$
$$b = 16 - 44$$
$$b = -28$$

The solution is $(22, -28)$.

9. Solve Equation 1 for s:

$$8r + s = 4$$
$$s = 4 - 8r$$

Substitute $4 - 8r$ for s into Equation 2:

$$2r + 7s = -8$$
$$2r + 7(4 - 8r) = -8$$
$$2r + 28 - 56r = -8$$
$$-54r + 28 = -8$$
$$-54r = -36$$
$$r = \frac{-36}{-54}$$
$$r = \frac{2}{3}$$

Substitute $\frac{2}{3}$ for r into:

$$s = 4 - 8r$$
$$s = 4 - 8\left(\frac{2}{3}\right)$$
$$s = 4 - \frac{16}{3}$$
$$s = \frac{12}{3} - \frac{16}{3}$$
$$s = \frac{-4}{3}$$

The solution is $\left(\frac{2}{3}, \frac{-4}{3}\right)$.

11. Solve Equation 1 for b:

$$36a = 6 - 3b$$
$$36a - 6 = -3b$$
$$\frac{36a - 6}{-3} = \frac{-3b}{-3}$$
$$\frac{36a}{-3} + \frac{-6}{-3} = b$$
$$-12a + 2 = b$$

Substitute $-12a + 2$ for b into Equation 2:

$$2b = 3a - 5$$
$$2(-12a + 2) = 3a - 5$$
$$-24a + 4 = 3a - 5$$
$$4 = 27a - 5$$
$$9 = 27a$$
$$\frac{9}{27} = a$$
$$\frac{1}{3} = a$$

Substitute $\frac{1}{3}$ for a into:

$$b = -12a + 2$$
$$b = -12\left(\frac{1}{3}\right) + 2$$
$$b = -4 + 2$$
$$b = -2$$

The solution is $\left(\frac{1}{3}, -2\right)$.

13. Calories in hamburger: h
Calories in shake: s

Equation 1: $h + s = 1030$
Equation 2: $2s + 3h = 2710$

Solve Equation 1 for s:
$$h + s = 1030$$
$$s = 1030 - h$$

Substitute $1030 - h$ for s into Equation 2:
$$2s + 3h = 2710$$
$$2(1030 - h) + 3h = 2710$$
$$2060 - 2h + 3h = 2710$$
$$2060 + h = 2710$$
$$h = 650$$

Substitute 650 for h into:
$$s = 1030 - h$$
$$s = 1030 - 650$$
$$s = 380$$

A hamburger has 650 calories, and a shake has 380 calories.

15. Number of chairs: c
Number of tables: t

Equation 1: $175c + 250t = 11{,}400$
Equation 2: $c = 4t$

Substitute $4t$ for c into Equation 1:
$$175c + 250t = 11{,}400$$
$$175(4t) + 250t = 11{,}400$$
$$700t + 250t = 11{,}400$$
$$950t = 11{,}400$$
$$t = 12$$

Substitute 12 for t into:
$$c = 4t$$
$$c = 4(12)$$
$$c = 48$$

He can buy 12 tables and 48 chairs.

17. Add the equations:
$$x + y = 5$$
$$\underline{x - y = 1}$$
$$2x = 6$$
$$x = 3$$

Substitute 3 for x into:
$$x + y = 5$$
$$3 + y = 5$$
$$y = 2$$
The solution is $(3, 2)$.

19. Multiply Equation 2 by -1, then add:

why

$$2a + b = 3 \quad \Rightarrow \quad 2a + b = 3$$
$$-1(a + b = 2) \quad \Rightarrow \quad \underline{-a - b = -2}$$
$$a = 1$$

Substitute 1 for a into:
$$a + b = 2$$
$$1 + b = 2$$
$$b = 1$$
The solution is $(1, 1)$.

21. Multiply Equation 2 by -2, then add:

why

$$3x + 2y = 7 \quad \Rightarrow \quad 3x + 2y = 7$$
$$-2(x + y = 3) \quad \Rightarrow \quad \underline{-2x - 2y = -6}$$
$$x = 1$$

Substitute 1 for x into:
$$x + y = 3$$
$$1 + y = 3$$
$$y = 2$$
The solution is $(1, 2)$.

23. Write each equation in general form:
$$3a = 1 + 5b \qquad\qquad 2b = 14 + 6a$$
$$3a - 5b = 1 \qquad\qquad -6a + 2b = 14$$

Multiply Equation 1 by 2, then add:
$$2(3a - 5b = 1) \quad \Rightarrow \quad 6a - 10b = 2$$
$$-6a + 2b = 14 \quad \Rightarrow \quad \underline{-6a + 2b = 14}$$
$$-8b = 16$$
$$b = -2$$

Substitute -2 for b into:
$$3a = 1 + 5b$$
$$3a = 1 + 5(-2)$$
$$3a = 1 + (-10)$$
$$3a = -9$$
$$a = -3$$
The solution is $(-3, -2)$.

25. Multiply Equation 1 by -3 and Equation 2 by 2:
$$-3(2x + 3y = -1) \quad \Rightarrow \quad -6x - 9y = 3$$
$$2(3x + 5y = -2) \quad \Rightarrow \quad \underline{6x + 10y = -4}$$
$$y = -1$$

Substitute -1 for y into:
$$2x + 3y = -1$$
$$2x + 3(-1) = -1$$
$$2x - 3 = -1$$
$$2x = 2$$
$$x = 1$$
The solution is $(1, -1)$.

27. Write each equation in general form:
$$5z = 1 - 3w \qquad\qquad 5w = 2 - 7z$$
$$3w + 5z = 1 \qquad\qquad 5w + 7z = 2$$

Multiply Equation 1 by 5 and Equation 2 by -3:
$$5(3w + 5z = 1) \quad \Rightarrow \quad 15w + 25z = 5$$
$$-3(5w + 7z = 2) \quad \Rightarrow \quad \underline{-15w - 21z = -6}$$
$$4z = -1$$
$$z = \frac{-1}{4}$$

Substitute $\frac{-1}{4}$ for z into:
$$5w = 2 - 7z$$
$$5w = 2 - 7\left(\frac{-1}{4}\right)$$
$$5w = \frac{8}{4} + \frac{7}{4}$$
$$5w = \frac{15}{4}$$
$$\frac{1}{5} \cdot 5w = \frac{15}{4} \cdot \frac{1}{5}$$
$$w = \frac{3}{4}$$
The solution is $\left(\frac{3}{4}, \frac{-1}{4}\right)$.

29. Length: L Equation 1: $2L + 2W = 42$
Width: W Equation 2: $W = L - 13$

Write Equation 2 in general form:
$$W = L - 13$$
$$-L + W = -13$$

Multiply Equation 2 by 2, then add:
$$2L + 2W = 42 \quad \Rightarrow \quad 2L + 2W = 42$$
$$2(-L + W = -13) \quad \Rightarrow \quad \underline{-2L + 2W = -26}$$
$$4W = 16$$
$$W = 4$$

Substitute 4 for W into:
$$W = L - 13$$
$$4 = L - 13$$
$$17 = L$$

The rectangle is 17 m by 4 m.

31. Cost of bacon: b Equation 1: $3b + 2c = 17.80$
Cost of coffee: c Equation 2: $2b + 5c = 32.40$

Multiply Equation 1 by -2 and Equation 2 by 3:
$$-2(3b + 2c = 17.80) \quad \Rightarrow \quad -6b - 4c = -35.60$$
$$3(2b + 5c = 32.40) \quad \Rightarrow \quad \underline{6b + 15c = 97.20}$$
$$11c = 61.60$$
$$c = 5.60$$

Substitute 5.60 for c into:
$$3b + 2c = 17.80$$
$$3b + 2(5.60) = 17.80$$
$$3b + 11.20 = 17.80$$
$$3b = 6.60$$
$$b = 2.20$$

Bacon costs \$2.20 per lb, and coffee costs \$5.60 per lb.

33. Multiply Equation 2 by -3, then add:
$$3x - 6y = 6 \quad \Rightarrow \quad 3x - 6y = 6$$
$$-3(x - 2y = 3) \quad \Rightarrow \quad \underline{-3x + 6y = -9}$$
$$0 = -3 \quad \text{False} \qquad \text{Therefore, the system is inconsistent.}$$

35. Write each equation in general form:

$$8a = 6 + 12b \qquad\qquad 4 = 6a - 9b$$
$$8a - 12b = 6 \qquad\qquad -6a + 9b = -4$$

Multiply Equation 1 by 3 and Equation 2 by 4:

$$3(8a - 12b = 6) \quad \Rightarrow \quad 24a - 36b = 18$$
$$4(-6a + 9b = -4) \quad \Rightarrow \quad \underline{-24a + 36b = -16}$$
$$0 = -2 \quad \text{False} \qquad \text{Therefore, the system is inconsistent.}$$

37. Write each equation in general form:

$$3p = 1 + 7q \qquad\qquad 21q = 9p - 3$$
$$3p - 7q = 1 \qquad\qquad -9p + 21q = -3$$

Multiply Equation 1 by 3, then add:

$$3(3p - 7q = 1) \quad \Rightarrow \quad 9p - 21q = 3$$
$$-9p + 21q = -3 \quad \Rightarrow \quad \underline{-9p + 21q = -3}$$
$$0 = 0 \quad \text{True} \qquad \text{Therefore, the system is dependent.}$$

39. Elimination   **41.** Elimination

Homework 3.8

1a. Amount in bonds: x **b.** Interest from bonds: $0.065x$
Amount in mutuals: y Interest from mutuals: $0.118y$
$x + y = 150,000$ **c.** $0.065x + 0.118y = 12,930$

d. $1000(0.065x + 0.118y = 12,930) \quad \Rightarrow \quad 65x + 118y = 12,930,000$ Substitute 60,000 for y into:
$-65(x + y = 150,000) \quad \Rightarrow \quad \underline{-65x - 65y = -9,750,000}$ $x + y = 150,000$
$53y = 3,180,000$ $x + 60,000 = 150,000$
$y = 60,000$ $x = 90,000$

They invested $90,000 in bonds and $60,000 in mutual funds.

3a.

	Principal	Rate	Interest
First Loan	x	0.12	$0.12x$
Second Loan	y	0.15	$0.15y$

b.
$$x + y = 30,000$$
$$0.12x + 0.15y = 3750$$

c. $100(0.12x + 0.15y = 3750) \quad \Rightarrow \quad 12x + 15y = 375,000$ Substitute 5,000 for y into:
$-12(x + y = 30,000) \quad \Rightarrow \quad \underline{-12x - 12y = -360,000}$ $x + y = 30,000$
$3y = 15,000$ $x + 5,000 = 30,000$
$y = 5,000$ $x = 25,000$

He borrowed $25,000 at 12% and $5,000 at 15%.

5.

	Principal	Rate	Interest
Car Loan	x	0.07	$0.07x$
Student Loan	y	0.04	$0.04y$

Substitute $2y$ for x into $0.07x = 0.04y + 500$ Substitute 5000 for y into $x = 2y$
$0.07(2y) = 0.04y + 500$ $x = 2(5000)$
$0.14y = 0.04y + 500$ $x = 10,000$
$(10)0.10y = 500$
$y = 5000$

$x = 2y$
$0.07x = 0.04y + 500$

He borrowed $10,000 on his car loan and $5000 on his student loan.

84

7a. $0.35(80) = 28$ women
$0.15(60) = 9$ women

b. $80 + 60 = 140$ students
$28 + 9 = 37$ women

c. $\dfrac{37}{140} \approx 26.4\%$

9a. $0.50(30) = 15$ mℓ
$0.15(12) = 1.8$ mℓ

b. $15 + 1.8 = 16.8$ mℓ
$30 + 12 = 42$ mℓ

c. $\dfrac{16.8}{42} = 40\%$

d.

	Number of Milliliters	Percent Acid	Amount of Acid
50% Solution	30	0.50	15
15% Solution	12	0.15	1.8
Mixture	42	0.40	16.8

11a.

	Number of Liters	Percent Salt	Amount of Salt
12% Solution	x	0.12	$0.12x$
30% Solution	y	0.30	$0.30y$
Mixture	45	0.24	$0.24(45)$

b.
$x + y = 45$
$0.12x + 0.30y = 0.24(45)$

c.
$100(0.12x + 0.30y = 0.24(45)) \Rightarrow 12x + 30y = 1080$
$-12(x + y = 45) \Rightarrow \underline{-12x - 12y = -540}$
$18y = 540$
$y = 30$

Substitute 30 for y into:
$x + y = 45$
$x + 30 = 45$
$x = 15$

He needs 15 liters of the 12% solution and 30 liters of the 30% solution.

13.

	Number of People	Percent in Favor	People in Favor
Men	x	0.50	$0.50x$
Women	y	0.70	$0.70y$
Mixture	400	0.58	$0.58(400)$

$x + y = 400$
$0.50x + 0.70y = 0.58(400)$

$100(0.50x + 0.70y = 0.58(400)) \Rightarrow 50x + 70y = 23200$
$-50(x + y = 400) \Rightarrow \underline{-50x - 50y = -20000}$
$20y = 3200$
$y = 160$

160 women were polled.

15a.

	Rate	Time	Distance
Delbert	x	6	$6x$
Francine	y	6	$6y$

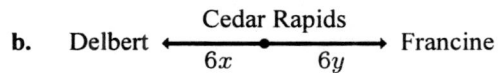

b.
Cedar Rapids
Delbert ←——— $6x$ —•— $6y$ ———→ Francine

c.
$y = x - 5$
$6x + 6y = 570$

d. Substitute $x - 5$ for y into
$6x + 6y = 570$
$6x + 6(x - 5) = 570$
$6x + 6x - 30 = 570$
$12x - 30 = 570$
$12x = 600$
$x = 50$

Substitute 50 for x into $y = x - 5$
$y = 50 - 5$
$y = 45$

Delbert's speed is 50 mph, and Francine's speed is 45 mph.

17a. Dallas

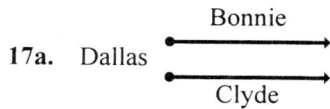

Bonnie

Clyde

b.

	Rate	Time	Distance
Bonnie	40	t	d
Clyde	70	$t-3$	d

c.

$$40t = d$$
$$70(t-3) = d$$

d. Substitute $40t$ for d into $70(t-3) = d$

$$70(t-3) = 40t$$
$$70t - 210 = 40t$$
$$-210 = -30t$$
$$7 = t$$

Substitute 7 for t into

$$40t = d$$
$$40(7) = d$$
$$280 = d$$

Bonnie drove for 7 hr and traveled 280 mi.

19. San Diego

Yacht

Cutter

	Rate	Time	Distance
Yacht	25	t	d
Cutter	40	$t-6$	d

$$25t = d$$
$$40(t-6) = d$$

Substitute $25t$ for d into $40(t-6) = d$

$$40(t-6) = 25t$$
$$40t - 240 = 25t$$
$$-240 = -15t$$
$$16 = t$$

Substitute 16 for t into

$$25t = d$$
$$25(16) = d$$
$$400 = d$$

The cutter's time is $t - 6 = 16 - 6 = 10$ hr, and it traveled 400 mi.

21.

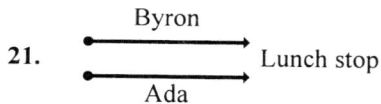

Byron

Lunch stop

Ada

	Rate	Time	Distance
Byron	10	t	d
Ada	12	$t-\frac{2}{3}$	d

$$10t = d$$
$$12(t-\tfrac{2}{3}) = d$$

Substitute $10t$ for d into $12(t-\frac{2}{3}) = d$

$$12(t-\tfrac{2}{3}) = 10t$$
$$12t - 8 = 10t$$
$$-8 = -2t$$
$$4 = t$$

Substitute 4 for t into

$$10t = d$$
$$10(4) = d$$
$$40 = d$$

The tour rode 40 mi.

23a. $g = 0.40t + 0.60(72) = 0.40t + 43.2$

b. $g = 0.40(65) + 43.2 = 26 + 43.2 = 69.2$
$g = 0.40(80) + 43.2 = 32 + 43.2 = 75.2$

c. $t = 92$

d. $g = 0.40(92) + 43.2 = 36.8 + 43.2 = 80$

Homework 3.9

1. $\dfrac{10-2}{2-8} = \dfrac{8}{-6} = \dfrac{-4}{3}$

3. $\dfrac{-5-(-5)}{2-9} = \dfrac{-5+5}{2-9} = \dfrac{0}{-7} = 0$

5.
$$\frac{3}{2}(4-7)+\frac{1}{2}$$
$$=\frac{3}{2}(-3)+\frac{1}{2}$$
$$=\frac{-9}{2}+\frac{1}{2}$$
$$=\frac{-8}{2}$$
$$=-4$$

7.
$$m=\frac{y_2-y_1}{x_2-x_1}$$
$$=\frac{7-2}{8-5}$$
$$=\frac{5}{3}$$

9.
$$m=\frac{y_2-y_1}{x_2-x_1}$$
$$=\frac{1-(-2)}{0-3}$$
$$=\frac{3}{-3}$$
$$=-1$$

11.
$$m=\frac{y_2-y_1}{x_2-x_1}$$
$$=\frac{-3-(-2)}{-3-6}$$
$$=\frac{-1}{-9}$$
$$=\frac{1}{9}$$

13.

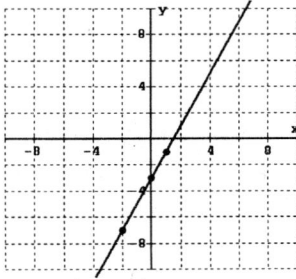

a. $(0,-3)$
b. $(-2,-7)$

15.

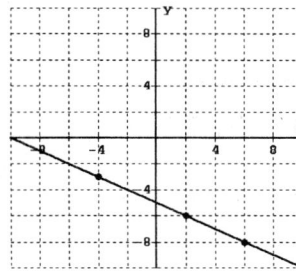

a. $(6,-8)$
b. $(-4,-3)$

17.
$$\frac{y-y_1}{x-x_1}=m \qquad \text{Substitute } m=-2,$$
$$y_1=4, \text{ and } x_1=-3$$
$$\frac{y-4}{x-(-3)}=-2 \qquad \text{Simplify}$$
$$\frac{y-4}{x+3}=\frac{-2}{1} \qquad \text{Cross-multiply}$$
$$1(y-4)=-2(x+3) \qquad \text{Distribute}$$
$$y-4=-2x-6 \qquad \text{Solve for } y$$
$$y=-2x-2$$

19.
$$\frac{y-y_1}{x-x_1}=m \qquad \text{Substitute } m=\tfrac{1}{2},$$
$$y_1=-3, \text{ and } x_1=4$$
$$\frac{y-(-3)}{x-4}=\frac{1}{2} \qquad \text{Simplify}$$
$$\frac{y+3}{x-4}=\frac{1}{2} \qquad \text{Cross-multiply}$$
$$2(y+3)=1(x-4) \qquad \text{Distribute}$$
$$2y+6=x-4 \qquad \text{Solve for } y$$
$$2y=x-10$$
$$\frac{2y}{2}=\frac{x}{2}-\frac{10}{2}$$
$$y=\frac{1}{2}x-5$$

21.
$$\frac{y-y_1}{x-x_1}=m \qquad \text{Substitute } m=\tfrac{-2}{3},$$
$$y_1=2, \text{ and } x_1=-6$$
$$\frac{y-2}{x-(-6)}=\frac{-2}{3} \qquad \text{Simplify}$$
$$\frac{y-2}{x+6}=\frac{-2}{3} \qquad \text{Cross-multiply}$$
$$3(y-2)=-2(x+6) \qquad \text{Distribute}$$
$$3y-6=-2x-12 \qquad \text{Solve for } y$$
$$3y=-2x-6$$
$$\frac{3y}{3}=\frac{-2x}{3}-\frac{6}{3}$$
$$y=\frac{-2}{3}x-2$$

87

23. $m = \dfrac{3}{5}; \ (0,-7)$

25. $m = 3; \ (-5,2)$

27. $m = \dfrac{4}{5}; \ (0,0)$

29a.

x	y
0	7
5	0

b.
$$m = \frac{y_2 - y_1}{x_2 - x_1}$$
$$= \frac{0 - 7}{5 - 0}$$
$$= \frac{-7}{5}$$

31a.

x	y
0	1.6
-2.4	0

b.
$$m = \frac{y_2 - y_1}{x_2 - x_1}$$
$$= \frac{0 - 1.6}{-2.4 - 0}$$
$$= \frac{-1.6}{-2.4}$$
$$= \frac{2}{3}$$

33a.

x	y
0	$\frac{7}{2}$
$\frac{1}{3}$	0

b.
$$m = \frac{y_2 - y_1}{x_2 - x_1}$$
$$= \frac{0 - \frac{7}{2}}{\frac{1}{3} - 0}$$
$$= \frac{\frac{-7}{2}}{\frac{1}{3}}$$
$$= \frac{-7}{2} \cdot \frac{3}{1}$$
$$= \frac{-21}{2}$$

Homework 3.10

1.
$$3x - 4y = 2$$
$$-4y = -3x + 2$$
$$\frac{-4y}{-4} = \frac{-3x}{-4} + \frac{2}{-4}$$
$$y = \frac{3}{4}x - \frac{1}{2}$$
$$\text{Slope} = \frac{3}{4}$$

$$8y - 6x = 6$$
$$8y = 6x + 6$$
$$\frac{8y}{8} = \frac{6x}{8} + \frac{6}{8}$$
$$y = \frac{3}{4}x + \frac{3}{4}$$
$$\text{Slope} = \frac{3}{4}$$

Since $m_1 = m_2$, the lines are parallel.

3.
$$2x = 4 - 5y$$
$$5y = -2x + 4$$
$$\frac{5y}{5} = \frac{-2x}{5} + \frac{4}{5}$$
$$y = \frac{-2}{5}x + \frac{4}{5}$$
$$\text{Slope} = \frac{-2}{5}$$

$$2y = 4x - 5$$
$$\frac{2y}{2} = \frac{4x}{2} - \frac{5}{2}$$
$$y = 2x - \frac{5}{2}$$
$$\text{Slope} = 2$$

The lines are neither.

5.
$$y + 3x - 2 = 0$$
$$y = -3x + 2$$
$$\text{Slope} = -3$$

$$x - 3y + 2 = 0$$
$$-3y = -x - 2$$
$$\frac{-3y}{-3} = \frac{-x}{-3} + \frac{-2}{-3}$$
$$y = \frac{1}{3}x + \frac{2}{3}$$
$$\text{Slope} = \frac{1}{3}$$

Since $m_1 m_2 = -1$, the lines are perpendicular.

7. $y = -5$

9. $x = 2$

11. $x = -8$

13. $y = 0$

15a. -3

b. -3

c.
$$\frac{y - y_1}{x - x_1} = m$$
$$\frac{y - 3}{x - 1} = -3$$
$$\frac{y - 3}{x - 1} = \frac{-3}{1}$$
$$1(y - 3) = -3(x - 1)$$
$$y - 3 = -3x + 3$$
$$y = -3x + 6$$

17a. -2

b. $\frac{1}{2}$

c.
$$\frac{y - y_1}{x - x_1} = m$$
$$\frac{y - 2}{x - 4} = \frac{1}{2}$$
$$2(y - 2) = 1(x - 4)$$
$$2y - 4 = x - 4$$
$$2y = x$$
$$\frac{2y}{2} = \frac{x}{2}$$
$$y = \frac{1}{2}x$$

19a.

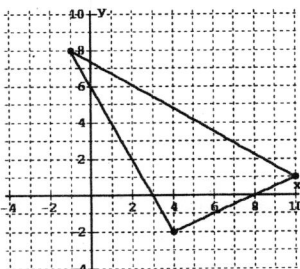

b. Using $B(-1, 8)$ and $C(4, -2)$:
$$m_{BC} = \frac{y_2 - y_1}{x_2 - x_1}$$
$$= \frac{-2 - 8}{4 - (-1)}$$
$$= \frac{-10}{5}$$
$$= -2$$

Using $C(4, -2)$ and $D(10, 1)$:
$$m_{CD} = \frac{y_2 - y_1}{x_2 - x_1}$$
$$= \frac{1 - (-2)}{10 - 4}$$
$$= \frac{3}{6}$$
$$= \frac{1}{2}$$

Since $m_{BC} m_{CD} = -1$, the lines are perpendicular.

21a.

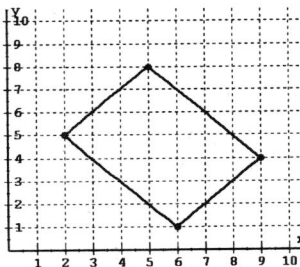

b. From the graph $Z(5, 8)$.

c. Using $W(2, 5)$ and $X(6, 1)$:
$$m_{WX} = \frac{y_2 - y_1}{x_2 - x_1}$$
$$= \frac{1 - 5}{6 - 2}$$
$$= \frac{-4}{4}$$
$$= -1$$

Using $Z(5, 8)$ and $Y(9, 4)$:
$$m_{ZY} = \frac{y_2 - y_1}{x_2 - x_1}$$
$$= \frac{4 - 8}{9 - 5}$$
$$= \frac{-4}{4}$$
$$= -1$$

Since $m_{WX} = m_{ZY}$, the lines are parallel.

Using $X(6, 1)$ and $Y(9, 4)$:
$$m_{XY} = \frac{y_2 - y_1}{x_2 - x_1}$$
$$= \frac{4 - 1}{9 - 6}$$
$$= \frac{3}{3}$$
$$= 1$$

Using $W(2, 5)$ and $Z(5, 8)$:
$$m_{WZ} = \frac{y_2 - y_1}{x_2 - x_1}$$
$$= \frac{8 - 5}{5 - 2}$$
$$= \frac{3}{3}$$
$$= 1$$

Since $m_{XY} = m_{WZ}$, the lines are parallel.

23.
$$m = \frac{y_2 - y_1}{x_2 - x_1}$$
$$= \frac{7 - 4}{1 - (-2)}$$
$$= \frac{3}{3}$$
$$= 1$$

$$\frac{y - y_1}{x - x_1} = m$$
$$\frac{y - 4}{x - (-2)} = 1$$
$$\frac{y - 4}{x + 2} = \frac{1}{1}$$
$$y - 4 = x + 2$$
$$y = x + 6$$

25.
$$m = \frac{y_2 - y_1}{x_2 - x_1}$$
$$= \frac{-5 - 5}{-3 - 3}$$
$$= \frac{-10}{-6}$$
$$= \frac{5}{3}$$

$$\frac{y - y_1}{x - x_1} = m$$
$$\frac{y - 5}{x - 3} = \frac{5}{3}$$
$$3y - 15 = 5x - 15$$
$$3y = 5x$$
$$y = \frac{5}{3}x$$

27.

$$m = \frac{y_2 - y_1}{x_2 - x_1}$$
$$= \frac{5 - 4}{-2 - 6}$$
$$= \frac{1}{-8}$$
$$= \frac{-1}{8}$$

$$\frac{y - y_1}{x - x_1} = m$$
$$\frac{y - 4}{x - 6} = \frac{-1}{8}$$
$$8y - 32 = -x + 6$$
$$8y = -x + 38$$
$$\frac{8y}{8} = \frac{-x}{8} + \frac{38}{8}$$
$$y = \frac{-1}{8}x + \frac{19}{4}$$

29a.

$$m = \frac{y_2 - y_1}{x_2 - x_1}$$
$$= \frac{77 - 65}{3200 - 8000}$$
$$= \frac{12}{-4800}$$
$$= \frac{-1}{400} \text{ or}$$
$$= -0.0025$$

$$\frac{y - y_1}{x - x_1} = m$$
$$\frac{y - 77}{x - 3200} = \frac{-0.0025}{1}$$
$$y - 77 = -0.0025x + 8$$
$$y = -0.0025x + 85$$

b. $m = \frac{-1°}{400 \text{ ft}}$; The temperature is dropping 1 degree every 400 feet.

c.
$$y = -0.0025x + 85$$
$$y = -0.0025(10,000) + 85$$
$$y = -25 + 85$$
$$y = 60°$$

d.
$$y = -0.0025x + 85$$
$$y = -0.0025(0) + 85$$
$$y = 0 + 85$$
$$y = 85°$$

31a.

$$m = \frac{y_2 - y_1}{x_2 - x_1}$$
$$= \frac{61,000 - 45,000}{180 - 100}$$
$$= \frac{16,000}{80}$$
$$= 200$$

$$\frac{y - y_1}{x - x_1} = m$$
$$\frac{y - 45,000}{x - 100} = \frac{200}{1}$$
$$y - 45,000 = 200x - 20,000$$
$$y = 200x + 25,000$$

b. $m = 200$ dollars per dryer; It costs $200 to produce one dryer.

c.
$$100,000 = 200x + 25,000$$
$$75,000 = 200x$$
$$375 = x$$

d.
$$y = 200x + 25,000$$
$$y = 200(0) + 25,000$$
$$y = 0 + 25,000$$
$$y = \$25,000$$

33. Using $(-2, 3)$ and $(5, -2)$:

$$m = \frac{y_2 - y_1}{x_2 - x_1}$$
$$= \frac{-2 - 3}{5 - (-2)}$$
$$= \frac{-5}{7}$$

$$\frac{y - y_1}{x - x_1} = m$$
$$\frac{y - 3}{x - (-2)} = \frac{-5}{7}$$
$$\frac{y - 3}{x + 2} = \frac{-5}{7}$$
$$7y - 21 = -5x - 10$$
$$7y = -5x + 11$$
$$y = \frac{-5}{7}x + \frac{11}{7}$$

35. Using $(-15, -4)$ and $(25, 6)$:

$$m = \frac{y_2 - y_1}{x_2 - x_1}$$
$$= \frac{6 - (-4)}{25 - (-15)}$$
$$= \frac{10}{40}$$
$$= \frac{1}{4}$$

$$\frac{y - y_1}{x - x_1} = m$$
$$\frac{y - (-4)}{x - (-15)} = \frac{1}{4}$$
$$\frac{y + 4}{x + 15} = \frac{1}{4}$$
$$4y + 16 = x + 15$$
$$4y = x - 1$$
$$y = \frac{1}{4}x - \frac{1}{4}$$

37.

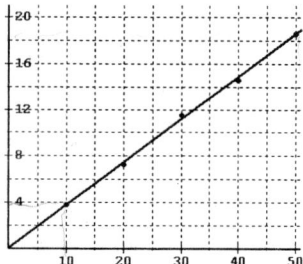

Using $(10, 3.8)$ and $(50, 18.6)$:

$$m = \frac{y_2 - y_1}{x_2 - x_1} \qquad \frac{y - y_1}{x - x_1} = m$$

$$= \frac{18.6 - 3.8}{50 - 10} \qquad \frac{y - 3.8}{x - 10} = \frac{0.37}{1}$$

$$= \frac{14.8}{40} \qquad y - 3.8 = 0.37x - 3.7$$

$$= 0.37 \qquad\qquad y = 0.37x + 0.1$$

a.
$$y = 0.37x + 0.1$$
$$y = 0.37(17) + 0.1$$
$$y = 6.29 + 0.1$$
$$y \approx 6.4 \text{ in.}$$

b.
$$y = 0.37x + 0.1$$
$$84 = 0.37x + 0.1$$
$$83.9 = 0.37x$$
$$227 \text{ in.} \approx x$$

39.

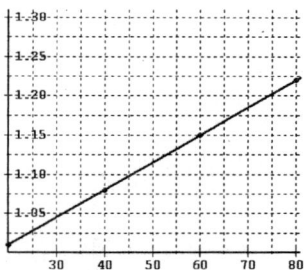

Using $(20, 1.01)$ and $(80, 1.22)$:

$$m = \frac{y_2 - y_1}{x_2 - x_1} \qquad \frac{y - y_1}{x - x_1} = m$$

$$= \frac{1.22 - 1.01}{80 - 20} \qquad \frac{y - 1.01}{x - 20} = \frac{0.0035}{1}$$

$$= \frac{0.21}{60} \qquad y - 1.01 = 0.0035x - 0.07$$

$$= 0.0035 \qquad\qquad y = 0.0035x + 0.94$$

a.
$$y = 0.0035x + 0.94$$
$$y = 0.0035(37) + 0.94$$
$$y = 0.1295 + 0.94$$
$$y \approx 1.07 \text{ atm.}$$

b.
$$y = 0.0035x + 0.94$$
$$1.5 = 0.0035x + 0.94$$
$$0.56 = 0.0035x$$
$$160° \text{ C} \approx x$$

c. The y-intercept represents the pressure when the temperature is $0°$ C.

d. The x-intercept represents the temperature when the pressure is 0 atm.

41a. Yes   **b.** $12.2 - 10.8 = 1.4$ sec   **c.** $10.94 - 9.84 = 1.1$ sec

d. 6; 5   **e.** $10.8 - 9.84 = 0.96$ sec; $12.2 - 10.94 = 1.26$ sec

f. Using $(24, 10.3)$ and $(56, 10)$ for men's times:

$$m = \frac{y_2 - y_1}{x_2 - x_1} \qquad \frac{y - y_1}{x - x_1} = m$$

$$= \frac{10 - 10.3}{56 - 24} \qquad \frac{y - 10.3}{x - 24} = \frac{-0.0094}{1}$$

$$= \frac{-0.3}{32} \qquad y - 10.3 = -0.0094x + 0.2256$$

$$\approx -0.0094 \qquad y \approx -0.0094x + 10.53$$

Using $(8, 11.8)$ and $(60, 10.9)$ for women's times:

$$m = \frac{y_2 - y_1}{x_2 - x_1} \qquad \frac{y - y_1}{x - x_1} = m$$

$$= \frac{10.9 - 11.8}{60 - 8} \qquad \frac{y - 11.8}{x - 8} = \frac{-0.0173}{1}$$

$$= \frac{-0.9}{52} \qquad y - 11.8 = -0.0173x + 0.1384$$

$$\approx -0.0173 \qquad y \approx -0.0173x + 11.94$$

g. The women's

h.
$$y = -0.0094x + 10.53$$
$$y = -0.0094(76) + 10.53$$
$$y = -0.7144 + 10.53$$
$$y \approx 9.82 \text{ sec}$$

i.
$$y = -0.0173x + 11.94$$
$$y = -0.0173(76) + 11.94$$
$$y = -1.3148 + 11.94$$
$$y \approx 10.63 \text{ sec}$$

j.
$$-0.0094x + 10.53 = -0.0173x + 11.94$$
$$0.0079x = 1.41$$
$$x \approx 178$$

$$y = -0.0094x + 10.53$$
$$y = -0.0094(178) + 10.53$$
$$y = -1.6732 + 10.53$$
$$y \approx 8.86 \text{ sec}$$

The intersection point, $(178, 8.86)$ predicts that the men's and the women's time will be the same, 8.86 sec, in the year 2106 $(178 + 1928 = 2106)$.

k. The x-intercept would represent a running time of zero seconds.

Chapter 3 Summary and Review

1. If $\dfrac{a}{b} = \dfrac{c}{d}$, then $ad = bc$.

2. $y = kx$

3. $m = \dfrac{\Delta y}{\Delta x}$

4. $y = mx + b$, where m = slope and b = y-intercept

5. $m = \dfrac{y_2 - y_1}{x_2 - x_1}$

6. $\dfrac{y - y_1}{x - x_1} = m$

7. If two lines are <u>parallel</u>, then $m_1 = m_2$.

8. If two lines are <u>perpendicular</u>, then $m_1 m_2 = -1$.

9. $x = a$ is the equation of a <u>vertical</u> line with undefined slope.

10. $y = b$ is the equation of a <u>horizontal</u> line with slope $= 0$.

11. Two variables are <u>proportional</u> if their ratios are always the same.

12. <u>To graph a line using the slope-intercept method:</u>
Step 1 Write the equation in the form, $y = mx + b$.
Step 2 Plot the y-intercept, $(0, b)$.
Step 3 Write the slope as a fraction, $m = \dfrac{\Delta y}{\Delta x}$
Step 4 Starting at the y-intercept, move Δy units in the y-direction, then Δx units in the x-direction to find a second point.
Step 5 Starting at the y-intercept, move $-\Delta y$ units in the y-direction, then $-\Delta x$ units in the x-direction to find a third point.
Step 6 Draw a line through the three points.

13. Solve the equation for y. The coefficient of x is the slope of the line.

14. <u>To solve a linear system by graphing:</u>
Graph both equations on the same axes.
The intersection point is the solution to the system.

15. Substitution and elimination

16. If combining the two equations results in a false statement, such as $0 = 3$, then the system is <u>inconsistent</u>.

If combining the two equations results in a true statement, such as $0 = 0$, then the system is <u>dependent</u>.

17.

Principal	Rate	Interest

18.

Total Amount	Percent of Important Ingredient	Amount of Important Ingredient

19.

Rate	Time	Distance

20. Use the point-slope formula, $\dfrac{y - y_1}{x - x_1} = m$

21a.

D	0.5	1	2	4
V	0.25	2	16	64
$\frac{V}{D}$	$\frac{0.25}{0.5} = 0.5$	$\frac{2}{1} = 2$	$\frac{16}{2} = 8$	$\frac{64}{4} = 16$

They are not proportional.

b.

s	0.2	0.8	1.6	2.5
M	0.08	0.32	0.64	1
$\frac{M}{s}$	$\frac{0.08}{0.2} = 0.4$	$\frac{0.32}{0.8} = 0.4$	$\frac{1.6}{0.64} = 0.4$	$\frac{1}{2.5} = 0.4$

They are proportional.

22.
$$\frac{16}{q} = \frac{52}{86.125}$$
$$1378 = 52q$$
$$\frac{1378}{52} = \frac{52q}{52}$$
$$26.5 = q$$

23.
$$\frac{2x - 1}{x + 3} = \frac{3}{2}$$
$$2(2x - 1) = 3(x + 3)$$
$$4x - 2 = 3x + 9$$
$$x - 2 = 9$$
$$x = 11$$

24.
$$\frac{36 \text{ frimbles}}{\$4.86} = \frac{x \text{ frimbles}}{\$6.75}$$
$$243 = 4.86x$$
$$50 = x$$

25.
$$\frac{70 \text{ g fat}}{200 \text{ g protein}} = \frac{112 \text{ g fat}}{x \text{ g protein}}$$
$$70x = 22400$$
$$x = 320 \text{ g protein}$$

$$\frac{70 \text{ g fat}}{1800 \text{ g carbohydrates}} = \frac{112 \text{ g fat}}{x \text{ g carbohydrates}}$$
$$70x = 201600$$
$$x = 2880 \text{ g carbohydrates}$$

26.
$$\frac{x}{100 - x} = \frac{7}{5}$$
$$5x = 7(100 - x)$$
$$5x = 700 - 7x$$
$$12x = 700$$
$$x = 58\tfrac{1}{3} \text{ lb}$$
$$100 - x = 41\tfrac{2}{3} \text{ lb}$$

27.
$$\frac{w}{w - 10} = \frac{30}{18}$$
$$18w = 30(w - 10)$$
$$18w = 30w - 300$$
$$-12w = -300$$
$$w = 25$$

28.
$$\frac{a}{a - 50} = \frac{108}{48}$$
$$48a = 108(a - 50)$$
$$48a = 108a - 5400$$
$$-60a = -5400$$
$$a = 90$$

29.
$$\frac{x}{6} = \frac{21}{9}$$
$$9x = 126$$
$$x = 14 \text{ ft}$$

30.
$$\frac{x}{160} = \frac{17}{8}$$
$$8x = 2720$$
$$x = 340 \text{ cm}$$

31a.
$$\frac{21}{28} = \frac{3}{4} = k;$$
$$B = \frac{3}{4}h$$

b.
$$B = \frac{3}{4}h$$
$$39 = \frac{3}{4}h$$
$$4 \cdot 39 = 4 \cdot \frac{3}{4}h$$
$$156 = 3h$$
$$\frac{156}{3} = \frac{3h}{3}$$
$$52 \text{ hr} = h$$

c.

h	B
28	21
52	39

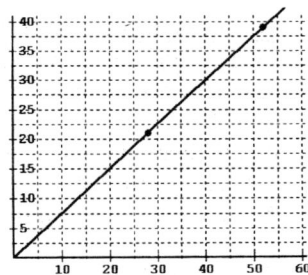

93

32. Since $\dfrac{100}{150} = \dfrac{2}{3}$ and $\dfrac{1600}{600} = \dfrac{8}{3}$, therefore, they are not proportional.

33. Because the number of defective chips varies directly with the size of the shipment, if the shipment is increased by 25%, then the number of defective chips also will increase by 25%. $0.25(12) = 3$ more defective chips and $12 + 3 = 15$ defective chips this month.

34. No. If the two variables H and L vary directly, they are related by the equation $H = kL$, where k is the constant of proportionality.

35.

x	y
0	-6
3	0

b.

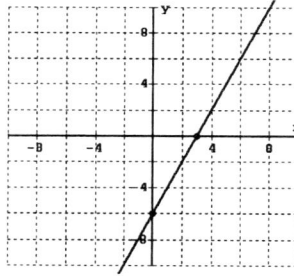

c. $\dfrac{\Delta y}{\Delta x} = \dfrac{6}{3} = 2$

36.

x	y
0	9
4	0

b.

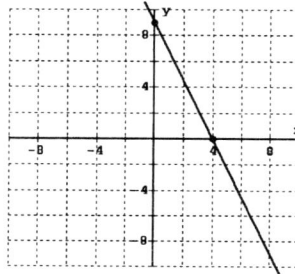

c. $\dfrac{\Delta y}{\Delta x} = \dfrac{-9}{4}$

37. $\dfrac{\Delta y}{\Delta x} = \dfrac{-12}{8} = \dfrac{-3}{2}$

38. $\dfrac{\Delta y}{\Delta x} = \dfrac{10}{6} = \dfrac{5}{3}$

39.
$$m = 8\% = \dfrac{8}{100}$$
$$\dfrac{\Delta y}{3600} = \dfrac{8}{100}$$
$$100 \cdot \Delta y = 28{,}800$$
$$\dfrac{100 \cdot \Delta y}{100} = \dfrac{28{,}800}{100}$$
$$\Delta y = 288 \text{ ft}$$

40.
$$m = 18.\overline{3}\% = \dfrac{18\frac{1}{3}}{100} = \dfrac{55}{3} \cdot \dfrac{1}{100} = \dfrac{11}{60}$$
$$\dfrac{44}{\Delta x} = \dfrac{11}{60}$$
$$2640 = 11 \cdot \Delta x$$
$$\dfrac{2640}{11} = \dfrac{11 \cdot \Delta x}{11}$$
$$240 \text{ in.} = \Delta x$$
$$\text{or } \Delta x = \dfrac{240 \text{ in.}}{12} = 20 \text{ ft}$$

41a.

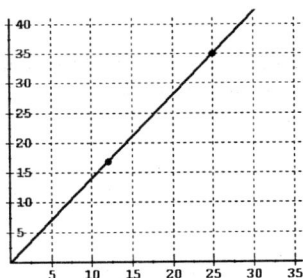

$$m = \frac{\Delta p}{\Delta g} = \frac{35 - 16.8}{25 - 12} = \frac{\$18.20}{13 \text{ gal}} = \$1.40/\text{gal}$$

b. The slope indicates the price per gallon.

42a. $\dfrac{1200}{8 \cdot 10} = \dfrac{2700}{12 \cdot 15} = 15 = k$

b. $C = 15A$

c. The slope indicates the price per square foot.
The tile cost \$15 per square foot.

43a.
$2y + 5x = -10$	Subtract $5x$ from both sides
$2y = -5x - 10$	Divide both sides by 2
$\dfrac{2y}{2} = \dfrac{-5x}{2} - \dfrac{10}{2}$	Simplify each quotient
$y = \dfrac{-5}{2}x - 5$	Slope-intercept form

The slope is $\dfrac{-5}{2}$ and the y-intercept is $(0, -5)$.

b.

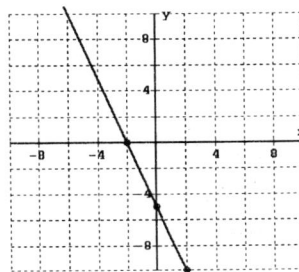

44a.
$y + x = 0$	Subtract x from both sides
$y = -x$	
$y = -x + 0$	Slope-intercept form

The slope is $\dfrac{-1}{1}$ and the y-intercept is $(0, 0)$.

b.

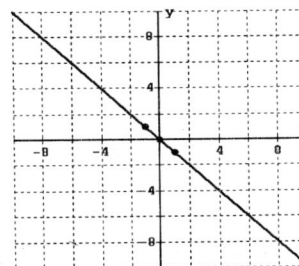

45. $y = 4$ is the equation of a horizontal line with slope is 0 and its y-intercept is $(0, 4)$.

46. The slope is 0.85 and the y-intercept is $(0, 1.35)$.
The slope represents the price per topping and the y-intercept represents the cost of the yogurt without toppings.

47. Check $(-2, 10)$:

$$x + y = 8$$
$$-2 + 10 \stackrel{?}{=} 8$$
$$8 = 8 \quad \text{True}$$

Check $(-2, 10)$:

$$x - y = 2$$
$$-2 - 10 \stackrel{?}{=} 2$$
$$-12 = 2 \quad \text{False}$$

Therefore, $(-2, 10)$ is not a solution.

48. Check $(-3, 1)$: $8x + 3y = 21$ Check $(-3, 1)$: $5x = y - 16$

$$8(-3) + 3(1) \stackrel{?}{=} 21 \qquad\qquad 5(-3) \stackrel{?}{=} 1 - 16$$

$$-24 + 3 \stackrel{?}{=} 21 \qquad\qquad\qquad -15 = -15 \quad \text{True}$$

$$-21 = 21 \quad \text{False}$$

Therefore, $(-3, 1)$ is not a solution.

49.

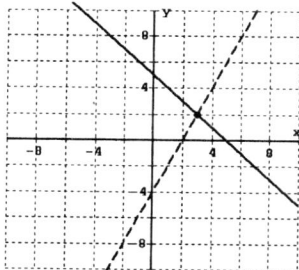

$(3, 2)$

50.

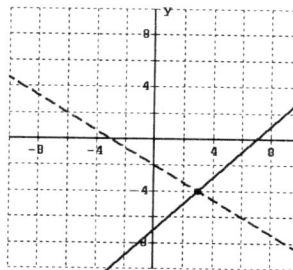

$(3, -4)$

51a. $y = 3x - 4$ $y = 3x + 4$

From their slope-intercept forms, both lines have slope 3, but the y-intercepts are different, so their graphs are parallel lines. The system is inconsistent.

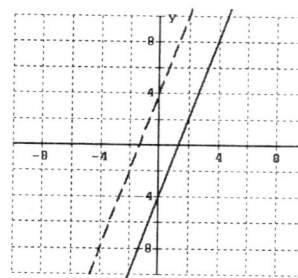

b. $y = \dfrac{2}{5}x - 2$

$$5y = 2x - 10$$

$$\frac{5y}{5} = \frac{2x}{5} - \frac{10}{5}$$

$$y = \frac{2}{5}x - 2$$

From their slope-intercept forms, both lines have slope $\frac{2}{5}$, and y-intercept -2, so their graphs are the same line. The system is dependent.

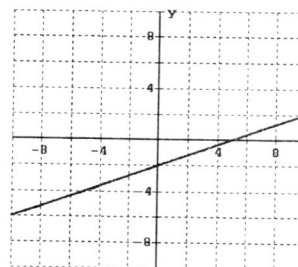

52a. $2y = 3x + 7$ $y = \dfrac{3}{2}x + 7$

$$\frac{2y}{2} = \frac{3x}{2} + \frac{7}{2}$$

$$y = \frac{3}{2}x + \frac{7}{2}$$

From their slope-intercept forms, both lines have slope $\frac{3}{2}$, but the y-intercepts are different, so their graphs are parallel lines. The system is inconsistent.

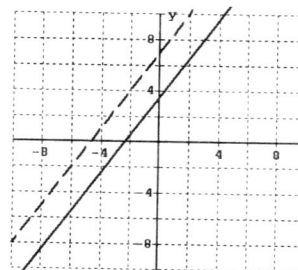

96

b.

$$3x - 4y = 6$$
$$-4y = -3x + 6$$
$$\frac{-4y}{-4} = \frac{-3x}{-4} + \frac{6}{-4}$$
$$y = \frac{3}{4}x - \frac{3}{2}$$

$$4x + 3y = 6$$
$$3y = -4x + 6$$
$$\frac{3y}{3} = \frac{-4x}{3} + \frac{6}{3}$$
$$y = \frac{-4}{3}x + 2$$

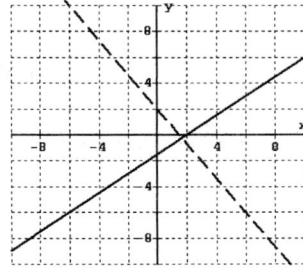

From their slope-intercept forms, the lines have different slopes, so their graphs are intersecting lines.
The system is consistent.

53. Substitute $2x + 1$ for y into Equation 2:

$$2x + 3y = -21$$
$$2x + 3(2x + 1) = -21$$
$$2x + 6x + 3 = -21$$
$$8x + 3 = -21$$
$$8x = -24$$
$$x = -3$$

Substitute -3 for x into:

$$y = 2x + 1$$
$$y = 2(-3) + 1$$
$$y = -6 + 1$$
$$y = -5$$

The solution is $(-3, -5)$.

54. Solve Equation 1 for x:

$$x + 4y = 1$$
$$x = 1 - 4y$$

Substitute $1 - 4y$ for x into Equation 2:

$$2x + 3y = -3$$
$$2(1 - 4y) + 3y = -3$$
$$2 - 8y + 3y = -3$$
$$2 - 5y = -3$$
$$-5y = -5$$
$$y = 1$$

Substitute 1 for y into:

$$x = 1 - 4y$$
$$x = 1 - 4(1)$$
$$x = 1 - 4$$
$$x = -3$$

The solution is $(-3, 1)$.

55. Multiply Equation 1 by 5 and Equation 2 by -2:

$$5(2x + 7y = -19) \Rightarrow 10x + 35y = -95$$
$$-2(5x - 3y = 14) \Rightarrow \underline{-10x + 6y = -28}$$
$$41y = -123$$
$$y = -3$$

Substitute -3 for y into:

$$2x + 7y = -19$$
$$2x + 7(-3) = -19$$
$$2x - 21 = -19$$
$$2x = 2$$
$$x = 1$$

The solution is $(1, -3)$.

56. Write Equation 2 in general form:

$$5x + 15 = -2y$$
$$5x + 2y = -15$$

Multiply Equation 1 by -2 and Equation 2 by 3:

$$-2(4x + 3y = -19) \Rightarrow -8x - 6y = 38$$
$$3(5x + 2y = -15) \Rightarrow \underline{15x + 6y = -45}$$
$$7x = -7$$
$$x = -1$$

Substitute -1 for x into:

$$4x + 3y = -19$$
$$4(-1) + 3y = -19$$
$$-4 + 3y = -19$$
$$3y = -15$$
$$y = -5$$

The solution is $(-1, -5)$.

57.

	Number of Pounds	Price per Pound	Total Cost
Cereal	x	65	$65x$
Dried Fruit	y	90	$90y$
Mixture	30	80	$80(30)$

$$x + y = 30$$
$$65x + 90y = 80(30)$$

Multiply Equation 1 by -65:

$$
\begin{aligned}
-65(x + y = 30) &\Rightarrow & -65x - 65y &= -1950 \\
65x + 90y = 80(30) &\Rightarrow & \underline{65x + 90y} &= \underline{2400} \\
& & 25y &= 450 \\
& & y &= 18
\end{aligned}
$$

Substitute 18 for y into:

$$
\begin{aligned}
x + y &= 30 \\
x + 18 &= 30 \\
x &= 12
\end{aligned}
$$

He needs 12 lb of cereal and 18 lb of dried fruit.

58.   Length: L Equation 1: $2L + 2W = 50$
Width: W Equation 2: $L = W + 9$

Substitute $W + 9$ for L into Equation 1:

$$
\begin{aligned}
2L + 2W &= 50 \\
2(W + 9) + 2W &= 50 \\
2W + 18 + 2W &= 50 \\
4W + 18 &= 50 \\
4W &= 32 \\
W &= 8
\end{aligned}
$$

Substitute 8 for W into:

$$
\begin{aligned}
L &= W + 9 \\
L &= 8 + 9 \\
L &= 17
\end{aligned}
$$

The rectangle is 17 yd by 8 yd.

59.   Amount in First Account: x Equation 1: $0.06x + 0.09y = 93$
Amount in Second Account: y Equation 2: $0.04x + 0.08y = 76$

Multiply Equation 1 by -200 and Equation 2 by 300:

$$
\begin{aligned}
-200(0.06x + 0.09y = 93) &\Rightarrow & -12x - 18y &= -18600 \\
300(0.04x + 0.08y = 76) &\Rightarrow & \underline{12x + 24y} &= \underline{22800} \\
& & 6y &= 4200 \\
& & y &= 700
\end{aligned}
$$

Substitute 700 for y into

$$
\begin{aligned}
0.04x + 0.08y &= 76 \\
0.04x + 0.08(700) &= 76 \\
0.04x + 56 &= 76 \\
0.04x &= 20 \\
x &= 500
\end{aligned}
$$

She invested \$500 in the first account and \$700 in the second account.

60.

	Principal	Rate	Interest
Amount at 6%	x	0.06	$0.06x$
Amount at 8%	y	0.08	$0.08y$

$$x = y + 300$$
$$0.06x + 0.08y = 242$$

Substitute $y + 300$ for x into Equation 2:

$$
\begin{aligned}
0.06x + 0.08y &= 242 \\
0.06(y + 300) + 0.08y &= 242 \\
0.06y + 18 + 0.08y &= 242 \\
0.14y + 18 &= 242 \\
0.14y &= 224 \\
y &= 1600
\end{aligned}
$$

Substitute 1600 for y into:

$$
\begin{aligned}
x &= y + 300 \\
x &= 1600 + 300 \\
x &= 1900
\end{aligned}
$$

He invested \$1900 at 6% and \$1600 at 8%.

61.

	Number of Pounds	Percent Copper	Amount of Copper
60% Alloy	x	0.60	$0.60x$
20% Alloy	y	0.20	$0.20y$
Mixture	8	0.30	$0.30(8)$

$$x + y = 8$$
$$0.60x + 0.20y = 0.30(8)$$

Multiply Equation 2 by 100 and Equation 1 by -20:   Substitute 2 for x into:

$$100(0.60x + 0.20y = 0.30(8)) \Rightarrow \quad 60x + 20y = \quad 240x + y = 8$$
$$-20(x + y = 8) \quad\quad\quad \Rightarrow \quad \underline{-20x - 20y = -160} \quad 2 + y = 8$$
$$40x \quad\quad = \quad 80 \quad\quad y = 6$$
$$x = \quad 2$$

He needs 2 lb of the 60% alloy and 6 lb of the 20% alloy.

62a. $0.15(20) = 3$ hits   **b.** $20 + 140 = 160$ times   **c.** $\dfrac{52}{160} = 32.5\%$ **d.** Joe; Joe; Jerry

$0.35(140) = 49$ hits $3 + 49 = 52$ hits

63.

Alida
Steve 93 (San Diego)

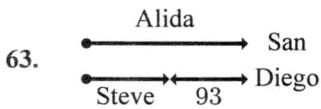

	Rate	Time	Distance
Alida	x	3	$3x$
Steve	y	3	$3y$

$$x = 2y$$
$$3x = 3y + 93$$

Substitute $2y$ for x into $3x = 3y + 93$ Substitute 31 for y into $x = 2y$
$$3(2y) = 3y + 93 \quad\quad\quad\quad\quad x = 2(31)$$
$$6y = 3y + 93 \quad\quad\quad\quad\quad\quad x = 62$$
$$3y = 93$$
$$y = 31$$

Alida is traveling at 62 mph, and Steve is traveling at 31 mph.

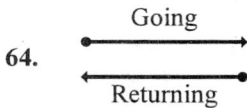

64.

Going
Returning

	Rate	Time	Distance
Going	r	6	d
Returning	$r - 8$	9	d

$$6r = d$$
$$9(r - 8) = d$$

Substitute $6r$ for d into $9(r - 8) = d$ Substitute 24 for r into $6r = d$
$$9(r - 8) = 6r \quad\quad\quad\quad\quad\quad\quad 6(24) = d$$
$$9r - 72 = 6r \quad\quad\quad\quad\quad\quad\quad 144 = d$$
$$-72 = -3r$$
$$24 = r$$

He rides $2d = 2(144) = 288$ mi.

65.
$$m = \frac{y_2 - y_1}{x_2 - x_1}$$
$$= \frac{3 - (-2)}{-7 - 1}$$
$$= \frac{5}{-8}$$
$$= \frac{-5}{8}$$

66.
$$y = \frac{1}{3}x - 2$$

The slope is $\dfrac{1}{3}$.

Then, $m_\perp = -3$

67. Using $(5, 0)$ and $(0, -1)$:

$$m = \frac{y_2 - y_1}{x_2 - x_1}$$
$$= \frac{-1 - 0}{0 - 5}$$
$$= \frac{-1}{-5}$$
$$= \frac{1}{5}$$

$$\frac{y - y_1}{x - x_1} = m$$

$$\frac{y - 0}{x - 5} = \frac{1}{5}$$ Cross-multiply

$$5y = x - 5$$ Solve for y

$$\frac{5y}{5} = \frac{x}{5} - \frac{5}{5}$$

$$y = \frac{1}{5}x - 1$$ Slope-intercept form

Substitute $m = \frac{1}{5}$, $y_1 = 0$, and $x_1 = 5$

68.
$$\frac{y - y_1}{x - x_1} = m$$

Substitute $m = \frac{-1}{4}$, $y_1 = -5$, and $x_1 = 3$

$$\frac{y - (-5)}{x - 3} = \frac{-1}{4}$$ Simplify

$$\frac{y + 5}{x - 3} = \frac{-1}{4}$$ Cross-multiply

$$4y + 20 = -x + 3$$ Solve for y

$$4y = -x - 17$$

$$\frac{4y}{4} = \frac{-x}{4} - \frac{17}{4}$$

$$y = \frac{-1}{4}x - \frac{17}{4}$$ Slope-intercept form

69. $x = -4$

70. $y = 8$

71. Using $(-3, 1)$ and $(-1, -1)$:

$$m = \frac{y_2 - y_1}{x_2 - x_1}$$
$$= \frac{-1 - 1}{-1 - (-3)}$$
$$= \frac{-2}{2}$$
$$= -1$$

$$\frac{y - y_1}{x - x_1} = m$$

$$\frac{y - 1}{x - (-3)} = -1$$ Simplify

$$\frac{y - 1}{x + 3} = \frac{-1}{1}$$ Cross-multiply

$$y - 1 = -x - 3$$ Solve for y

$$y = -x - 2$$ Slope-intercept form

Substitute $m = -1$, $y_1 = 1$, and $x_1 = -3$

72.
$$2x + 3y = 1$$
$$3y = -2x + 1$$
$$\frac{3y}{3} = \frac{-2x}{3} + \frac{1}{3}$$
$$y = \frac{-2}{3}x + \frac{1}{3}$$

The slope $= \frac{-2}{3}$

Then, $m_\perp = \frac{3}{2}$

$$\frac{y - y_1}{x - x_1} = m$$

$$\frac{y - 1}{x - (-4)} = \frac{3}{2}$$ Simplify

$$\frac{y - 1}{x + 4} = \frac{3}{2}$$ Cross-multiply

$$2y - 2 = 3x + 12$$ Solve for y

$$2y = 3x + 14$$

$$\frac{2y}{2} = \frac{3x}{2} + \frac{14}{2}$$

$$y = \frac{3}{2}x + 7$$ Slope-intercept form

Substitute $m = \frac{3}{2}$, $y_1 = 1$, and $x_1 = -4$

100

73.

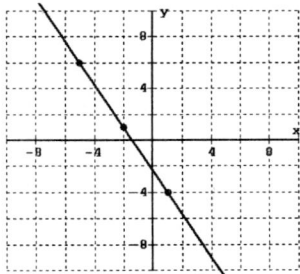

74a.

$$5x + 4y = 3 \qquad\qquad 3y - 5x = 4$$
$$\qquad 4y = -5x + 3 \qquad\qquad 3y = 5x + 4$$
$$\qquad \frac{4y}{4} = \frac{-5x}{4} + \frac{3}{4} \qquad\qquad \frac{3y}{3} = \frac{5x}{3} + \frac{4}{3}$$
$$\qquad y = \frac{-5}{4}x + \frac{3}{4} \qquad\qquad y = \frac{5}{3}x + \frac{4}{3}$$

Slope $= \dfrac{-5}{4}$ \qquad\qquad Slope $= \dfrac{5}{3}$

The lines are neither.

b.

$$5x + 3y = 1 \qquad\qquad 5y - 3x = 6$$
$$\qquad 3y = -5x + 1 \qquad\qquad 5y = 3x + 6$$
$$\qquad \frac{3y}{3} = \frac{-5x}{3} + \frac{1}{3} \qquad\qquad \frac{5y}{5} = \frac{3x}{5} + \frac{6}{5}$$
$$\qquad y = \frac{-5}{3}x + \frac{1}{3} \qquad\qquad y = \frac{3}{5}x + \frac{6}{5}$$

Slope $= \dfrac{-5}{3}$ \qquad\qquad Slope $= \dfrac{3}{5}$

Since $m_1 m_2 = -1$, the lines are perpendicular.

c.

$$4y - 5x = 1 \qquad\qquad 2\left(2y - \frac{5}{2}x = 2\right)$$
$$\qquad 4y = 5x + 1 \qquad\qquad 4y - 5x = 4$$
$$\qquad \frac{4y}{4} = \frac{5x}{4} + \frac{1}{4} \qquad\qquad 4y = 5x + 4$$
$$\qquad y = \frac{5}{4}x + \frac{1}{4} \qquad\qquad \frac{4y}{4} = \frac{5x}{4} + \frac{4}{4}$$

Slope $= \dfrac{5}{4}$ \qquad\qquad $y = \frac{5}{4}x + 1$

\qquad\qquad\qquad Slope $= \dfrac{5}{4}$

Since $m_1 = m_2$, the lines are parallel.

101

CHAPTER 4

Homework 4.1

1a. $4^3 = 64$ **b.** $5^3 = 125$ **c.** $5^4 = 625$

3a. $\left(\dfrac{2}{3}\right)^4 = \dfrac{16}{81}$ **b.** $\left(\dfrac{4}{5}\right)^3 = \dfrac{64}{125}$ **c.** $\left(\dfrac{11}{9}\right)^2 = \dfrac{121}{81}$

5a. $(3.1)^3 \approx 29.79$ **b.** $(2.6)^4 \approx 45.70$ **c.** $(0.8)^4 \approx 0.41$

7a. $-5^2 = -(5 \cdot 5)^2 = -25$ **b.** $-5^3 = -(5 \cdot 5 \cdot 5)^3 = -125$
c. $(-5)^2 = (-5)(-5) = 25$ **d.** $(-5)^3 = (-5)(-5)(-5) = -125$

9a. $-(-2)^2 = -4$ **b.** $-(-2)^3 = -(-8) = 8$
c. $-2^3 - 2^2 = -8 - 4 = -12$ **d.** $-(2^3 - 2)^2 = -(8 - 2)^2 = -6^2 = -36$

11. $(3 + 4)^2 = 7^2 = 49$ **13.** $5 + 2.3^3 = 5 + 12.167 = 17.167 \approx 17.17$

15. $0.25 \cdot 6^3 = 0.25 \cdot 216 = 54$

17a. $5x^3$ **b.** $5x^2$ **c.** $5 - x^2$ **d.** $5 - x^3$
$5(-2)^3$ $5(-2)^2$ $5 - (-2)^2$ $5 - (-2)^3$
$= 5(-8)$ $= 5(4)$ $= 5 - 4$ $= 5 - (-8)$
$= -40$ $= 20$ $= 1$ $= 13$

19a. ab^3 **b.** $a - b^3$ **c.** $(a - b^2)^2$ **d.** $ab(a^2 - b^2)$
$(-3)(-4)^3$ $-3 - (-4)^3$ $[-3 - (-4)^2]^2$ $(-3)(-4)[(-3)^2 - (-4)^2]$
$= (-3)(-64)$ $= -3 - (-64)$ $= [-3 - 16]^2$ $= (-3)(-4)[9 - 16[$
$= 192$ $= 61$ $= [-19]^2$ $= (-3)(-4)[-7]$
$= 361$ $= -84$

21a. $x + x + x = 3x$ **b.** $x \cdot x \cdot x = x^3$ **23a.** $2a \cdot 2a = 4a^2$ **b.** $2a + 2a = 4a$

25a. $-q - q - q = -3q$ **b.** $-q(-q)(-q) = -q^3$

27a. $-3m - 3m = -6m$ **b.** $(-3m)(-3m) = 9m^2$

29a. If $z = 2$: $\left.\begin{array}{l} 3z^2 = 3(2)^2 = 3 \cdot 4 = 12 \\ (3z)^2 = (3 \cdot 2)^2 = 6^2 = 36 \end{array}\right\}$ Not the same **b.** $(3z)^2 = (3z)(3z) = 9z^2$

31a. For $x = 0$: $2x = 2 \cdot 0 = 0$ **b.** For $x = 1$: $2x = 2 \cdot 1 = 2$ For $x = 4$: $2x = 2 \cdot 4 = 8$
$x^2 = 0^2 = 0$ $x^2 = 1^2 = 1$ $x^2 = 4^2 = 16$
For $x = 2$: $2x = 2 \cdot 2 = 4$ For $x = 3$: $2x = 2 \cdot 3 = 6$ For $x = 5$: $2x = 2 \cdot 5 = 10$
$x^2 = 2^2 = 4$ $x^2 = 3^2 = 9$ $x^2 = 5^2 = 25$
c. No, $2x$ is not equivalent to x^2.

102

33. $A = (4w)^2 - (6t)^2$
$= 16w^2 - 36t^2$

35. $A = \pi(2w)^2 - 5^2$
$= \pi(4w^2) - 25$
$= 4\pi w^2 - 25$

37. $V = \dfrac{4}{3}\pi r^3$
$= \dfrac{4}{3}\pi(2h)^3$
$= \dfrac{4}{3}\pi(8h^3)$
$= \dfrac{32\pi h^3}{3}$

39. $V = lwh$
$= (2c)(3c)(7c)$
$= 42c^3$

41a. $V = \dfrac{1}{2} \cdot \dfrac{4}{3}\pi r^3$
$= \dfrac{1}{2} \cdot \dfrac{4}{3}\pi(40)^3$
$= \dfrac{1}{2} \cdot \dfrac{4}{3}\pi(64{,}000)$
$= \dfrac{128{,}000\pi}{3}$ cu ft

b. $S = \dfrac{1}{2} \cdot 4\pi r^2$
$= \dfrac{1}{2} \cdot 4\pi(40)^2$
$= \dfrac{1}{2} \cdot 4\pi(1600)$
$= 3200\pi$ sq ft

c.

40 ft

43a. $S = 2\pi r^2 + 2\pi rh$
$= 2\pi(5)^2 + 2\pi(5)(16)$
$= 2\pi(25) + 2\pi(80)$
$= 50\pi + 160\pi$
$= 210\pi$ sq cm

b. $S = 2\pi rh$
$848.25 = 2\pi(1.5)h$
$848.25 = 3\pi h$
$\dfrac{848.25}{3\pi} = \dfrac{3\pi h}{3\pi}$
$90 \text{ ft} \approx h$

c.
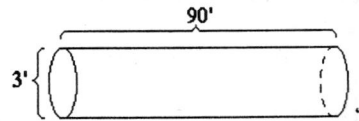
90'
3'

45a. $y = 16$ **b.** $x = -3,\ x = 3$ **c.** $y \approx 6.3$ **d.** $y \approx 12.3$
e. $x \approx -4.5,\ x \approx 4.5$ **f.** There is no solution since the graph never has a y-coordinate of -5.

47. $v = lwh$ Divide both sides by lh
$\dfrac{v}{lh} = \dfrac{lwh}{lh}$
$\dfrac{v}{lh} = w$

49. $E = \dfrac{mv^2}{2}$ Multiply both sides by 2
$2 \cdot E = 2 \cdot \dfrac{mv^2}{2}$
$2E = mv^2$ Divide both sides by v^2
$\dfrac{2E}{v^2} = \dfrac{mv^2}{v^2}$
$\dfrac{2E}{v^2} = m$

51. $A = \dfrac{h}{2}(b+c)$ Multiply both sides by 2
$2 \cdot A = 2 \cdot \dfrac{h}{2}(b+c)$
$2A = h(b+c)$ Divide both sides by $(b+c)$
$\dfrac{2A}{b+c} = \dfrac{h(b+c)}{b+c}$
$\dfrac{2A}{b+c} = h$

53.

$$F = \frac{9}{5}C + 32 \qquad \text{Subtract 32 from both sides}$$

$$F - 32 = \frac{9}{5}C + 32 - 32$$

$$F - 32 = \frac{9}{5}C \qquad \text{Multiply both sides by 5}$$

$$5(F - 32) = 5 \cdot \frac{9}{5}C$$

$$5(F - 32) = 9C \qquad \text{Divide both sides by 9}$$

$$\frac{5(F - 32)}{9} = \frac{9C}{9}$$

$$\frac{5}{9}(F - 32) = C$$

55.

$$A = \pi rh + 2\pi r^2 \qquad \text{Subtract } 2\pi r^2 \text{ from both sides}$$

$$A - 2\pi r^2 = \pi rh + 2\pi r^2 - 2\pi r^2$$

$$A - 2\pi r^2 = \pi rh \qquad \text{Divide both sides by } \pi r$$

$$\frac{A - 2\pi r^2}{\pi r} = \frac{\pi rh}{\pi r}$$

$$\frac{A - 2\pi r^2}{\pi r} = h$$

57. $24 \times 10^2 = 2400$ **59.** $0.003 \times 10^4 = 30$ **61.** $2 \cdot 10^2 + 3 \cdot 10 + 4 = 234$

Homework 4.2

1a. p^2 **b.** $\pm\sqrt{k}$

3a. The square root of a negative number is undefined because the square of any number is positive or zero (never negative).

b. $\sqrt{-x}$ is the square root of a negative number, which is undefined.

$-\sqrt{x}$ indicates the negative of the square root of the positive number x, which is defined.

5a.

$$4 - 2\sqrt{64} \qquad \text{Simplify root}$$
$$= 4 - 2 \cdot 8 \qquad \text{Multiply before subtract}$$
$$= 4 - 16 \qquad \text{Subtract}$$
$$= -12$$

b.

$$\frac{4 - \sqrt{64}}{2} \qquad \text{Simplify root}$$
$$= \frac{4 - 8}{2} \qquad \text{Subtract}$$
$$= \frac{-4}{2} \qquad \text{Divide}$$
$$= -2$$

73.

$3t^2 - 16 = 16t^2$ Subtract $3t^2$ from both sides

$-16 = 13t^2$ Divide both sides by 13

$\dfrac{-16}{13} = t^2$ Take square roots

$\pm\sqrt{\dfrac{-16}{13}} = t$ Note: The square root of a negative number is undefined.

No solution

75.

$A = 4\pi r^2$ Divide both sides by 4π

$\dfrac{A}{4\pi} = \dfrac{4\pi r^2}{4\pi}$

$\dfrac{A}{4\pi} = r^2$ Take the square root of both sides

$\pm\sqrt{\dfrac{A}{4\pi}} = r$

77.

$V = \dfrac{1}{3}\pi r^2 h$ Multiply both sides by 3

$3 \cdot V = 3 \cdot \dfrac{1}{3}\pi r^2 h$

$3V = \pi r^2 h$ Divide both sides by πh

$\dfrac{3V}{\pi h} = \dfrac{\pi r^2 h}{\pi h}$

$\dfrac{3V}{\pi h} = r^2$ Take the square root of both sides

$\pm\sqrt{\dfrac{3V}{\pi h}} = r$

Homework 4.3

1. A <u>rational number</u> can be written as a fraction $\frac{a}{b}$ where a and b are both integers and $b \neq 0$. For example, $2 = \frac{2}{1}$ is rational.

An <u>irrational number</u> cannot be written as a fraction $\frac{a}{b}$ where a and b are both integers and $b \neq 0$. For example, $\sqrt{2}$ is irrational.

3. Rational numbers: $\sqrt{1}, \sqrt{4}, \sqrt{9}$

Irrational numbers: $\sqrt{2}, \sqrt{3}, \sqrt{5}$

5. True   **7.** False   **9.** False

11. $\sqrt{6}$ is irrational.

13. $\sqrt{16} = 4 = \frac{4}{1}$ is rational.

15. $6.008 = 6\frac{8}{1000} = \frac{6008}{1000}$ is rational.

17. $\sqrt{250}$ is irrational

19. $\frac{\sqrt{81}}{4} = \frac{9}{4}$ is rational.

21. $2 + \sqrt{5}$ is irrational

23.

a. $x \approx \pm 3.5$ **b.** $-2.4 \leq x \leq 2.4$

25.

a. $x \approx 6.3$ **b.** $1 < x \leq 9$

27.

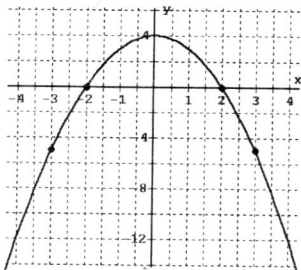

a. $x = \pm 3$ **b.** $x < -2$ or $x > 2$

29.

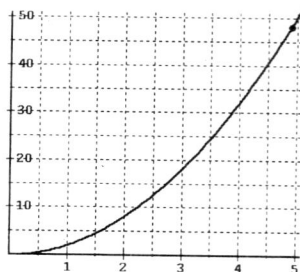

a. $x = 7$ **b.** $x > 2$

31b. $A = (2w)w = 2w^2$

c.

w	A
0	0
1	2
2	8
3	18
4	32
5	50

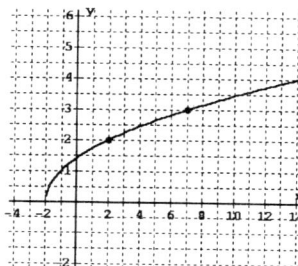

d. $w \approx 4.9$ in.

33. $v = \sqrt{2.5r}$

a.

r	v
0	0
100	15.8
200	22.4
300	27.4
400	31.6
500	35.4
600	38.7
700	41.8

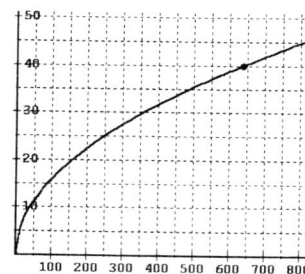

b. $r \approx 640$ ft

35. (b)   **37.** (c)   **39.** (b)

41a. $\sqrt{x^2}$

For $x = 3$: $\sqrt{3^2} = \sqrt{9} = 3$

For $x = 5$: $\sqrt{5^2} = \sqrt{25} = 5$

For $x = 8$: $\sqrt{8^2} = \sqrt{64} = 8$

For $x = 12$: $\sqrt{12^2} = \sqrt{144} = 12$

b. If $a \geq 0$, then $\sqrt{a^2} = a$

43. $(\sqrt{16})^2 = 16$

45. $(\sqrt{7})(\sqrt{7}) = (\sqrt{7})^2 = 7$

47. $(\sqrt[3]{5})^3 = 5$

49. $(\sqrt[3]{9})(\sqrt[3]{9})(\sqrt[3]{9}) = (\sqrt[3]{9})^3 = 9$

51. $\dfrac{3}{\sqrt{3}} = \dfrac{\sqrt{3}\sqrt{3}}{\sqrt{3}} = \sqrt{3}$

53. $\dfrac{-11}{\sqrt{11}} = \dfrac{-\sqrt{11}\sqrt{11}}{\sqrt{11}} = -\sqrt{11}$

55. $(\sqrt{2b})(\sqrt{2b}) = (\sqrt{2b})^2 = 2b$

57. $\dfrac{2m}{\sqrt{m}} = \dfrac{2\sqrt{m}\sqrt{m}}{\sqrt{m}} = 2\sqrt{m}$

59. $\sqrt{x}\sqrt{x} = (\sqrt{x})^2 = x$ when $x \geq 0$

61. 64 because 64 is 8^2 and 64 is 4^3.

Homework 4.4

In this section, because distance is a positive number, when extracting roots, we will take only the positive root.

1.
$$a^2 + b^2 = c^2 \qquad \text{Substitute } x \text{ for } a, \text{ 6 for } b, \text{ and 10 for } c$$
$$x^2 + 6^2 = 10^2$$
$$x^2 + 36 = 100 \qquad \text{Subtract 36 from both sides}$$
$$x^2 = 64 \qquad \text{Take square roots}$$
$$x = \sqrt{64} \qquad \text{Taking only the positive root,}$$
$$\text{because distance is positive.}$$
$$x = 8$$

3.
$$a^2 + b^2 = c^2$$
$$12^2 + 35^2 = z^2$$
$$144 + 1225 = z^2$$
$$1369 = z^2$$
$$\sqrt{1369} = z$$
$$37 = z$$

5.
$$a^2 + b^2 = c^2$$
$$p^2 + 16^2 = 18^2$$
$$p^2 + 256 = 324$$
$$p^2 = 68$$
$$p = \sqrt{68}$$

7.
$$a^2 + b^2 = c^2$$
$$r^2 + (\sqrt{10})^2 = 5^2$$
$$r^2 + 10 = 25$$
$$r^2 = 15$$
$$r = \sqrt{15}$$

9.
$$a^2 + b^2 = c^2$$
$$k^2 + 20^2 = (3k)^2$$
$$k^2 + 400 = 9k^2$$
$$400 = 8k^2$$
$$50 = k^2$$
$$\sqrt{50} = k$$

11.
$$a^2 + b^2 = c^2$$
$$9^2 + 16^2 \overset{?}{=} 25^2$$
$$81 + 256 \overset{?}{=} 625$$
$$337 = 625 \quad \text{False}$$
Therefore, this is not a right triangle.

13.
$$a^2 + b^2 = c^2$$
$$5^2 + 12^2 \overset{?}{=} 13^2$$
$$25 + 144 \overset{?}{=} 169$$
$$169 = 169 \quad \text{True}$$
Therefore, it is a right triangle.

15.
$$a^2 + b^2 = c^2$$
$$(5^2)^2 + (8^2)^2 \overset{?}{=} (13^2)^2$$
$$25^2 + 64^2 \overset{?}{=} 169^2$$
$$625 + 4096 \overset{?}{=} 28{,}561$$
$$4721 = 28{,}561 \qquad \text{False}$$
Therefore, it is not a right triangle.

17.

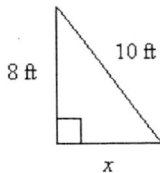

$$a^2 + b^2 = c^2$$
$$x^2 + 8^2 = 10^2$$
$$x^2 + 64 = 100$$
$$x^2 = 36$$
$$x = \sqrt{36}$$
$$x = 6; \ \text{Yes}$$

19.

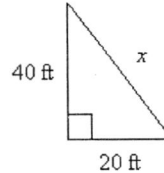

$$a^2 + b^2 = c^2$$
$$20^2 + 40^2 = x^2$$
$$400 + 1600 = x^2$$
$$2000 = x^2$$
$$\sqrt{2000} = x$$
$$44.72 \text{ ft} \approx x$$

21.

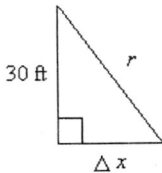

$$m = 1.2 = 1\frac{2}{10} = \frac{12}{10}$$
$$\frac{30}{\Delta x} = \frac{12}{10}$$
$$300 = 12 \cdot \Delta x$$
$$\frac{300}{12} = \frac{12 \cdot \Delta x}{12}$$
$$25 \text{ ft} = \Delta x$$

$$a^2 + b^2 = c^2$$
$$25^2 + 30^2 = r^2$$
$$625 + 900 = r^2$$
$$1525 = r^2$$
$$\sqrt{1525} = r$$
$$39.05 \text{ ft} \approx r$$

23.
$$a^2 + b^2 = c^2$$
$$x^2 + 3^2 = 4^2$$
$$x^2 + 9 = 16$$
$$x^2 = 7$$
$$x = \sqrt{7}$$
$$x \approx 2.65 \text{ mi}$$

25.

a	b	$a+b$	$(a+b)^2$	a^2	b^2	a^2+b^2
2	3	5	25	4	9	13
3	4	7	49	9	16	25
1	5	6	36	1	25	26
−2	6	4	16	4	36	40

No, $(a+b)^2 \neq a^2 + b^2$.

27.

a	b	$a+b$	$\sqrt{a+b}$	$\sqrt{a}$	$\sqrt{b}$	$\sqrt{a}+\sqrt{b}$
2	7	9	3	$\sqrt{2}$	$\sqrt{7}$	4.1
4	9	13	3.6	2	3	5
1	5	6	2.4	1	$\sqrt{5}$	3.2
9	16	25	5	3	4	7

No, $\sqrt{a+b} \neq \sqrt{a} + \sqrt{b}$.

29.

$\sqrt{x^2-4}$
$\sqrt{3^2-4}$
$= \sqrt{9-4}$
$= \sqrt{5}$
≈ 2.236

$\sqrt{x^2-4}$
$\sqrt{(\sqrt{5})^2-4}$
$= \sqrt{5-4}$
$= \sqrt{1}$
$= 1$

$\sqrt{x^2-4}$
$\sqrt{(-2)^2-4}$
$= \sqrt{4-4}$
$= \sqrt{0}$
$= 0$

31.

$\sqrt{x} - \sqrt{x+3}$
$\sqrt{1} - \sqrt{1+3}$
$= \sqrt{1} - \sqrt{4}$
$= 1 - 2$
$= -1$

$\sqrt{x} - \sqrt{x+3}$
$\sqrt{4} - \sqrt{4+3}$
$= \sqrt{4} - \sqrt{7}$
$= 2 - \sqrt{7}$
≈ -0.646

$\sqrt{x} - \sqrt{x+3}$
$\sqrt{100} - \sqrt{100+3}$
$= \sqrt{100} - \sqrt{103}$
$= 10 - \sqrt{103}$
≈ -0.149

33.

$2x^2 - 4x$
$2(-3)^2 - 4(-3)$
$= 2(9) - 4(-3)$
$= 18 + 12$
$= 30$

$2x^2 - 4x$
$2(\sqrt{2})^2 - 4(\sqrt{2})$
$= 2(2) - 4\sqrt{2}$
$= 4 - 4\sqrt{2}$
≈ -1.657

$2x^2 - 4x$
$2\left(\dfrac{1}{4}\right)^2 - 4\left(\dfrac{1}{4}\right)$
$= 2\left(\dfrac{1}{16}\right) - 4\left(\dfrac{1}{4}\right)$
$= \dfrac{1}{8} - 1$
$= \dfrac{1}{8} - \dfrac{8}{8}$
$= \dfrac{-7}{8}$

35. $\dfrac{5}{4}$, 2, 2.3, $\sqrt{8}$

37. $\sqrt{6}$, 3, $2\sqrt{3}$, $\dfrac{23}{6}$

39. $t = \pm 100$

41. $u = \pm\dfrac{1}{7}$

43. $n = \pm 0.5$

45a. To find height:
$a^2 + b^2 = c^2$
$h^2 + 3^2 = 6^2$
$h^2 + 9 = 36$
$h^2 = 27$
$h = \sqrt{27}$ ft

b. To find area:
$A = \dfrac{1}{2}bh$
$= \dfrac{1}{2}(6)(\sqrt{27})$
$= 3\sqrt{27}$ sq ft

47a. To find diagonal of base:

$$a^2 + b^2 = c^2$$
$$2^2 + 2^2 = d^2$$
$$4 + 4 = d^2$$
$$8 = d^2$$
$$\sqrt{8} \text{ cm} = d$$

b. To find height of pyramid:

$$a^2 + b^2 = c^2$$
$$h^2 + \left(\frac{\sqrt{8}}{2}\right)^2 = 4^2$$
$$h^2 + \frac{8}{4} = 16$$
$$h^2 + 2 = 16$$
$$h^2 = 14$$
$$h = \sqrt{14}$$

c. To find volume:

$$V = \frac{1}{3}s^2 h$$
$$= \frac{1}{3}(2)^2(\sqrt{14})$$
$$= \frac{1}{3}(4)\sqrt{14}$$
$$= \frac{4\sqrt{14}}{3} \text{ cu cm}$$

49. To find shorter leg:

$$A = s^2$$
$$2 = a^2$$
$$\sqrt{2} = a$$

To find longer leg:

$$A = s^2$$
$$16 = b^2$$
$$\sqrt{16} = b$$
$$4 = b$$

To find hypotenuse:

$$a^2 + b^2 = c^2$$
$$(\sqrt{2})^2 + 4^2 = c^2$$
$$2 + 16 = c^2$$
$$18 = c^2$$
$$\sqrt{18} = c$$

To find area of largest square:

$$A = s^2$$
$$= (\sqrt{18})^2$$
$$= 18$$

51. True; $\dfrac{15}{\sqrt{5}} = \dfrac{3 \cdot 5}{\sqrt{5}} = \dfrac{3 \cdot \sqrt{5}\sqrt{5}}{\sqrt{5}} = 3\sqrt{5}$

53. True; $(\sqrt{13})^3 = \sqrt{13}\sqrt{13}\sqrt{13} = 13\sqrt{13}$

55. False; Can add only like terms that have the same variable with the same exponent, $5x + 3x^2 \neq 8x^3$

57. False; $(\sqrt{2}m)(\sqrt{2}m) = (\sqrt{2}\sqrt{2})(m \cdot m) = 2m^2 \neq 2m$

59. True; $(\sqrt[3]{b})(2\sqrt[3]{b})(3\sqrt[3]{b}) = (2 \cdot 3)(\sqrt[3]{b}\sqrt[3]{b}\sqrt[3]{b}) = 6b$

Homework 4.5

1a.

	$5a$	3
$2a$	$10a^2$	$6a$

$A = 10a^2 + 6a$

b. $2a(5a + 3)$
$= 2a(5a) + 2a(3)$
$= 10a^2 + 6a$

3a.

	$2x$	-5	y
$3xy$	$6x^2y$	$-15xy$	$3xy^2$

$A = 6x^2y - 15xy + 3xy^2$

b. $3xy(2x - 5 + y)$
$= 3xy(2x) + 3xy(-5) + 3xy(y)$
$= 6x^2y - 15xy + 3xy^2$

5. $-2b(6b - 2) = -12b^2 + 4b$

7. $-4x^2(2x + 3y) = -8x^3 - 12x^2y$

9. $(y^3 + 3y - 2)(2y) = 2y^4 + 6y^2 - 4y$

11. $-xy(2x^2 - xy + 3y^2) = -2x^3y + x^2y^2 - 3xy^3$

13. $2a(x + 3) - 3a(x - 3)$
$= 2ax + 6a - 3ax + 9a$
$= -ax + 15a$

15. $2x(3 - x) + 2(x^2 + 1) - 2x$
$= 6x - 2x^2 + 2x^2 + 2 - 2x$
$= 4x + 2$

17. -3

19. $-3m$

21. $4; \ -3h$

23.

	40	2
30	1200	60
6	240	12

36×42
$= 1200 + 60 + 240 + 12$
$= 1512$

25.

	10	6
80	800	480
2	20	12

82×16
$= 800 + 480 + 20 + 12$
$= 1312$

27a. $x^2 + 4x + 3x + 12$ **b.** $(x+3)(x+4)$

29a. $4x^2 - 4x - 2x + 2$ **b.** $(4x-2)(x-1)$

31a.

	a	-3
a	a^2	$-3a$
-5	$-5a$	15

b. $(a-5)(a-3)$
$= a^2 - 3a - 5a + 15$
$= a^2 - 8a + 15$

33a.

	$3y$	-2
y	$3y^2$	$-2y$
1	$3y$	-2

b. $(y+1)(3y-2)$
$= 3y^2 - 2y + 3y - 2$
$= 3y^2 + y - 2$

35a.

	$4x$	3
$5x$	$20x^2$	$15x$
-2	$-8x$	-6

b. $(5x-2)(4x+3)$
$= 20x^2 + 15x - 8x - 6$
$= 20x^2 + 7x - 6$

37. $(x+2y)(x-y)$
$= x^2 - xy + 2xy - 2y^2$
$= x^2 + xy - 2y^2$

39. $(3s+t)(2s+3t)$
$= 6s^2 + 9st + 2st + 3t^2$
$= 6s^2 + 11st + 3t^2$

41. $(2x-a)(x-3a)$
$= 2x^2 - 6ax - ax + 3a^2$
$= 2x^2 - 7ax + 3a^2$

43a. $-9x + 6x = -3x$

b.

	x	-9
x	x^2	$-9x$
6	$6x$	-54

45a. $8x - 5x = 3x$

b.

	x	4
$2x$	$2x^2$	$8x$
-5	$-5x$	-20

47. $2(3x-1)(x-3)$
$= 2(3x^2 - 9x - x + 3)$
$= 2(3x^2 - 10x + 3)$
$= 6x^2 - 20x + 6$

49. $-3(x+4)(x-1)$
$= -3(x^2 - x + 4x - 4)$
$= -3(x^2 + 3x - 4)$
$= -3x^2 - 9x + 12$

51. $-(4x+3)(x-2)$
$= -(4x^2 - 8x + 3x - 6)$
$= -(4x^2 - 5x - 6)$
$= -4x^2 + 5x + 6$

53. $(x+3)(x-3)$
$= x^2 - 3x + 3x - 9$
$= x^2 - 9$

55. $(x-2a)(x+2a)$
$= x^2 + 2ax - 2ax - 4a^2$
$= x^2 - 4a^2$

57. $(3x+1)(3x-1)$
$= 9x^2 - 3x + 3x - 1$
$= 9x^2 - 1$

In **53-57**, there is no linear (middle) term in the simplified form because the inner and the outer terms are opposites and add to zero.

59. $(w+4)(w+4)$
$= w^2 + 4w + 4w + 16$
$= w^2 + 8w + 16$

61. $(z-6)(z-6)$
$= z^2 - 6z - 6z + 36$
$= z^2 - 12z + 36$

63. $(3a-2c)(3a-2c)$
$= 9a^2 - 6ac - 6ac + 4c^2$
$= 9a^2 - 12ac + 4c^2$

65.

a	b	$a-b$	$(a-b)^2$	a^2	b^2	a^2-b^2
5	3	2	4	25	9	16
2	6	-4	16	4	36	-32
-4	-3	-1	1	16	9	7

No, $(a-b)^2 \neq a^2 - b^2$.

67a. $x^2 + 7x + 7x + 49$ **b.** $(x+7)^2$ **69a.** $x^2 - 5x - 5x + 25$ **b.** $(x-5)^2$

71. No; $(x+4)^2$ will have a linear (middle) term because both the inner and the outer terms are $4x$, and their sum will be $8x$

For $x = 1$: $(x+4)^2 = (1+4)^2 = 5^2 = 25$
$x^2 + 4^2 = 1^2 + 4^2 = 1 + 16 = 17$

73.
$(x-2)^2$
$= (x-2)(x-2)$
$= x^2 - 2x - 2x + 4$
$= x^2 - 4x + 4$

75.
$(2x+1)^2$
$= (2x+1)(2x+1)$
$= 4x^2 + 2x + 2x + 1$
$= 4x^2 + 4x + 1$

77.
$(3x-4y)^2$
$= (3x-4y)(3x-4y)$
$= 9x^2 - 12xy - 12xy + 16y^2$
$= 9x^2 - 24xy + 16y^2$

79.
$a^2 + b^2 = c^2$
$h^2 + (\sqrt{7})^2 = (h+1)^2$

81.
$a^2 + b^2 = c^2$
$12^2 + (x-1)^2 = (2x+1)^2$

Midchapter 4 Review

1. -3^2 means the negative of the square of 3: $-3^2 = -(3 \cdot 3) = -9$. The exponent applies only to 3
$(-3)^2$ means the square of negative 3: $(-3)^2 = (-3)(-3) = 9$. The exponent applies to (-3)

2. square root

3. cube root

4. "The square of 4" means 4^2 and "the square root of 4" means $\sqrt{4}$.

5. $x^2 = 9$ is true for $x = \pm 3$, but $x = \sqrt{9}$ is true only for $x = 3$.

6. $-\sqrt{25}$ means -5, but $\sqrt{-25}$ is undefined.

7. The Pythagorean theorem only applies to <u>right</u> triangles, and c must be the <u>hypotenuse</u>.

8. No, a rational number can be written as a fraction $\frac{a}{b}$ where a and b are both integers, and $\sqrt[3]{2}$ is not an integer.

9. $-\sqrt{2}$ (other answers possible)

10. $(\sqrt[3]{5})^2(\sqrt[3]{5}) = (\sqrt[3]{5})(\sqrt[3]{5})(\sqrt[3]{5}) = (\sqrt[3]{5})^3 = 5$

11a. $3(-7)^2 = 3(49) = 147$ **b.** $3^2(-7^2) = 9(-49) = -441$

12a. $3^2 - 7^2 = 9 - 49 = -40$ **b.** $(3-7)^2 = (-4)^2 = 16$

13a. $\sqrt{13^2 - 5^2} = \sqrt{169 - 25} = \sqrt{144} = 12$ **b.** $\sqrt{13^2} - \sqrt{5^2} = \sqrt{169} - \sqrt{25} = 13 - 5 = 8$

14a. $\sqrt{13^2}\sqrt{5^2} = \sqrt{169}\sqrt{25} = 13 \cdot 5 = 65$ **b.** $\sqrt{(13-5)^2} = \sqrt{8^2} = \sqrt{64} = 8$

15a. $\sqrt{\left(\dfrac{4}{5}\right)^2 + \left(\dfrac{3}{5}\right)^2}$

$= \sqrt{\dfrac{16}{25} + \dfrac{9}{25}}$

$= \sqrt{\dfrac{25}{25}}$

$= \sqrt{1}$

$= 1$

b. $\sqrt{\left(\dfrac{4}{5}\right)^2} + \sqrt{\left(\dfrac{3}{5}\right)^2}$

$= \sqrt{\dfrac{16}{25}} + \sqrt{\dfrac{9}{25}}$

$= \dfrac{4}{5} + \dfrac{3}{5}$

$= \dfrac{7}{5}$

16a. $\sqrt{\left(\dfrac{4}{5}\right)^2 \left(\dfrac{3}{5}\right)^2}$

$= \sqrt{\dfrac{16}{25} \cdot \dfrac{9}{25}}$

$= \sqrt{\dfrac{144}{625}}$

$= \dfrac{12}{25}$

b. $\sqrt{\left(\dfrac{4}{5} + \dfrac{3}{5}\right)^2}$

$= \sqrt{\left(\dfrac{7}{5}\right)^2}$

$= \sqrt{\dfrac{49}{25}}$

$= \dfrac{7}{5}$

17a. $2 - \sqrt[3]{64} = 2 - 4 = -2$

b. $2\sqrt[3]{-64} = 2(-4) = -8$

18a. $\left(\sqrt[3]{101}\right)^3 = 101$

b. $\sqrt[3]{7} \cdot \sqrt[3]{7} \cdot \sqrt[3]{7} = \left(\sqrt[3]{7}\right)^3 = 7$

19a. $\left(\sqrt[3]{8}\right)^2 = 2^2 = 4$

b. $\sqrt[3]{8^2} = \sqrt[3]{64} = 4$

20a. $\left(\sqrt{9}\right)^3 = 3^3 = 27$

b. $\sqrt{9^3} = \sqrt{729} = 27$

21a. $\sqrt[3]{-5}$ is irrational.

c. $\sqrt{(-5)^2} = \sqrt{25} = 5$ is rational.

b. $\sqrt{-5^2} = \sqrt{-25}$ is undefined.

d. $(\sqrt[3]{-5})^2$ is irrational.

22a. $-3.1\overline{6}$ is rational.

c. $-\sqrt{10}$ is irrational.

b. -3.16 is rational.

d. $-\pi$ is irrational.

23a. $5 - a^2$

$5 - 3^2$

$= 5 - 9$

$= -4$

b. $5 - a^2$

$5 - (-3)^2$

$= 5 - 9$

$= -4$

24a. $(5 - a)^2$

$(5 - 3)^2$

$= 2^2$

$= 4$

b. $(5 - a)^2$

$[5 - (-3)]^2$

$= 8^2$

$= 64$

25a. $-w^2$

$-(-2)^2$

$= -4$

b. $(-w)^2$

$[-(-2)]^2$

$= 2^2$

$= 4$

26a. $\sqrt{-w}$

$\sqrt{-(-2)}$

$= \sqrt{2}$

≈ 1.414

b. $\sqrt[3]{-w}$

$\sqrt[3]{-(-2)}$

$= \sqrt[3]{2}$

≈ 1.260

27a. $\sqrt{x^2 - y^2}$

$\sqrt{13^2 - (-5)^2}$

$= \sqrt{169 - 25}$

$= \sqrt{144}$

$= 12$

b. $\left(\sqrt{x}\right)^2 + \left(\sqrt{-y}\right)^2$

$\left(\sqrt{13}\right)^2 + \left(\sqrt{-(-5)}\right)^2$

$= \left(\sqrt{13}\right)^2 + \left(\sqrt{5}\right)^2$

$= 13 + 5$

$= 18$

28a. $\sqrt{b^2 - 4ac}$
$\sqrt{(-3)^2 - 4(2)(1)}$
$= \sqrt{9 - 8}$
$= \sqrt{1}$
$= 1$

b. $\dfrac{-b + \sqrt{b^2 - 4ac}}{2a}$
$\dfrac{-(-3) + \sqrt{(-3)^2 - 4(2)(1)}}{2(2)}$
$= \dfrac{3 + 1}{4}$
$= \dfrac{4}{4}$
$= 1$

29a. $\sqrt[3]{x^3 + y^3 + z^3}$
$\sqrt[3]{3^3 + 4^3 + 5^3}$
$= \sqrt[3]{27 + 64 + 125}$
$= \sqrt[3]{216}$
$= 6$

b. $\sqrt[3]{x^3} + \sqrt[3]{y^3} + \sqrt[3]{z^3}$
$\sqrt[3]{3^3} + \sqrt[3]{4^3} + \sqrt[3]{5^3}$
$= 3 + 4 + 5$
$= 12$

30a. $m^3 + n^3$
$3^3 + 2^3$
$= 27 + 8$
$= 35$

b. $(m + n)(m^2 - mn + n^2)$
$(3 + 2)(3^2 - 3 \cdot 2 + 2^2)$
$= (5)(9 - 6 + 4)$
$= (5)(7)$
$= 35$

31. $V = lwh$
$= (3\sqrt{7})(3)(2\sqrt{7})$
$= (3 \cdot 3 \cdot 2)(\sqrt{7}\sqrt{7})$
$= (18)(7)$
$= 126$

$S = 2lw + 2wh + 2lh$
$= 2(3\sqrt{7})(3) + 2(3)(2\sqrt{7}) + 2(3\sqrt{7})(2\sqrt{7})$
$= (2 \cdot 3 \cdot 3)\sqrt{7} + (2 \cdot 3 \cdot 2)\sqrt{7} + (2 \cdot 3 \cdot 2)(\sqrt{7}\sqrt{7})$
$= 18\sqrt{7} + 12\sqrt{7} + (12)(7)$
$= 30\sqrt{7} + 84$

32. $V = lwh$
$= (8)(\sqrt{5})(\sqrt{5})$
$= (8)(5)$
$= 40$

$S = 2lw + 2wh + 2lh$
$= 2(8)(\sqrt{5}) + 2(\sqrt{5})(\sqrt{5}) + 2(8)(\sqrt{5})$
$= (2 \cdot 8)\sqrt{5} + 2(5) + (2 \cdot 8)\sqrt{5}$
$= 16\sqrt{5} + 10 + 16\sqrt{5}$
$= 32\sqrt{5} + 10$

33. $9x^2 - 4 = 0$
$9x^2 = 4$
$x^2 = \dfrac{4}{9}$
$x = \pm\sqrt{\dfrac{4}{9}}$
$x = \pm\dfrac{2}{3}$

34. $3 + 2t^2 = 9$
$2t^2 = 6$
$t^2 = 3$
$t = \pm\sqrt{3}$

35. $5y^2 - 12 = 2y^2$
$3y^2 - 12 = 0$
$3y^2 = 12$
$y^2 = 4$
$y = \pm\sqrt{4}$
$y = \pm 2$

36. $q^2 - 3 = 5 - q^2$
$2q^2 - 3 = 5$
$2q^2 = 8$
$q^2 = 4$
$q = \pm\sqrt{4}$
$q = \pm 2$

37.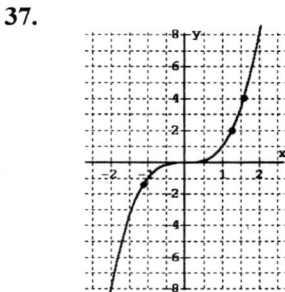

a. $x \approx 1.3$
b. $x \approx -1.1$
c. $x \approx 1.6$

38.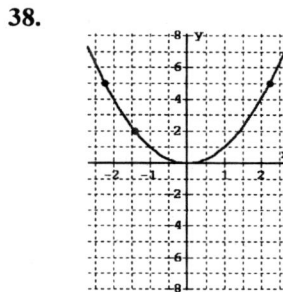

a. $x \approx \pm 2.2$
b. No solution
c. $x \approx -1.4$

39.
$$s = vt + \frac{1}{2}at^2 \qquad \text{Multiply both sides by 2}$$

$$2 \cdot s = 2\left(vt + \frac{1}{2}at^2\right) \qquad \text{Distribute 2 to each term}$$

$$2 \cdot s = 2 \cdot vt + 2 \cdot \frac{1}{2}at^2$$

$$2s = 2vt + at^2 \qquad \text{Subtract } at^2 \text{ from both sides}$$

$$2s - at^2 = 2vt + at^2 - at^2$$

$$2s - at^2 = 2vt \qquad \text{Divide both sides by } 2t$$

$$\frac{2s - at^2}{2t} = \frac{2vt}{2t}$$

$$\frac{2s}{2t} - \frac{at^2}{2t} = v \qquad \text{Reduce each fraction}$$

$$\frac{s}{t} - \frac{at}{2} = v$$

40.
$$A = \frac{h}{2}(b + c) \qquad \text{Multiply both sides by 2}$$

$$2 \cdot A = 2 \cdot \frac{h}{2}(b + c)$$

$$2A = h(b + c) \qquad \text{Divide both sides by } h$$

$$\frac{2A}{h} = \frac{h(b + c)}{h}$$

$$\frac{2A}{h} = b + c \qquad \text{Subtract } b \text{ from both sides}$$

$$\frac{2A}{h} - b = b + c - b$$

$$\frac{2A}{h} - b = c$$

41.
$$V = \frac{s^2 h}{3} \qquad \text{Multiply both sides by 3}$$

$$3 \cdot V = 3 \cdot \frac{s^2 h}{3}$$

$$3V = s^2 h \qquad \text{Divide both sides by } h$$

$$\frac{3V}{h} = \frac{s^2 h}{h}$$

$$\frac{3V}{h} = s^2 \qquad \text{Take the square root of both sides}$$

$$\pm\sqrt{\frac{3V}{h}} = s$$

42.

$$A = \frac{\pi d^2}{4}$$ Multiply both sides by 4

$$4 \cdot A = 4 \cdot \frac{\pi d^2}{4}$$

$$4A = \pi d^2$$ Divide both sides by π

$$\frac{4A}{\pi} = \frac{\pi d^2}{\pi}$$

$$\frac{4A}{\pi} = d^2$$ Take the square root of both sides

$$\pm\sqrt{\frac{4A}{\pi}} = d$$

43a.
$$d = 16t^2$$
$$144 = 16t^2$$
$$9 = t^2$$
$$\sqrt{9} = t$$
$$3 \text{ sec} = t$$

b.
$$d = 16t^2$$
$$d = 16(1.5)^2$$
$$d = 16(2.25)$$
$$d = 36 \text{ ft}$$

44a.
$$V = \frac{4}{3}\pi r^3$$
$$V = \frac{4}{3}\pi(3)^3$$
$$V = \frac{4}{3}\pi(27)$$
$$V = 36\pi$$
$$V \approx 113 \text{ cu ft}$$

b.
$$V = \frac{4}{3}\pi r^3$$
$$7000 = \frac{4}{3}\pi r^3$$
$$21{,}000 = 4\pi r^3$$
$$\frac{21{,}000}{4\pi} = r^3$$
$$\sqrt[3]{\frac{21{,}000}{4\pi}} = r$$
$$\sqrt[3]{1671.126\ldots} = r$$
$$11.87 \text{ in.} \approx r$$

45.
$$V = \frac{1}{3}s^2 h$$
$$V = \frac{1}{3}(4b)^2(3b)$$
$$V = \frac{1}{3}(16b^2)(3b)$$
$$V = 16b^3$$

46.
$$V = \frac{1}{3}\pi r^2 h$$
$$V = \frac{1}{3}\pi(3c)^2(5c)$$
$$V = \frac{1}{3}\pi(9c^2)(5c)$$
$$V = 15\pi c^3$$

47a.
$$a^2 + b^2 = c^2$$
$$15^2 + 8^2 = d^2$$
$$225 + 64 = d^2$$
$$289 = d^2$$
$$\sqrt{289} = d$$
$$17 = d$$

b.
$$P = 15 + 8 + 17$$
$$= 40$$

c.
$$A = \frac{1}{2}bh$$
$$= \frac{1}{2}(8)(15)$$
$$= 60$$

48a.
$$a^2 + b^2 = c^2$$
$$m^2 + 5^2 = 9^2$$
$$m^2 + 25 = 81$$
$$m^2 = 56$$
$$m = \sqrt{56}$$

b.
$$P = 5 + 9 + \sqrt{56}$$
$$= 14 + \sqrt{56}$$

c.
$$A = \frac{1}{2}bh$$
$$= \frac{1}{2}(5)(\sqrt{56})$$
$$= \frac{5\sqrt{56}}{2}$$

49a.
$$a^2 + b^2 = c^2$$
$$r^2 + (\sqrt{3})^2 = 2^2$$
$$r^2 + 3 = 4$$
$$r^2 = 1$$
$$r = \sqrt{1}$$
$$r = 1$$

b.
$$P = 1 + 2 + \sqrt{3}$$
$$= 3 + \sqrt{3}$$

c.
$$A = \frac{1}{2}bh$$
$$= \frac{1}{2}(1)(\sqrt{3})$$
$$= \frac{\sqrt{3}}{2}$$

50a.
$$a^2 + b^2 = c^2$$
$$s^2 + s^2 = 2^2$$
$$2s^2 = 4$$
$$s^2 = 2$$
$$s = \sqrt{2}$$

b.
$$P = 2 + \sqrt{2} + \sqrt{2}$$
$$= 2 + 2\sqrt{2}$$

c.
$$A = \frac{1}{2}bh$$
$$= \frac{1}{2}(\sqrt{2})(\sqrt{2})$$
$$= \frac{1}{2}(2)$$
$$= 1$$

51.

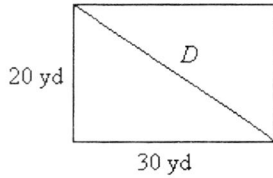

$$a^2 + b^2 = c^2$$
$$20^2 + 30^2 = D^2$$
$$400 + 900 = D^2$$
$$1300 = D^2$$
$$\sqrt{1300} = D$$
$$36.1 \text{ yd} \approx D$$

$$F = 20 + 30 = 50 \text{ yd}$$

$$\text{Difference} = F - D$$
$$= 50 - 36.1$$
$$= 13.9 \text{ yd}$$

52.

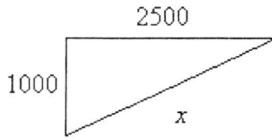

$$a^2 + b^2 = c^2$$
$$2500^2 + 1000^2 = x^2$$
$$6{,}250{,}000 + 1{,}000{,}000 = x^2$$
$$7{,}250{,}000 = x^2$$
$$\sqrt{7{,}250{,}000} = x$$
$$2692.6 \text{ m} \approx x$$

$$\text{Total} = 2500 + 1000 + 2692.6$$
$$= 6192.6 \text{ m}$$

53. $-5x(4 - 3x)$
$= -5x(4) - 5x(-3x)$
$= -20x + 15x^2$

54. $(3xy - 4x^2 + 2)(-2xy)$
$= 3xy(-2xy) - 4x^2(-2xy) + 2(-2xy)$
$= -6x^2y^2 + 8x^3y - 4xy$

55. $6a(2a - 1) - (3a^2 - 3a)$
$= 12a^2 - 6a - 3a^2 + 3a$
$= 9a^2 - 3a$

56. $4b - 2(3 - b^2) - b(b - 3)$
$= 4b - 6 + 2b^2 - b^2 + 3b$
$= b^2 + 7b - 6$

57. 1 **58.** $-7b$ **59a.** $5x^2 - 2x + 5x - 2$ **b.** $(x + 1)(5x - 2)$

60a. $6a^2 + 21a - 8a - 28$ **b.** $(3a - 4)(2a + 7)$

61. $(x - 2)(x + 4)$
$= x^2 + 4x - 2x - 8$
$= x^2 + 2x - 8$

62. $(2y + 1)(3y - 2)$
$= 6y^2 - 4y + 3y - 2$
$= 6y^2 - y - 2$

63. $(5a + 1)(5a - 1)$
$= 25a^2 - 5a + 5a - 1$
$= 25a^2 - 1$

64. $(n - 7)(n - 7)$
$= n^2 - 7n - 7n + 49$
$= n^2 - 14n + 49$

65. $(2q + 5)(2q + 5)$
$= 4q^2 + 10q + 10q + 25$
$= 4q^2 + 20q + 25$

66. $(3c - 8)(3c + 8)$
$= 9c^2 + 24c - 24c - 64$
$= 9c^2 - 64$

Homework 4.6

1a. $y = x^2 + 1$

x	y
-3	10
-2	5
-1	2
0	1
1	2
2	5
3	10

b. $y = x^2 - 3$

x	y
-3	6
-2	1
-1	-2
0	-3
1	-2
2	1
3	6

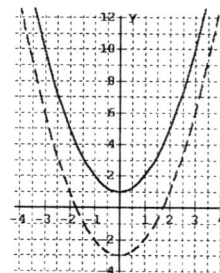

118

3a. $y = -x^2 - 2$

x	y
-3	-11
-2	-6
-1	-3
0	-2
1	-3
2	-6
3	-11

b. $y = 5 - x^2$

x	y
-3	-4
-2	1
-1	4
0	5
1	4
2	1
3	-4

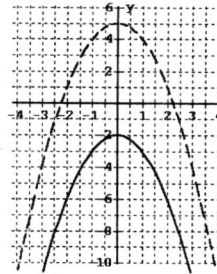

5.
$y = x^2 + 1$	vertex $= (0, 1)$
$y = x^2 - 3$	vertex $= (0, -3)$
$y = x^2 - 1$	vertex $= (0, -1)$
$y = x^2 + 2$	vertex $= (0, 2)$
$y = -x^2 - 2$	vertex $= (0, -2)$
$y = 5 - x^2$	vertex $= (0, 5)$
$y = -x^2 + 3$	vertex $= (0, 3)$
$y = 9 - x^2$	vertex $= (0, 9)$

7a. $y = (x + 2)^2$

x	y
-4	4
-3	1
-2	0
-1	1
0	4
1	9
2	16

b. $y = (x - 1)^2$

x	y
-3	16
-2	9
-1	4
0	1
1	0
2	1
3	4

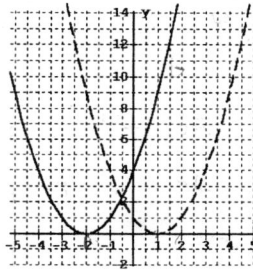

9a. $y = -(x + 1)^2$

x	y
-3	-4
-2	-1
-1	0
0	-1
1	-4
2	-9
3	-16

b. $y = -(x - 4)^2$

x	y
0	-16
1	-9
2	-4
3	-1
4	0
5	-1
6	-4

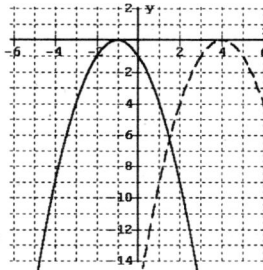

11.
$y = (x + 2)^2$	vertex $= (-2, 0)$
$y = (x - 1)^2$	vertex $= (1, 0)$
$y = (x - 2)^2$	vertex $= (2, 0)$
$y = (x + 3)^2$	vertex $= (-3, 0)$
$y = -(x + 1)^2$	vertex $= (-1, 0)$
$y = -(x - 4)^2$	vertex $= (4, 0)$
$y = -(x - 3)^2$	vertex $= (3, 0)$
$y = -(x + 4)^2$	vertex $= (-4, 0)$

119

13a.

x	y
0	11
1	6
2	3
3	2
4	3
5	6
6	11

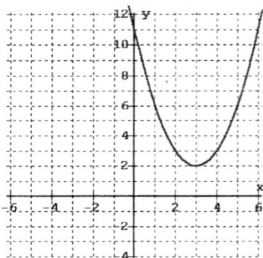

b. $y = 2 + (x-3)^2$ vertex $= (3,2)$

c. $y = 2 + (x-3)(x-3)$
$y = 2 + x^2 - 3x - 3x + 9$
$y = x^2 - 6x + 11$

15a.

x	y
-3	0
-2	3
-1	4
0	3
1	0
2	-5
3	-12

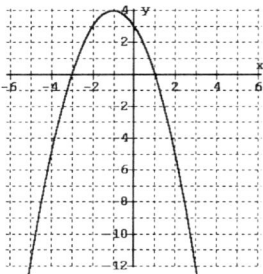

b. $y = 4 - (x+1)^2$ vertex $= (-1, 4)$

c. $y = 4 - (x+1)(x+1)$
$y = 4 - (x^2 + x + x + 1)$
$y = 4 - x^2 - x - x - 1$
$y = -x^2 - 2x + 3$

17.

x	-5	-4	-3	-2	-1	0	1	2	3
y	7	0	-5	-8	-9	-8	-5	0	7

a. $x = -4,\ x = 2$

b.
Check $x = -4$:

$x^2 + 2x - 8 = 0$
$(-4)^2 + 2(-4) - 8 \overset{?}{=} 0$
$16 - 8 - 8 \overset{?}{=} 0$
$0 = 0$

Check $x = 2$:

$x^2 + 2x - 8 = 0$
$(2)^2 + 2(2) - 8 \overset{?}{=} 0$
$4 + 4 - 8 \overset{?}{=} 0$
$0 = 0$

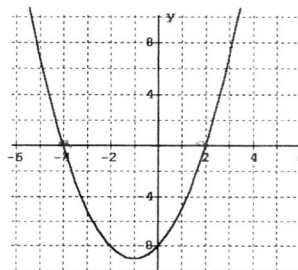

c.
$(x+4)(x-2)$
$= x^2 - 2x + 4x - 8$
$= x^2 + 2x - 8$

d.
When $x = -4$:
$(x+4)(x-2)$
$= (-4+4)(-4-2)$
$= (0)(-6)$
$= 0$

When $x = 2$:
$(x+4)(x-2)$
$= (2+4)(2-2)$
$= (6)(0)$
$= 0$

19.

x	-1	0	1	2	3	4	5	6	7
y	-7	0	5	8	9	8	5	0	-7

a. $x = 0,\ x = 6$

b.
Check $x = 0$:

$-x^2 + 6x = 0$
$-(0)^2 + 6(0) \overset{?}{=} 0$
$0 + 0 \overset{?}{=} 0$
$0 = 0$

Check $x = 6$:

$-x^2 + 6x = 0$
$-(6)^2 + 6(6) \overset{?}{=} 0$
$-36 + 36 \overset{?}{=} 0$
$0 = 0$

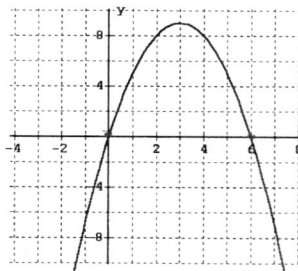

120

c. $-x(x-6) = -x^2 + 6x$

d.
When $x = 0$:
$$-x(x-6)$$
$$= -0(0-6)$$
$$= 0(-6)$$
$$= 0$$

When $x = 6$:
$$-x(x-6)$$
$$= -6(6-6)$$
$$= -6(0)$$
$$= 0$$

21.

x	-1	0	1	2	3	4	5	6	7
y	16	9	4	1	0	1	4	9	16

a. $x = 3$

b. Check $x = 3$:
$$x^2 - 6x + 9 = 0$$
$$(3)^2 - 6(3) + 9 \overset{?}{=} 0$$
$$9 - 18 + 9 \overset{?}{=} 0$$
$$0 = 0$$

c.
$$(x-3)^2$$
$$= (x-3)(x-3)$$
$$= x^2 - 3x - 3x + 9$$
$$= x^2 - 6x + 9$$

d.
When $x = 3$:
$$(x-3)^2$$
$$= (3-3)^2$$
$$= 0^2$$
$$= 0$$

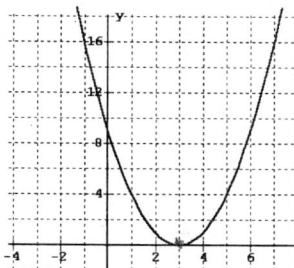

23.

x	-1	0	1	2	3	4	5	6	7
y	-24	-10	0	6	8	6	0	-10	-24

a. $x = 1$ or $x = 5$

b.
Check $x = 1$:
$$-2x^2 + 12x - 10 = 0$$
$$-2(1)^2 + 12(1) - 10 \overset{?}{=} 0$$
$$-2 + 12 - 10 \overset{?}{=} 0$$
$$0 = 0$$

Check $x = 5$:
$$-2x^2 + 12x - 10 = 0$$
$$-2(5)^2 + 12(5) - 10 \overset{?}{=} 0$$
$$-50 + 60 - 10 \overset{?}{=} 0$$
$$0 = 0$$

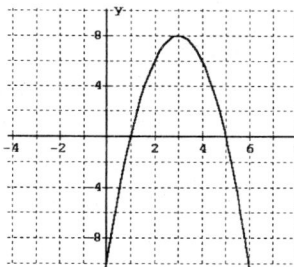

c.
$$-2(x-1)(x-5)$$
$$= -2(x^2 - 5x - x + 5)$$
$$= -2(x^2 - 6x + 5)$$
$$= -2x^2 + 12x - 10$$

d.
When $x = 1$:
$$-2(x-1)(x-5)$$
$$= -2(1-1)(1-5)$$
$$= -2(0)(-4)$$
$$= 0$$

When $x = 5$:
$$-2(x-1)(x-5)$$
$$= -2(5-1)(5-5)$$
$$= -2(4)(0)$$
$$= 0$$

25. To find the solutions to a quadratic equation, set each factor equal to zero and solve for x.

27a.

t	h
0	48
0.5	60
1	64
1.5	60
2	48
2.5	28
3	0

b.

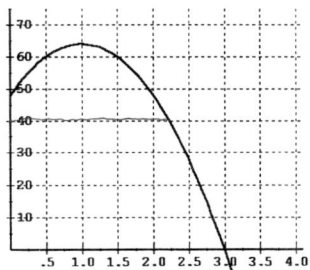

c. ≈ 55 ft;
$$h = 48 + 32t - 16t^2$$
$$= 48 + 32(1.75) - 16(1.75)^2$$
$$= 48 + 56 - 49$$
$$= 55$$

d. ≈ 2.2 sec

e. $40 = 48 + 32t - 16t^2$

f. 1 sec

g. 2 sec

121

29a.

x	C
10	330
20	280
30	250
40	240
50	250
60	280
70	330
80	400
90	490
100	600

b.

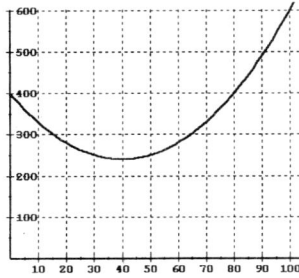

c. $\approx \$260$

d. 10 or 70 toasters

e. $330 = 0.1x^2 - 8x + 400$

f. 40 toasters

g. 90 or less

Homework 4.7

1. $(x+1)(x-4) = 0$
$x + 1 = 0 \quad$ or $\quad x - 4 = 0$
$\qquad x = -1 \qquad\qquad x = 4$

3. $0 = p(p + 7)$
$p = 0 \quad$ or $\quad p + 7 = 0$
$\qquad\qquad\qquad p = -7$

5. $(2v+3)(4v-1) = 0$
$2v + 3 = 0 \quad$ or $\quad 4v - 1 = 0$
$\quad 2v = -3 \qquad\qquad 4v = 1$
$\quad v = \dfrac{-3}{2} \qquad\qquad v = \dfrac{1}{4}$

7a. $(x-9)(x+3)$
$= x^2 + 3x - 9x - 27$
$= x^2 - 6x - 27$

b. $(x-9)(x+3) = 0$
$x - 9 = 0 \quad$ or $\quad x + 3 = 0$
$\quad x = 9 \qquad\qquad x = -3$

c. Check $x = 9$:
$x^2 - 6x - 27 = 0$
$(9)^2 - 6(9) - 27 \overset{?}{=} 0$
$81 - 54 - 27 \overset{?}{=} 0$
$0 = 0$

Check $x = -3$:
$x^2 - 6x - 27 = 0$
$(-3)^2 - 6(-3) - 27 \overset{?}{=} 0$
$9 + 18 - 27 \overset{?}{=} 0$
$0 = 0$

9a. $-3x(x-16) = -3x^2 + 48x$

b. $-3x(x-16) = 0$
$-3x = 0 \quad$ or $\quad x - 16 = 0$
$\quad x = 0 \qquad\qquad x = 16$

c. Check $x = 0$:
$-3x^2 + 48x = 0$
$-3(0)^2 + 48(0) \overset{?}{=} 0$
$0 + 0 \overset{?}{=} 0$
$0 = 0$

Check $x = 16$:
$-3x^2 + 48x = 0$
$-3(16)^2 + 48(16) \overset{?}{=} 0$
$-768 + 768 \overset{?}{=} 0$
$0 = 0$

11. $6a^2 - 8a = 2a(3a - 4)$

13. $-18v^2 - 6v = 6v(-3v - 1)$ or
$-18v^2 - 6v = -6v(3v + 1)$

15. $4h - 9h^2 = h(4 - 9h)$

17. $d = k - kat$
$d = k(1 - at)$

19. $A = 2rh - \pi r^2$
$A = r(2h - \pi r)$

21. $V = \pi r^2 h - \frac{1}{3}s^2 h$
$V = h(\pi r^2 - \frac{1}{3}s^2)$

23. $10x^2 - 15x = 0 \quad$ Factor out $5x$
$5x(2x - 3) = 0 \quad$ Set each factor equal to 0
$5x = 0 \quad$ or $\quad 2x - 3 = 0 \quad$ Solve each
$\quad x = 0 \qquad\qquad 2x = 3$
$\qquad\qquad\qquad\qquad x = \dfrac{3}{2}$

25. $20x^2 = x \quad$ Write in standard form
$20x^2 - x = 0 \quad$ Factor out x
$x(20x - 1) = 0 \quad$ Set each factor equal to 0
$x = 0 \quad$ or $\quad 20x - 1 = 0 \quad$ Solve each
$\qquad\qquad\qquad 20x = 1$
$\qquad\qquad\qquad x = \dfrac{1}{20}$

27. $0 = 144x + 3x^2$ Factor out $3x$
$0 = 3x(48 + x)$ Set each factor equal to 0
$3x = 0$ or $48 + x = 0$ Solve each
$x = 0$ $x = -48$

29. The result is not a value for x that is a solution to the original equation; it only gives another (equivalent) equation with x's on both sides.

31a. $R = -2n^2 + 80n$
$R = -2n(n - 40)$

b.

n	R
0	0
5	350
10	600
20	800
30	600
40	0

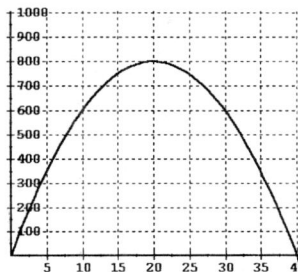

c. $800; 20 bracelets

33a. $h = -16t^2 + 80t$
$h = -16t(t - 5)$

b.

t	h
0	0
1	64
2	96
2.5	100
3	96
4	64
5	0

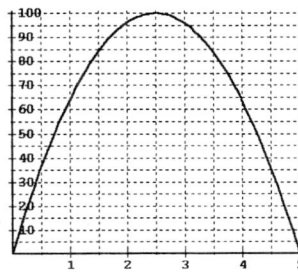

c. 100 ft; 2.5 sec

35. $(x - 3)(2x + 5) = 0$
$x - 3 = 0$ or $2x + 5 = 0$
$x = 3$ $2x = -5$
$x = \dfrac{-5}{2}$
x-intercepts: $(3, 0)$ and $\left(\dfrac{-5}{2}, 0\right)$

37. $2x^2 - 6x = 0$ Factor out $2x$
$2x(x - 3) = 0$ Set each factor equal to 0
$2x = 0$ or $x - 3 = 0$
$x = 0$ $x = 3$
x-intercepts: $(0, 0)$ and $(3, 0)$

39. $8x - 3x^2 = 0$ Factor out x
$x(8 - 3x) = 0$ Set each factor equal to 0
$x = 0$ or $8 - 3x = 0$
$-3x = -8$
$x = \dfrac{8}{3}$
x-intercepts: $(0, 0)$ and $\left(\dfrac{8}{3}, 0\right)$

41a. $(y + 4)(y + 2)$
$= y^2 + 2y + 4y + 8$
$= y^2 + 6y + 8$

b.

	y	2
y	y^2	$2y$
4	$4y$	8

43a. $(w - 6)(w + 3)$
$= w^2 + 3w - 6w - 18$
$= w^2 - 3w - 18$

b.

	w	3
w	w^2	$3w$
-6	$-6w$	-18

45a. $(t + 5)(2t + 3)$
$= 2t^2 + 3t + 10t + 15$
$= 2t^2 + 13t + 15$

b.

	$2t$	3
t	$2t^2$	$3t$
5	$10t$	15

123

Homework 4.8

1a.

	x	6
x	x^2	$6x$
5	$5x$	30

b. $(x+5)(x+6)$
$= x^2 + 6x + 5x + 30$
$= x^2 + 11x + 30$

3a.

	x	-3
x	x^2	$-3x$
-9	$-9x$	27

b. $(x-9)(x-3)$
$= x^2 - 3x - 9x + 27$
$= x^2 - 12x + 27$

5a.

	x	2
x	x^2	$2x$
8	$8x$	16

b. $(x+8)(x+2)$
$= x^2 + 2x + 8x + 16$
$= x^2 + 10x + 16$

7. $n^2 + 10n + 16 = (n+2)(n+8)$

9. $h^2 + 26h + 48 = (h+2)(h+24)$

11. $a^2 - 8a + 12 = (a-2)(a-6)$

13. $t^2 - 15t + 36 = (t-3)(t-12)$

15. $x^2 - 3x - 10 = (x+2)(x-5)$

17. $a^2 + 8a - 20 = (a-2)(a+10)$

19. $x^2 - 17x + 30 = (x-2)(x-15)$

21. $x^2 + 4x + 2$ cannot be factored

23. $y^2 - 44y - 45 = (y+1)(y-45)$

25. $t^2 - 9t - 20$ cannot be factored

27. $q^2 - 5q - 6 = (q+1)(q-6)$

29. $n^2 - 5n + 6 = (n-2)(n-3)$

31. $x^2 - 9 = (x+3)(x-3)$

33. $4 - w^2 = (2+w)(2-w)$

35. $-121 + b^2 = b^2 - 121 = (b+11)(b-11)$

37.
$x^2 + 3x - 10 = 0$ Factor
$(x-2)(x+5) = 0$ Set each factor equal to 0
$x - 2 = 0$ or $x + 5 = 0$ Solve each
$x = 2$ $x = -5$

39.
$t^2 + t = 42$ Write in standard form
$t^2 + t - 42 = 0$ Factor
$(t-6)(t+7) = 0$ Set each factor equal to 0
$t - 6 = 0$ or $t + 7 = 0$ Solve each
$t = 6$ $t = -7$

41.
$2x^2 - 10x = 12$ Write in standard form
$2x^2 - 10x - 12 = 0$ Factor out 2
$2(x^2 - 5x - 6) = 0$ Divide both sides by 2
$\dfrac{2(x^2 - 5x - 6)}{2} = \dfrac{0}{2}$
$x^2 - 5x - 6 = 0$ Factor
$(x+1)(x-6) = 0$ Set each factor equal to 0
$x + 1 = 0$ or $x - 6 = 0$ Solve each
$x = -1$ $x = 6$

43.
$0 = n^2 - 14n + 49$ Factor
$0 = (n-7)(n-7)$ Set each factor equal to 0
$n - 7 = 0$ Note: The other factor is the same,
$n = 7$ and will give the same solution

124

45.
$$5q^2 = 10q \quad \text{Write in standard form}$$
$$5q^2 - 10q = 0 \quad \text{Factor}$$
$$5q(q-2) = 0 \quad \text{Set each factor equal to 0}$$
$$5q = 0 \quad \text{or} \quad q - 2 = 0 \quad \text{Solve each}$$
$$q = 0 \qquad\qquad q = 2$$

47.
$$x(x-4) = 21 \quad \text{Distribute } x$$
$$x^2 - 4x = 21 \quad \text{Write in standard form}$$
$$x^2 - 4x - 21 = 0 \quad \text{Factor}$$
$$(x+3)(x-7) = 0 \quad \text{Set each factor equal to 0}$$
$$x + 3 = 0 \quad \text{or} \quad x - 7 = 0 \quad \text{Solve each}$$
$$x = -3 \qquad\qquad x = 7$$

49.
$$(x-2)(x+1) = 4 \quad \text{Multiply using FOIL}$$
$$x^2 + x - 2x - 2 = 4 \quad \text{Combine like terms}$$
$$x^2 - x - 2 = 4 \quad \text{Write in standard form}$$
$$x^2 - x - 6 = 0 \quad \text{Factor}$$
$$(x+2)(x-3) = 0 \quad \text{Set each factor equal to 0}$$
$$x + 2 = 0 \quad \text{or} \quad x - 3 = 0 \quad \text{Solve each}$$
$$x = -2 \qquad\qquad x = 3$$

51a. To find x-intercepts, set $y = 0$.   To find y-intercept, set $x = 0$.
$$x^2 + 2x = 0 \qquad\qquad y = x^2 + 2x$$
$$x(x+2) = 0 \qquad\qquad y = 0^2 + 2(0) = 0$$
$$x = 0 \quad \text{or} \quad x + 2 = 0 \qquad y\text{-intercept: } (0,0)$$
$$x = -2$$
x-intercepts: $(0,0)$ and $(-2,0)$

b. To find vertex, find average of x-intercepts.   **c.**
$$x = \frac{0 + (-2)}{2} = \frac{-2}{2} = -1$$
$$y = x^2 + 2x$$
$$y = (-1)^2 + 2(-1) = 1 - 2 = -1$$
vertex: $(-1, -1)$

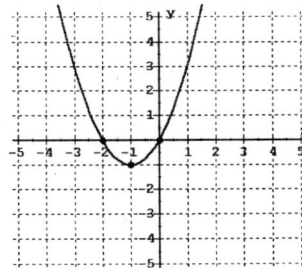

53a. To find x-intercepts, set $y = 0$.   To find y-intercept, set $x = 0$.
$$x^2 - 2x - 3 = 0 \qquad\qquad y = x^2 - 2x - 3$$
$$(x+1)(x-3) = 0 \qquad\qquad y = 0^2 - 2(0) - 3 = -3$$
$$x + 1 = 0 \quad \text{or} \quad x - 3 = 0 \qquad y\text{-intercept: } (0,-3)$$
$$x = -1 \qquad\qquad x = 3$$
x-intercepts: $(-1, 0)$ and $(3, 0)$

b. To find vertex, find average of x-intercepts.

$x = \dfrac{-1+3}{2} = \dfrac{2}{2} = 1$

$y = x^2 - 2x - 3$

$y = 1^2 - 2(1) - 3 = 1 - 2 - 3 = -4$

vertex: $(1, -4)$

c.

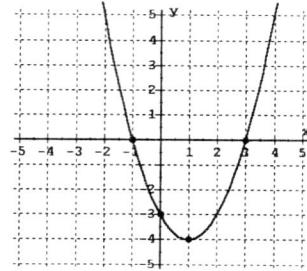

55a. To find x-intercepts, set $y = 0$.

$9 - x^2 = 0$

$(3 + x)(3 - x) = 0$

$3 + x = 0 \quad$ or $\quad 3 - x = 0$

$\quad\quad x = -3 \quad\quad\quad -x = -3$

$\quad\quad\quad\quad\quad\quad\quad\quad x = 3$

x-intercepts: $(-3, 0)$ and $(3, 0)$

To find y-intercept, set $x = 0$.

$y = 9 - x^2$

$y = 9 - 0^2 = 9$

y-intercept: $(0, 9)$

b. To find vertex, find average of x-intercepts.

$x = \dfrac{-3+3}{2} = \dfrac{0}{2} = 0$

$y = 9 - x^2$

$y = 9 - 0^2 = 9$

vertex: $(0, 9)$

c.

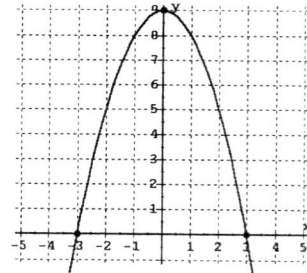

57a.

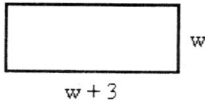

b. $A = w(w + 3)$

c. $w(w + 3) = 54$

d.

$w(w + 3) = 54 \quad$ Distribute w

$w^2 + 3w = 54 \quad$ Write in standard form

$w^2 + 3w - 54 = 0 \quad$ Factor

$(w - 6)(w + 9) = 0 \quad$ Set each factor equal to 0

$w - 6 = 0 \quad$ or $\quad w + 9 = 0$

$\quad w = 6 \quad\quad\quad\quad w = -9 \leftarrow$ Not possible

The width $w = 6$ yd.

59. $C = 8t^2 - 32t - 16$

$80 = 8t^2 - 32t - 16 \quad$ Write in standard form

$0 = 8t^2 - 32t - 96 \quad$ Factor out 8

$0 = 8(t^2 - 4t - 12) \quad$ Divide both sides by 8

$\dfrac{0}{8} = \dfrac{8(t^2 - 4t - 12)}{8}$

$0 = t^2 - 4t - 12 \quad$ Factor

$0 = (t - 6)(t + 2) \quad$ Set each factor equal to 0

$t - 6 = 0 \quad$ or $\quad t + 2 = 0$

$\quad t = 6 \quad\quad\quad\quad t = -2 \leftarrow$ Not possible

It takes 6 hr.

126

Homework 4.9

1. $2x^2 + 11x + 5 = (2x + 1)(x + 5)$

3. $5t^2 + 7t + 2 = (5t + 2)(t + 1)$

5. $3x^2 - 8x + 5 = (3x - 5)(x - 1)$

7. $D = 2x^2 \cdot 18 = 36x^2$
$(-4x)(-9x) = 36x^2$
$-4x - 9x = -13x$

	x	-2
$2x$	$2x^2$	$-4x$
-9	$-9x$	18

$2x^2 - 13x + 18 = (2x - 9)(x - 2)$

9. $D = 5x^2(-16) = -80x^2$
$(-4x)(20x) = -80x^2$
$-4x + 20x = 16x$

	$5x$	-4
x	$5x^2$	$-4x$
4	$20x$	-16

$5x^2 + 16x - 16 = (x + 4)(5x - 4)$

11. $D = 6h^2 \cdot 2 = 12h^2$
$(3h)(4h) = 12h^2$
$3h + 4h = 7h$

	$2h$	1
$3h$	$6h^2$	$3h$
2	$4h$	2

$6h^2 + 7h + 2 = (3h + 2)(2h + 1)$

13. $D = 9n^2(-1) = -9n^2$
$(n)(-9n) = -9n^2$
$n - 9n = -8n$

	$9n$	1
n	$9n^2$	n
-1	$-9n$	-1

$9n^2 - 8n - 1 = (n - 1)(9n + 1)$

15. $D = 6t^2(-25) = -150t^2$
$(10t)(-15t) = -150t^2$
$10t - 15t = -5t$

	$3t$	5
$2t$	$6t^2$	$10t$
-5	$-15t$	-25

$6t^2 - 5t - 25 = (2t - 5)(3t + 5)$

17. $D = 5x^2(-24) = -120x^2$
$(6x)(-20x) = -120x^2$
$6x - 20x = -14x$

	$5x$	6
x	$5x^2$	$6x$
-4	$-20x$	-24

$5x^2 - 14x - 24 = (x - 4)(5x + 6)$

19. $9x^2 - 8 - 21x$
$= 9x^2 - 21x - 8$
$= (3x - 8)(3x + 1)$

21. $-5 - 2z + 16z^2$
$= 16z^2 - 2z - 5$
$= (8z - 5)(2z + 1)$

23. $23a + 4a^2 - 6$
$= 4a^2 + 23a - 6$
$= (4a - 1)(a + 6)$

25.
$3n^2 - n = 4$ Write in standard form
$3n^2 - n - 4 = 0$ Factor
$(3n - 4)(n + 1) = 0$ Set each factor equal to 0
$3n - 4 = 0$ or $n + 1 = 0$ Solve each
$3n = 4$ $n = -1$
$n = \dfrac{4}{3}$

27. $\quad 11t = 6t^2 + 3 \qquad$ Write in standard form

$\qquad 0 = 6t^2 - 11t + 3 \qquad$ Factor

$\qquad 0 = (3t - 1)(2t - 3) \qquad$ Set each factor equal to 0

$\qquad 3t - 1 = 0 \quad$ or $\quad 2t - 3 = 0 \qquad$ Solve each

$\qquad\qquad 3t = 1 \qquad\qquad 2t = 3$

$\qquad\qquad t = \dfrac{1}{3} \qquad\qquad t = \dfrac{3}{2}$

29. $\qquad\qquad\qquad 1 = 4y - 4y^2 \qquad$ Write in standard form

$\qquad\qquad 4y^2 - 4y + 1 = 0 \qquad\qquad$ Factor

$\qquad (2y - 1)(2y - 1) = 0 \qquad\qquad$ Set the factor equal to 0

$\qquad\qquad 2y - 1 = 0 \qquad\qquad$ Note: The other factor is the same,

$\qquad\qquad\qquad 2y = 1 \qquad\qquad\qquad$ and will give the same solution

$\qquad\qquad\qquad y = \dfrac{1}{2}$

31. $\qquad\qquad 12z^2 + 26z = 10 \qquad$ Write in standard form

$\qquad 12z^2 + 26z - 10 = 0 \qquad$ Factor out 2

$\qquad 2(6z^2 + 13z - 5) = 0 \qquad$ Divide both sides by 2

$\qquad \dfrac{2(6z^2 + 13z - 5)}{2} = \dfrac{0}{2}$

$\qquad\quad 6z^2 + 13z - 5 = 0 \qquad$ Factor

$\qquad (2z + 5)(3z - 1) = 0 \qquad$ Set each factor equal to 0

$\qquad 2z + 5 = 0 \qquad$ or $\quad 3z - 1 = 0 \qquad$ Solve each

$\qquad\qquad 2z = -5 \qquad\qquad 3z = 1$

$\qquad\qquad z = \dfrac{-5}{2} \qquad\qquad z = \dfrac{1}{3}$

33. $\qquad y(3y + 4) = 4 \qquad$ Distribute y

$\qquad\quad 3y^2 + 4y = 4 \qquad$ Write in standard form

$\qquad 3y^2 + 4y - 4 = 0 \qquad$ Factor

$\qquad (3y - 2)(y + 2) = 0 \qquad$ Set each factor equal to 0

$\qquad 3y - 2 = 0 \quad$ or $\quad y + 2 = 0 \qquad$ Solve each

$\qquad\qquad 3y = 2 \qquad\qquad y = -2$

$\qquad\qquad y = \dfrac{2}{3}$

35. $\quad (2x - 1)(x - 2) = -1 \qquad$ FOIL

$\qquad\quad 2x^2 - 5x + 2 = -1 \qquad$ Write in standard form

$\qquad\quad 2x^2 - 5x + 3 = 0 \qquad$ Factor

$\qquad (2x - 3)(x - 1) = 0 \qquad$ Set each factor equal to 0

$\qquad 2x - 3 = 0 \quad$ or $\quad x - 1 = 0 \qquad$ Solve each

$\qquad\qquad 2x = 3 \qquad\qquad x = 1$

$\qquad\qquad x = \dfrac{3}{2}$

37.

$$-16t^2 + 8t + 24 = 0 \qquad \text{Factor out } -8$$
$$-8(2t^2 - t - 3) = 0 \qquad \text{Divide both sides by } -8$$
$$\frac{-8(2t^2 - t - 3)}{-8} = \frac{0}{-8}$$
$$2t^2 - t - 3 = 0 \qquad \text{Factor}$$
$$(2t - 3)(t + 1) = 0 \qquad \text{Set each factor equal to 0}$$
$$2t - 3 = 0 \quad \text{or} \quad t + 1 = 0$$
$$2t = 3 \qquad\qquad t = -1 \ \leftarrow \text{Not possible}$$
$$t = \frac{3}{2}$$

She has $\frac{3}{2} = 1\frac{1}{2}$ sec.

39a.

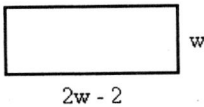

b.
$$w(2w - 2) = 60 \qquad \text{Distribute } w$$

c.
$$2w^2 - 2w = 60 \qquad \text{Write in standard form}$$
$$2w^2 - 2w - 60 = 0 \qquad \text{Factor out 2}$$
$$2(w^2 - w - 30) = 0 \qquad \text{Divide both sides by 2}$$
$$w^2 - w - 30 = 0 \qquad \text{Factor}$$
$$(w - 6)(w + 5) = 0 \qquad \text{Set each factor equal to 0}$$
$$w - 6 = 0 \quad \text{or} \quad w + 5 = 0$$
$$w = 6 \qquad\qquad w = -5 \ \leftarrow \text{Not possible}$$

The width is $w = 6$ cm, and length is $2w - 2 = 2(6) - 2 = 10$ cm.

41a.

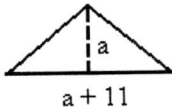

b.
$$\tfrac{1}{2}a(a + 11) = 40 \qquad \text{Multiply both sides by 2}$$

c.
$$2 \cdot \tfrac{1}{2}a(a + 11) = 2 \cdot 40$$
$$a(a + 11) = 80 \qquad \text{Distribute } a$$
$$a^2 + 11a = 80 \qquad \text{Write in standard form}$$
$$a^2 + 11a - 80 = 0 \qquad \text{Factor}$$
$$(a - 5)(a + 16) = 0 \qquad \text{Set each factor equal to 0}$$
$$a - 5 = 0 \quad \text{or} \quad a + 16 = 0$$
$$a = 5 \qquad\qquad a = -16 \ \leftarrow \text{Not possible}$$

The altitude is $a = 5$ in., and the base is $a + 11 = 5 + 11 = 16$ in.

43a.

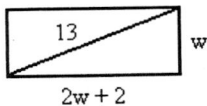

b.
$$(2w + 2)^2 + w^2 = 13^2$$

c.
$$(2w + 2)(2w + 2) + w^2 = 13^2 \qquad \text{FOIL}$$
$$4w^2 + 4w + 4w + 4 + w^2 = 169 \qquad \text{Write in standard form}$$
$$5w^2 + 8w - 165 = 0 \qquad \text{Factor}$$
$$(5w + 33)(w - 5) = 0 \qquad \text{Set each factor equal to 0}$$
$$5w + 33 = 0 \quad \text{or} \quad w - 5 = 0$$
$$5w = -33 \qquad\qquad w = 5$$
$$w = \frac{-33}{5} \ \leftarrow \text{Not possible}$$

The width is $w = 5$ ft, and the length is $2w + 2 = 2(5) + 2 = 12$ ft.

45a. For $y = x^2 - 2x - 15$

$x^2 - 2x - 15 = 0$

$(x - 5)(x + 3) = 0$

$x - 5 = 0 \quad$ or $\quad x + 3 = 0$

$\qquad x = 5 \qquad\qquad x = -3$

x-intercepts: $(5, 0)$ and $(-3, 0)$

$y = x^2 - 2x - 15$

$y = 0^2 - 2(0) - 15 = -15$

y-intercept: $(0, -15)$

For $y = x^2 + 2x - 15$

$x^2 + 2x - 15 = 0$

$(x + 5)(x - 3) = 0$

$x + 5 = 0 \quad$ or $\quad x - 3 = 0$

$\qquad x = -5 \qquad\qquad x = 3$

x-intercepts: $(-5, 0)$ and $(3, 0)$

$y = x^2 + 2x - 15$

$y = 0^2 + 2(0) - 15 = -15$

y-intercept: $(0, -15)$

b. For $y = x^2 - 2x - 15$

$x = \dfrac{5 + (-3)}{2} = \dfrac{2}{2} = 1$

$y = x^2 - 2x - 15$

$y = (1)^2 - 2(1) - 15 = 1 - 2 - 15 = -16$

vertex: $(1, -16)$

For $y = x^2 + 2x - 15$

$x = \dfrac{-5 + 3}{2} = \dfrac{-2}{2} = -1$

$y = x^2 + 2x - 15$

$y = (-1)^2 + 2(-1) - 15 = 1 - 2 - 15 = -16$

vertex: $(-1, -16)$

c.

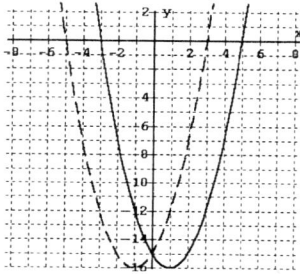

47a. For $y = x^2 - x - 2$

$x^2 - x - 2 = 0$

$(x - 2)(x + 1) = 0$

$x - 2 = 0 \quad$ or $\quad x + 1 = 0$

$\qquad x = 2 \qquad\qquad x = -1$

x-intercepts: $(2, 0)$ and $(-1, 0)$

$y = x^2 - x - 2$

$y = 0^2 - 0 - 2 = -2$

y-intercept: $(0, -2)$

For $y = -x^2 + x + 2$

$-x^2 + x + 2 = 0$

$-(x^2 - x - 2) = 0$

$x^2 - x - 2 = 0$

$(x - 2)(x + 1) = 0$

$x - 2 = 0 \quad$ or $\quad x + 1 = 0$

$\qquad x = 2 \qquad\qquad x = -1$

x-intercepts: $(2, 0)$ and $(-1, 0)$

$y = -x^2 + x + 2$

$y = -(0)^2 + 0 + 2 = 2$

y-intercept: $(0, 2)$

b. For $y = x^2 - x - 2$

$x = \dfrac{2 + (-1)}{2} = \dfrac{1}{2}$

$y = x^2 - x - 2$

$y = \left(\tfrac{1}{2}\right)^2 - \left(\tfrac{1}{2}\right) - 2$

$y = \dfrac{1}{4} - \dfrac{1}{2} - 2$

$y = \dfrac{1}{4} - \dfrac{2}{4} - \dfrac{8}{4} = \dfrac{-9}{4}$

vertex: $\left(\tfrac{1}{2}, \tfrac{-9}{4}\right)$

For $y = -x^2 + x + 2$

$x = \dfrac{2 + (-1)}{2} = \dfrac{1}{2}$

$y = -x^2 + x + 2$

$y = -\left(\tfrac{1}{2}\right)^2 + \left(\tfrac{1}{2}\right) + 2$

$y = -\dfrac{1}{4} + \dfrac{1}{2} + 2$

$y = -\dfrac{1}{4} + \dfrac{2}{4} + \dfrac{8}{4} = \dfrac{9}{4}$

vertex: $\left(\tfrac{1}{2}, \tfrac{9}{4}\right)$

c.

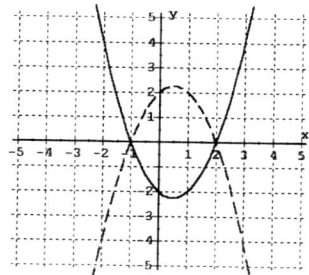

130

49a. **For** $y = 3x^2 - 6x + 3$

$$3x^2 - 6x + 3 = 0$$
$$3(x^2 - 2x + 1) = 0$$
$$x^2 - 2x + 1 = 0$$
$$(x-1)(x-1) = 0$$
$$x - 1 = 0$$
$$x = 1$$

x-intercept: $(1, 0)$

$$y = 3x^2 - 6x + 3$$
$$y = 3(0)^2 - 6(0) + 3 = 3$$

y-intercept: $(0, 3)$

b. **For** $y = 3x^2 - 6x + 3$

$$x = \frac{1+1}{2} = \frac{2}{2} = 1$$
$$y = 3x^2 - 6x + 3$$
$$y = 3(1)^2 - 6(1) + 3 = 3 - 6 + 3 = 0$$

vertex: $(1, 0)$

For $y = 3x^2 + 6x + 3$

$$3x^2 + 6x + 3 = 0$$
$$3(x^2 + 2x + 1) = 0$$
$$x^2 + 2x + 1 = 0$$
$$(x+1)(x+1) = 0$$
$$x + 1 = 0$$
$$x = -1$$

x-intercept: $(-1, 0)$

$$y = 3x^2 + 6x + 3$$
$$y = 3(0)^2 + 6(0) + 3 = 3$$

y-intercept: $(0, 3)$

For $y = 3x^2 + 6x + 3$

$$x = \frac{-1 + (-1)}{2} = \frac{-2}{2} = -1$$
$$y = 3x^2 + 6x + 3$$
$$y = 3(-1)^2 + 6(-1) + 3 = 3 - 6 + 3 = 0$$

vertex: $(-1, 0)$

c.

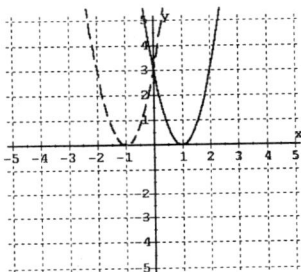

51.

$$\frac{p-3}{2} = \frac{7}{p+2} \qquad \text{Cross-multiply}$$
$$(p-3)(p+2) = 14 \qquad \text{FOIL}$$
$$p^2 - p - 6 = 14 \qquad \text{Write in standard form}$$
$$p^2 - p - 20 = 0 \qquad \text{Factor}$$
$$(p-5)(p+4) = 0 \qquad \text{Set each factor equal to 0}$$
$$p - 5 = 0 \quad \text{or} \quad p + 4 = 0 \qquad \text{Solve each}$$
$$p = 5 \qquad\qquad p = -4$$

53.

$$\frac{m-4}{2m+1} = \frac{m+1}{2m-1} \qquad \text{Cross-multiply}$$
$$(m-4)(2m-1) = (2m+1)(m+1) \qquad \text{FOIL}$$
$$2m^2 - 9m + 4 = 2m^2 + 3m + 1 \qquad \text{Subtract } 2m^2 \text{ from both sides}$$
$$-9m + 4 = 3m + 1 \qquad \text{Subtract } 3m \text{ from both sides}$$
$$-12m + 4 = 1 \qquad \text{Subtract 4 from both sides}$$
$$-12m = -3 \qquad \text{Divide both sides by } -12$$
$$m = \frac{-3}{-12} \qquad \text{Reduce fraction}$$
$$m = \frac{1}{4}$$

131

55. $x^2 + 15x + 56 = (x + 7)(x + 8)$

57. $b^2 + 8b - 240 = (b - 12)(b + 20)$

59. $n^2 - 97n - 300 = (n - 100)(n + 3)$

61. $3u^2 - 17u - 6 = (u - 6)(3u + 1)$

63. $2t^2 - 21t + 54 = (t - 6)(2t - 9)$

Homework 4.10

1a. $x^2 - 8x + 4 = 0 \qquad a = 1, \, b = -8, \, c = 4$

$$x = \frac{-b \pm \sqrt{b^2 - 4ac}}{2a}$$

$$= \frac{-(-8) \pm \sqrt{(-8)^2 - 4(1)(4)}}{2(1)}$$

$$= \frac{8 \pm \sqrt{64 - 16}}{2}$$

$$= \frac{8 \pm \sqrt{48}}{2}$$

b. $x = \dfrac{8 + \sqrt{48}}{2} \approx 7.46$

$x = \dfrac{8 - \sqrt{48}}{2} \approx 0.54$

3a. $3s^2 + 2s = 2 \quad$ Write in standard form

$3s^2 + 2s - 2 = 0 \quad a = 3, \, b = 2, \, c = -2$

$$s = \frac{-b \pm \sqrt{b^2 - 4ac}}{2a}$$

$$= \frac{-2 \pm \sqrt{2^2 - 4(3)(-2)}}{2(3)}$$

$$= \frac{-2 \pm \sqrt{4 + 24}}{6}$$

$$= \frac{-2 \pm \sqrt{28}}{6}$$

b. $s = \dfrac{-2 + \sqrt{28}}{6} \approx 0.55$

$s = \dfrac{-2 - \sqrt{28}}{6} \approx -1.22$

5a. $n^2 = n + 1 \quad$ Write in standard form

$n^2 - n - 1 = 0 \qquad a = 1, \, b = -1, \, c = -1$

$$n = \frac{-b \pm \sqrt{b^2 - 4ac}}{2a}$$

$$= \frac{-(-1) \pm \sqrt{(-1)^2 - 4(1)(-1)}}{2(1)}$$

$$= \frac{1 \pm \sqrt{1 + 4}}{2}$$

$$= \frac{1 \pm \sqrt{5}}{2}$$

b. $n = \dfrac{1 + \sqrt{5}}{2} \approx 1.62$

$n = \dfrac{1 - \sqrt{5}}{2} \approx -0.62$

7a. $-4z^2 + 2z + 1 = 0 \quad a = -4, \, b = 2, \, c = 1$

$$z = \frac{-b \pm \sqrt{b^2 - 4ac}}{2a}$$

$$= \frac{-2 \pm \sqrt{2^2 - 4(-4)(1)}}{2(-4)}$$

$$= \frac{-2 \pm \sqrt{4 + 16}}{-8}$$

$$= \frac{-2 \pm \sqrt{20}}{-8}$$

b. $z = \dfrac{-2 + \sqrt{20}}{-8} \approx -0.31$

$z = \dfrac{-2 - \sqrt{20}}{-8} \approx 0.81$

9a. $3t^2 - 5 = 0 \qquad a = 3,\ b = 0,\ c = -5$

$$t = \frac{-b \pm \sqrt{b^2 - 4ac}}{2a}$$

$$= \frac{-0 \pm \sqrt{0^2 - 4(3)(-5)}}{2(3)}$$

$$= \frac{0 \pm \sqrt{0 + 60}}{6}$$

$$= \frac{\pm\sqrt{60}}{6}$$

$$t \approx \pm 1.29$$

b. $3t^2 - 5 = 0 \qquad$ Add 5 to both sides

$\qquad 3t^2 = 5 \qquad$ Divide both sides by 3

$\qquad t^2 = \dfrac{5}{3} \qquad$ Take the square root

$\qquad\qquad\qquad$ of both sides

$$t = \pm\sqrt{\frac{5}{3}}$$

$$t \approx \pm 1.29$$

11a. $z = 3z^2 \qquad$ Write in standard form

$0 = 3z^2 - z \qquad a = 3,\ b = -1,\ c = 0$

$$z = \frac{-b \pm \sqrt{b^2 - 4ac}}{2a}$$

$$= \frac{-(-1) \pm \sqrt{(-1)^2 - 4(3)(0)}}{2(3)}$$

$$= \frac{1 \pm \sqrt{1 - 0}}{6}$$

$$= \frac{1 \pm \sqrt{1}}{6}$$

$$= \frac{1 \pm 1}{6}$$

$$z = \frac{1 + 1}{6} = \frac{2}{6} = \frac{1}{3}$$

$$z = \frac{1 - 1}{6} = \frac{0}{6} = 0$$

b. $z = 3z^2 \qquad$ Write in standard form

$0 = 3z^2 - z \qquad$ Factor out z

$0 = z(3z - 1) \qquad$ Set each factor equal to 0

$z = 0 \quad$ or $\quad 3z - 1 = 0 \qquad$ Solve each

$\qquad\qquad\qquad 3z = 1$

$$z = \frac{1}{3}$$

13a. $2w^2 + 6 = 7w \qquad$ Write in standard form

$2w^2 - 7w + 6 = 0 \qquad a = 2,\ b = -7,\ c = 6$

$$w = \frac{-b \pm \sqrt{b^2 - 4ac}}{2a}$$

$$= \frac{-(-7) \pm \sqrt{(-7)^2 - 4(2)(6)}}{2(2)}$$

$$= \frac{7 \pm \sqrt{49 - 48}}{4}$$

$$= \frac{7 \pm \sqrt{1}}{4}$$

$$= \frac{7 \pm 1}{4}$$

$$w = \frac{7 + 1}{4} = \frac{8}{4} = 2$$

$$w = \frac{7 - 1}{4} = \frac{6}{4} = \frac{3}{2}$$

b. $2w^2 + 6 = 7w \qquad$ Write in standard form

$2w^2 - 7w + 6 = 0 \qquad$ Factor

$(2w - 3)(w - 2) = 0 \qquad$ Set each equal to 0

$2w - 3 = 0 \quad$ or $\quad w - 2 = 0$

$\qquad 2w = 3 \qquad\qquad\quad w = 2$

$$w = \frac{3}{2}$$

15.
$$\frac{x^2}{6} + x = \frac{2}{3} \quad \text{Multiply both sides by LCD} = 6$$

$$6 \cdot \left(\frac{x^2}{6} + x\right) = 6 \cdot \frac{2}{3} \quad \begin{array}{l}\text{Distribute 6 to}\\ \text{each term}\end{array}$$

$$6 \cdot \frac{x^2}{6} + 6 \cdot x = 6 \cdot \frac{2}{3}$$

$$x^2 + 6x = 4 \quad \text{Write in standard form}$$

$$x^2 + 6x - 4 = 0 \quad a = 1,\, b = 6,\, c = -4$$

$$x = \frac{-b \pm \sqrt{b^2 - 4ac}}{2a}$$

$$= \frac{-6 \pm \sqrt{6^2 - 4(1)(-4)}}{2(1)}$$

$$= \frac{-6 \pm \sqrt{36 + 16}}{2}$$

$$= \frac{-6 \pm \sqrt{52}}{2}$$

17.
$$v^2 + 3v - 2 = 9v^2 - 12v + 5 \quad \begin{array}{l}\text{Write in}\\ \text{standard form}\end{array}$$

$$0 = 8v^2 - 15v + 7 \quad \text{Factor}$$

$$0 = (8v - 7)(v - 1) \quad \begin{array}{l}\text{Set each}\\ \text{equal to 0}\end{array}$$

$$8v - 7 = 0 \quad \text{or} \quad v - 1 = 0 \quad \text{Solve each}$$
$$8v = 7 \qquad\qquad v = 1$$
$$v = \frac{7}{8}$$

19.
$$m^2 - 3m = \frac{1}{3}(m^2 - 1) \quad \text{Multiply both sides by 3}$$

$$3 \cdot (m^2 - 3m) = 3 \cdot \frac{1}{3}(m^2 - 1) \quad \text{Distribute 3 to each term}$$

$$3 \cdot m^2 - 3 \cdot 3m = 3 \cdot \frac{1}{3}(m^2 - 1)$$

$$3m^2 - 9m = m^2 - 1 \quad \text{Write in standard form}$$

$$2m^2 - 9m + 1 = 0 \quad a = 2,\, b = -9,\, c = 1$$

$$m = \frac{-b \pm \sqrt{b^2 - 4ac}}{2a}$$

$$= \frac{-(-9) \pm \sqrt{(-9)^2 - 4(2)(1)}}{2(2)}$$

$$= \frac{9 \pm \sqrt{81 - 8}}{4}$$

$$= \frac{9 \pm \sqrt{73}}{4}$$

21.
$$-0.2x^2 + 3.6x - 9 = 0 \quad \text{Multiply both sides by 10}$$

$$10 \cdot \left(-0.2x^2 + 3.6x - 9\right) = 10 \cdot 0 \quad \text{Distribute 10 to each term}$$

$$10(-0.2x^2) + 10(3.6x) + 10(-9) = 10 \cdot 0$$

$$-2x^2 + 36x - 90 = 0 \quad \text{Factor out } -2$$

$$-2(x^2 - 18x + 45) = 0 \quad \text{Divide both sides by } -2$$

$$\frac{-2(x^2 - 18x + 45)}{-2} = \frac{0}{-2}$$

$$x^2 - 18x + 45 = 0 \quad \text{Factor}$$

$$(x - 15)(x - 3) = 0 \quad \text{Set each factor equal to 0}$$

$$x - 15 = 0 \quad \text{or} \quad x - 3 = 0 \quad \text{Solve each equation}$$
$$x = 15 \qquad\qquad x = 3$$

134

23a. $x^2 - 2x - 2 = 0 \quad a = 1, b = -2, c = -2$

$$x = \frac{-b \pm \sqrt{b^2 - 4ac}}{2a}$$

$$= \frac{-(-2) \pm \sqrt{(-2)^2 - 4(1)(-2)}}{2(1)}$$

$$= \frac{2 \pm \sqrt{4 + 8}}{2}$$

$$= \frac{2 \pm \sqrt{12}}{2}$$

$$x = \frac{2 + \sqrt{12}}{2} \approx 2.73 \quad \text{or} \quad x = \frac{2 - \sqrt{12}}{2} \approx -0.73$$

x-intercepts: $(2.73, 0)$ and $(-0.73, 0)$

b.

x	y
-1	1
0	-2
1	-3
2	-2
3	1

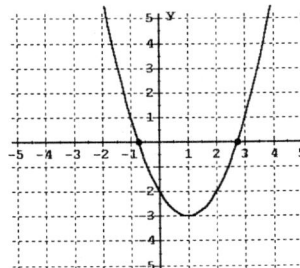

25a. $-3x^2 + 2x - 1 = 0 \quad a = -3, b = 2, c = -1$

$$x = \frac{-b \pm \sqrt{b^2 - 4ac}}{2a}$$

$$= \frac{-2 \pm \sqrt{2^2 - 4(-3)(-1)}}{2(-3)}$$

$$= \frac{-2 \pm \sqrt{4 - 12}}{-6}$$

$$= \frac{-2 \pm \sqrt{-8}}{-6} \leftarrow \text{Undefined}$$

No x-intercepts

b.

x	y
-1	-6
0	-1
1	-2
2	-9

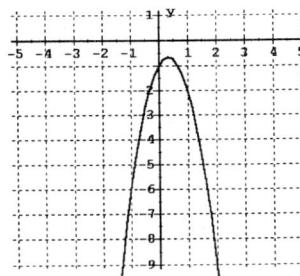

27a. $x^2 - 4x - 1 = 0 \quad a = 1, b = -4, c = -1$

$$x = \frac{-b \pm \sqrt{b^2 - 4ac}}{2a}$$

$$= \frac{-(-4) \pm \sqrt{(-4)^2 - 4(1)(-1)}}{2(1)}$$

$$= \frac{4 \pm \sqrt{16 + 4}}{2}$$

$$= \frac{4 \pm \sqrt{20}}{2}$$

$$x = \frac{4 + \sqrt{20}}{2} \approx 4.24 \quad \text{or} \quad x = \frac{4 - \sqrt{20}}{2} \approx -0.24$$

x-intercepts: $(4.24, 0)$ and $(-0.24, 0)$

b.

x	y
-1	4
0	-1
1	-4
2	-5
3	-4
4	-1
5	4

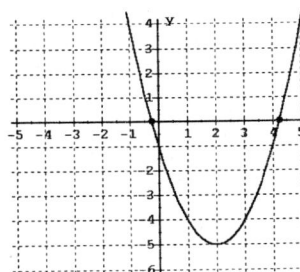

29a.

$$h = -16t^2 + 8t + 380 \qquad \text{Substitute } h = 300$$
$$300 = -16t^2 + 8t + 380 \qquad \text{Write in standard form}$$
$$16t^2 - 8t - 80 = 0 \qquad \text{Factor out 8}$$
$$8(2t^2 - t - 10) = 0 \qquad \text{Divide both sides by 8}$$
$$2t^2 - t - 10 = 0 \qquad \text{Factor}$$
$$(2t - 5)(t + 2) = 0 \qquad \text{Set each factor equal to 0}$$
$$2t - 5 = 0 \quad \text{or} \quad t + 2 = 0$$
$$2t = 5 \qquad\qquad t = -2 \leftarrow \text{Not possible}$$
$$t = \frac{5}{2}$$

It passes the window in $\frac{5}{2} = 2\frac{1}{2}$ sec.

b.

$$h = -16t^2 + 8t + 380 \qquad \text{Substitute } h = 0$$
$$0 = -16t^2 + 8t + 380 \qquad \text{Write in standard form}$$
$$16t^2 - 8t - 380 = 0 \qquad \text{Factor out 4}$$
$$4(4t^2 - 2t - 95) = 0 \qquad \text{Divide both sides by 4}$$
$$4t^2 - 2t - 95 = 0 \qquad a = 4,\ b = -2,\ c = -95$$
$$t = \frac{-b \pm \sqrt{b^2 - 4ac}}{2a}$$
$$= \frac{-(-2) \pm \sqrt{(-2)^2 - 4(4)(-95)}}{2(4)}$$
$$= \frac{2 \pm \sqrt{4 + 1520}}{8}$$
$$= \frac{2 \pm \sqrt{1524}}{8}$$

If $t = \dfrac{2 - \sqrt{1524}}{8} \approx -4.63$ (a negative number) which is not possible.

It takes $\dfrac{2 + \sqrt{1524}}{8} \approx 5.13$ sec.

31.

Width: w
Length: $3w - 5$

$$V = lwh \qquad\qquad\qquad \text{Substitute}$$
$$12,000 = (3w - 5)(w)(20) \qquad \text{Multiply 20 times } w$$
$$12,000 = (3w - 5)20w \qquad \text{Distribute } 20w$$
$$12,000 = 60w^2 - 100w \qquad \text{Write in standard form}$$
$$0 = 60w^2 - 100w - 12,000 \qquad \text{Factor out 20}$$
$$0 = 20(3w^2 - 5w - 600) \qquad \text{Divide both sides by 20}$$
$$0 = 3w^2 - 5w - 600 \qquad \text{Factor}$$
$$0 = (3w + 40)(w - 15) \qquad \text{Set each factor equal to 0}$$
$$3w + 40 = 0 \quad \text{or} \quad w - 15 = 0$$
$$3w = -40 \qquad\qquad w = 15$$
$$w = \frac{-40}{3} \quad \leftarrow \text{Not possible}$$

The width $w = 15$ in., the length $3w - 5 = 40$ in., and the height $= 20$ in.

33.

$$P = 2l + 2w \quad \text{Substitute } P = 42 \text{ and solve for } l$$
$$42 = 2l + 2w \quad \text{Factor out 2}$$
$$42 = 2(l + w) \quad \text{Divide both sides by 2}$$
$$21 = l + w \quad \text{Subtract } w \text{ from both sides}$$
$$21 - w = l$$

Width: w
Length: $21 - w$

$$a^2 + b^2 = c^2 \quad \text{Substitute}$$
$$(21 - w)^2 + w^2 = 15^2$$
$$(21 - w)(21 - w) + w = 15^2 \quad \text{FOIL}$$
$$441 - 42w + w^2 + w^2 = 225 \quad \text{Write in standard form}$$
$$2w^2 - 42w + 216 = 0 \quad \text{Factor out 2}$$
$$2(w^2 - 21w + 108) = 0 \quad \text{Divide both sides by 2}$$
$$w^2 - 21w + 108 = 0 \quad \text{Factor}$$
$$(w - 9)(w - 12) = 0 \quad \text{Set each factor equal to 0}$$
$$w - 9 = 0 \quad \text{or} \quad w - 12 = 0$$
$$w = 9 \qquad\qquad w = 12$$

If width $w = 9$, then length $21 - w = 12$, and if width $w = 12$, then length $21 - w = 9$. Therefore, the rectangle is 9 in. by 12 in.

35.

$$a^2 + b^2 = c^2 \quad \text{Substitute}$$
$$(2x - 1)^2 + x^2 = 17^2$$
$$(2x - 1)(2x - 1) + x^2 = 17^2 \quad \text{FOIL}$$
$$4x^2 - 4x + 1 + x^2 = 289 \quad \text{Write in standard form}$$
$$5x^2 - 4x - 288 = 0 \quad \text{Factor}$$
$$(5x + 36)(x - 8) = 0 \quad \text{Set each factor equal to 0}$$
$$5x + 36 = 0 \quad \text{or} \quad x - 8 = 0$$
$$5x = -36 \qquad\qquad x = 8$$
$$x = \frac{-36}{5} \quad \leftarrow \text{Not possible}$$

The sides are $x = 8$ and $2x - 1 = 15$.

37.

$$a^2 + b^2 = c^2 \qquad\qquad \text{Substitute}$$
$$(\sqrt{13})^2 + k^2 = (2k - 5)^2$$
$$13 + k^2 = (2k - 5)(2k - 5) \quad \text{FOIL}$$
$$13 + k^2 = 4k^2 - 20k + 25 \quad \text{Write in standard form}$$
$$0 = 3k^2 - 20k + 12 \quad \text{Factor}$$
$$0 = (3k - 2)(k - 6) \quad \text{Set each factor equal to 0}$$
$$3k - 2 = 0 \quad \text{or} \quad k - 6 = 0$$
$$3k = 2 \qquad\qquad k = 6$$
$$k = \frac{2}{3}$$

If $k = \frac{2}{3}$, then $2k - 5 = 2\left(\frac{2}{3}\right) - 5 = -3\frac{2}{3} \leftarrow$ Not possible
The sides are $k = 6$ and $2k - 5 = 7$.

Chapter 4 Summary and Review

1. Pythagorean theorem: $a^2 + b^2 = c^2$
where c is the hypotenuse of a right triangle
and a and b are the two legs. This theorem
gives us a formula for finding the third side of
a right triangle if we know the other two sides.

2. Zero factor principal:
If $a \cdot b = 0$, then either $a = 0$ or $b = 0$.
If the product of two numbers is zero, then one
(or both) of the numbers is zero.

3. Property of binomial products: When we
represent the product of two binomials by
the area of a rectangle, the product of the
entries on the two diagonals are equal.

4. Quadratic formula:
For the quadratic equation $ax^2 + bx + c = 0$,
the solutions are $x = \dfrac{-b \pm \sqrt{b^2 - 4ac}}{2a}$.
The formula is used to calculate the solutions
of a quadratic equation.

5. Order of operations:
1. Perform any operations inside parentheses, under a radical, or above or
   below a fraction bar.
2. Compute all powers and roots.
3. Perform all multiplications and divisions
   in order from left to right.
4. Perform all additions and subtractions
   in order from left to right.

6. A quadratic equation is in standard form when
the trinomial is written in descending powers
of the variable and the other side of the
equation is equal to zero.

7. To solve a quadratic equation by extraction
of roots:
1. Isolate the square of the variable.
2. Take the square root of both sides.
Note: since every positive number has two
square roots, use $\pm$.

8. The decimal representation of a rational number
either a) terminates, such as $\frac{3}{4} = 0.75$
or   b) repeats, such as $\frac{2}{3} = 0.66666... = 0.\overline{6}$
The decimal representation of an irrational number
never ends and does not repeat a pattern, such
as $\sqrt{2} = 1.414213562...$.

9. To multiply two binomials, use FOIL.
Multiply each term of the first binomial
by each term of the second binomial.

10. To solve a quadratic equation by factoring:
1. Write the equation in standard form.
   $ax^2 + bx + c = 0$
2. Factor the equation.
3. Set each factor equal to zero.
4. Solve each equation to obtain two solutions.

11. To find the x-coordinate of the vertex,
find the average of the x-intercepts.
To find the y-coordinate of the vertex,
substitute the x-coordinate into the
equation for the parabola.

138

12. To factor a quadratic trinomial using rectangles: $ax^2 + bx + c$
   1. Write the quadratic term ax^2 in the upper left sub-rectangle, and the constant term c in the lower right.
   2. Multiply these terms to find the diagonal product D.
   3. List all possible factors px and qx of D, and choose the pair whose sum is the linear term bx.
   4. Write the factors px and qx in the remaining sub-rectangles.
   5. Factor each row of the rectangle, writing the factors on the outside. These are the factors of the quadratic trinomial.

13.
$$4 - 2 \cdot 3^2$$
$$= 4 - 2 \cdot 9$$
$$= 4 - 18$$
$$= -14$$

14.
$$-2 - 3(-3)^3 - 2$$
$$= -2 - 3(-27) - 2$$
$$= -2 + 81 - 2$$
$$= 79 - 2$$
$$= 77$$

15.
$$\frac{-6 - \sqrt{6^2 - 4(2)(4)}}{2(2)}$$
$$= \frac{-6 - \sqrt{36 - 32}}{4}$$
$$= \frac{-6 - \sqrt{4}}{4}$$
$$= \frac{-6 - 2}{4}$$
$$= \frac{-8}{4}$$
$$= -2$$

16.
$$18 - 2\sqrt[3]{\frac{4}{3}(48)}$$
$$= 18 - 2\sqrt[3]{\frac{4}{3} \cdot \frac{48}{1}}$$
$$= 18 - 2\sqrt[3]{64}$$
$$= 18 - 2(4)$$
$$= 18 - 8$$
$$= 10$$

17a.
$$3 - x^2$$
$$3 - (-4)^2$$
$$= 3 - 16$$
$$= -13$$

b.
$$3(-x)^2$$
$$3[-(-4)]^2$$
$$= 3[4]^2$$
$$= 3[16]$$
$$= 48$$

18a.
$$\frac{3b^3}{6} - \frac{4 - a^2}{8}$$
$$\frac{3(-2)^3}{6} - \frac{4 - 6^2}{8}$$
$$= \frac{3(-8)}{6} - \frac{4 - 36}{8}$$
$$= \frac{-24}{6} - \frac{-32}{8}$$
$$= -4 - (-4)$$
$$= -4 + 4$$
$$= 0$$

b.
$$\frac{(a - b)^2}{ab^2}$$
$$\frac{[6 - (-2)]^2}{6(-2)^2}$$
$$= \frac{[8]^2}{6(4)}$$
$$= \frac{64}{24}$$
$$= \frac{8}{3}$$

19. The <u>coefficient</u> indicates how many times the power is added together.
For example, $2x^3$ means $x^3 + x^3$
and $3x^2$ means $x^2 + x^2 + x^2$.

The <u>exponent</u> indicates how many times the base is multiplied together.
For example, x^3 means $x \cdot x \cdot x$
and x^2 means $x \cdot x$.

20. No. $\sqrt{81} = \sqrt{9^2} = 9$

21a. $6t^2 - 8t^2 = -2t^2$ **b.** $6t - 8t^2$ cannot be simplified   **c.** $6t(-8t^2) = -48t^3$

22a. $w^2 + w^2 = 2w^2$ **b.** $-w - w = -2w$ **c.** $w^2 - w$ cannot be simplified

23. $-8 - 5\sqrt{6} \approx -20.247$ **24.** $\dfrac{3 + \sqrt[3]{3}}{3} \approx 1.481$

25.

$$V = lwh$$
$$= (3p)(5p)(10p)$$
$$= 150p^3$$

$$S = 2lw + 2wh + 2lh$$
$$= 2(3p)(5p) + 2(5p)(10p) + 2(3p)(10p)$$
$$= 30p^2 + 100p^2 + 60p^2$$
$$= 190p^2$$

26.

$$V = \pi r^2 h$$
$$= \pi(2a)^2(5a)$$
$$= \pi(4a^2)(5a)$$
$$= 20\pi a^3$$

$$S = 2\pi r^2 + 2\pi rh$$
$$= 2\pi(2a)^2 + 2\pi(2a)(5a)$$
$$= 2\pi(4a^2) + 2\pi(10a^2)$$
$$= 8\pi a^2 + 20\pi a^2$$
$$= 28\pi a^2$$

27.

$$9k^2 + 21 = 25$$
$$9k^2 = 4$$
$$k^2 = \frac{4}{9}$$
$$k = \pm\sqrt{\frac{4}{9}}$$
$$k = \pm\frac{2}{3}$$

28.

$$6a^2 + 3 = 4a^2 + 19$$
$$2a^2 + 3 = 19$$
$$2a^2 = 16$$
$$a^2 = 8$$
$$a = \pm\sqrt{8}$$

29a.

$$V = \frac{4}{3}\pi r^3$$
$$65.45 = \frac{4}{3}\pi r^3$$
$$196.35 = 4\pi r^3$$
$$\frac{196.35}{4\pi} = r^3$$
$$\sqrt[3]{\frac{196.35}{4\pi}} = r$$
$$\sqrt[3]{15.6250...} = r$$
$$2.5 \text{ ft} \approx r$$

b.

$$S = 4\pi r^2$$
$$= 4\pi(2.5)^2$$
$$= 4\pi(6.25)$$
$$= 25\pi$$
$$\approx 78.54 \text{ sq ft}$$

30a.

$$V = \pi r^2 h$$
$$= \pi(3.36)^2(10)$$
$$= \pi(11.2896)(10)$$
$$= 112.896\pi$$
$$\approx 355 \text{ cu cm}$$

b.

$$V = 355 \div 29.56$$
$$\approx 12 \text{ fl oz}$$

c.

$$V = 24(29.56)$$
$$= 709.44 \text{ cu cm}$$
$$V = \pi r^2 h$$
$$709.44 = \pi r^2(10)$$
$$70.944 = \pi r^2$$
$$\frac{70.944}{\pi} = r^2$$
$$\sqrt{\frac{70.944}{\pi}} = r$$
$$4.75 \text{ cm} \approx r$$

31.

$$S = \frac{n}{2}(a+f) \qquad \text{Multiply both sides by 2}$$
$$2 \cdot S = 2 \cdot \frac{n}{2}(a+f)$$
$$2S = n(a+f) \qquad \text{Divide both sides by } n$$
$$\frac{2S}{n} = \frac{n(a+f)}{n}$$
$$\frac{2S}{n} = a+f \qquad \text{Subtract } f \text{ from both sides}$$
$$\frac{2S}{n} - f = a+f-f$$
$$\frac{2S}{n} - f = a$$

32.

$$s = vt + \frac{1}{2}at^2 \qquad \text{Multiply both sides by 2}$$

$$2 \cdot s = 2\left(vt + \frac{1}{2}at^2\right) \qquad \text{Distribute 2 to each term}$$

$$2s = 2 \cdot vt + 2 \cdot \frac{1}{2}at^2$$

$$2s = 2vt + at^2 \qquad \text{Subtract } 2vt \text{ from both sides}$$

$$2s - 2vt = 2vt + at^2 - 2vt$$

$$2s - 2vt = at^2 \qquad \text{Divide both sides by } t^2$$

$$\frac{2s - 2vt}{t^2} = \frac{at^2}{t^2}$$

$$\frac{2s - 2vt}{t^2} = a$$

33.

$$C = bh^2r \qquad \text{Divide both sides by } br$$

$$\frac{C}{br} = \frac{bh^2r}{br}$$

$$\frac{C}{br} = h^2 \qquad \text{Take square root of both sides}$$

$$\pm\sqrt{\frac{C}{br}} = h$$

34.

$$G = \frac{np}{r^2} \qquad \text{Multiply both sides by } r^2$$

$$r^2 \cdot G = r^2 \cdot \frac{np}{r^2}$$

$$r^2G = np \qquad \text{Divide both sides by } G$$

$$\frac{r^2G}{G} = \frac{np}{G}$$

$$r^2 = \frac{np}{G} \qquad \text{Take square root of both sides}$$

$$r = \pm\sqrt{\frac{np}{G}}$$

35a. Irrational **b.** $\sqrt{300} \approx 17.32$

36a. Rational **b.** $\sqrt[3]{512} = 8$

37a. Irrational **b.** $5 + \sqrt[3]{15} \approx 7.47$

38a. Rational **b.** $\dfrac{7}{\sqrt{81}} = \dfrac{7}{9} = 0.\overline{7}$

39a.

h	d
0	0
1000	110
2000	155
3000	190
4000	219
5000	245

b. ≈ 225 mi

c. $d = \sqrt{12h}$
$= \sqrt{12(4149)}$
$= \sqrt{49{,}788}$
≈ 223.13 mi

e. ≈ 800 m

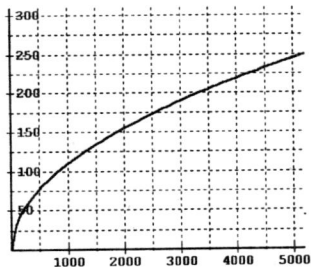

40a.

x	-1	0	1	2	3	4	5
y	$\sqrt{-2}$ is undefined	$\sqrt{-1}$ is undefined	0	1	1.4	1.7	2

b.

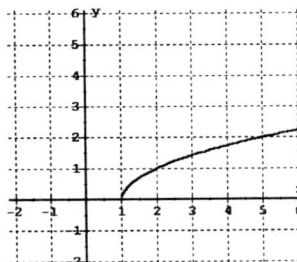

141

41. $(\sqrt{3x})(\sqrt{3x}) = (\sqrt{3x})^2 = 3x$ **42.** $3\sqrt{x}(3\sqrt{x}) = (3 \cdot 3)(\sqrt{x})^2 = 9x$

43. $(5\sqrt{a})^2 = (5\sqrt{a})(5\sqrt{a}) = (5 \cdot 5)(\sqrt{a})^2 = 25a$

44. $(\sqrt[3]{B})^2(\sqrt[3]{B}) = (\sqrt[3]{B})(\sqrt[3]{B})(\sqrt[3]{B}) = (\sqrt[3]{B})^3 = B$

45.
$a^2 + b^2 = c^2$
$7^2 + 9^2 = D^2$
$49 + 81 = D^2$
$130 = D^2$
$\sqrt{130} = D$
$11.40 \text{ m} \approx D$

46.
$a^2 + b^2 = c^2$
$12^2 + 16^2 = D^2$
$144 + 256 = D^2$
$400 = D^2$
$\sqrt{400} = D$
$20 \text{ in.} = D$

47.
$a^2 + b^2 = c^2$
$x^2 + 24^2 = 26^2$
$x^2 + 576 = 676$
$x^2 = 100$
$x = \sqrt{100}$
$x = 10 \text{ m}$

48.
$b = \frac{1}{2}(8) = 4$
$a^2 + b^2 = c^2$
$a^2 + 4^2 = 8^2$
$a^2 + 16 = 64$
$a^2 = 48$
$a = \sqrt{48}$
$a \approx 6.93 \text{ cm}$

49. $3xy(2x - 4 - y) = 6x^2y - 12xy - 3xy^2$ **50.** $-2a(-a^2 - 2a + 4) = 2a^3 + 4a^2 - 8a$

51.
$5a(2a - 3) - 4(3a^2 - 2) + 6a$
$= 10a^2 - 15a - 12a^2 + 8 + 6a$
$= -2a^2 - 9a + 8$

52.
$2x(x - 3y) - 3y(x - 3y)$
$= 2x^2 - 6xy - 3xy + 9y^2$
$= 2x^2 - 9xy + 9y^2$

53.
$(u - 5)(u - 2)$
$= u^2 - 2u - 5u + 10$
$= u^2 - 7u + 10$

54.
$(3r + 2)(r - 4)$
$= 3r^2 - 12r + 2r - 8$
$= 3r^2 - 10r - 8$

55.
$(a + 6)^2$
$= (a + 6)(a + 6)$
$= a^2 + 6a + 6a + 36$
$= a^2 + 12a + 36$

56.
$(6y + 5)(6y - 5)$
$= 36y^2 - 30y + 30y - 25$
$= 36y^2 - 25$

57.
$(2a - 5c)(3a + 2c)$
$= 6a^2 + 4ac - 15ac - 10c^2$
$= 6a^2 - 11ac - 10c^2$

58.
$-3(x - 4)(2x + 5)$
$= -3(2x^2 + 5x - 8x - 20)$
$= -3(2x^2 - 3x - 20)$
$= -6x^2 + 9x + 60$

59a.

t	h
0	100
0.5	96
1	84
1.25	75
1.5	64
1.75	51
2	36
2.25	19
2.5	0

b.

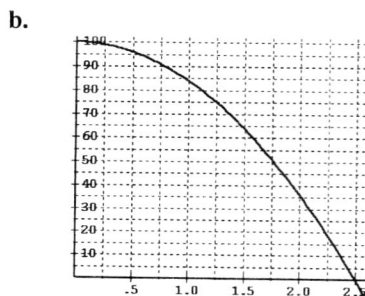

c. 36 ft;
$h = 100 - 16t^2$
$= 100 - 16(2)^2$
$= 100 - 16(4)$
$= 100 - 64$
$= 36$

d. 1.25 sec;
$h = 100 - 16t^2$
$75 = 100 - 16t^2$
$-25 = -16t^2$
$\dfrac{-25}{-16} = t^2$
$\sqrt{\dfrac{25}{16}} = t$
$\dfrac{5}{4} = t$

142

60a. $R = p(180 - 3p)$
$= 180p - 3p^2$

b.

p	R
0	0
10	1500
20	2400
30	2700
40	2400
50	1500
60	0

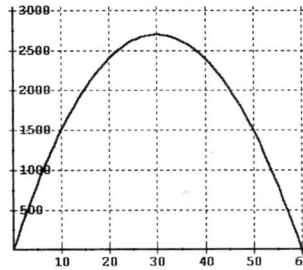

c. $0 and $60

d. $30; $2700

61a.

x	y
-2	3
-1	0
0	-1
1	0
2	3

b.

x	y
-2	-3
-1	0
0	1
1	0
2	-3

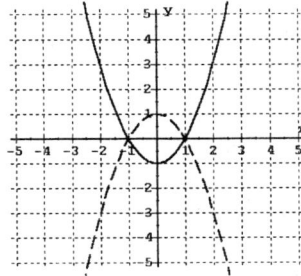

62a.

x	y
0	4
1	1
2	0
3	1
4	4

b.

x	y
-4	4
-3	1
-2	0
-1	1
0	4

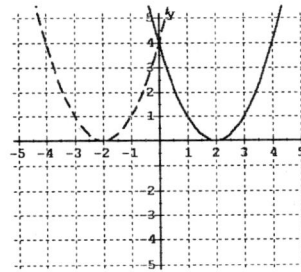

63. $24x^2 - 18x$
$= 6x(4x - 3)$

64. $-32y^2 + 24y$
$= -8y(4y - 3)$

65. $a^2 - 18a + 45$
$= (a - 15)(a - 3)$

66. $x^2 - 14x - 51$
$= (x - 17)(x + 3)$

67. $4y^2 + 16y + 15$
$= (2y + 3)(2y + 5)$

68. $8b^2 - 18b + 9$
$= (4b - 3)(2b - 3)$

69. $14w - 5 + 3w^2$
$= 3w^2 + 14w - 5$
$= (3w - 1)(w + 5)$

70. $-3 + 2p^2 - 5p$
$= 2p^2 - 5p - 3$
$= (2p + 1)(p - 3)$

71. $z^2 - 121$
$= (z + 11)(z - 11)$

72. $81 - 4t^2$
$= (9 + 2t)(9 - 2t)$

73. $6x^2 + 21x + 9$
$= 3(2x^2 + 7x + 3)$
$= 3(2x + 1)(x + 3)$

74. $8y^2 - 6y - 2$
$= 2(4y^2 - 3y - 1)$
$= 2(4y + 1)(y - 1)$

143

75. $0 = m^2 + 10m + 25$ Factor
$0 = (m + 5)(m + 5)$ Set the factor equal to 0
$m + 5 = 0$ Note: The other factor is the same,
$m = -5$ and will give the same solution

76. $b^2 - 25 = 0$ Factor
$(b + 5)(b - 5) = 0$ Set each factor equal to 0
$b + 5 = 0$ or $b - 5 = 0$ Solve each
$b = -5$ $b = 5$

77. $4p^2 = 16p$ Write in standard form
$4p^2 - 16p = 0$ Factor out $4p$
$4p(p - 4) = 0$ Set each factor equal to 0
$4p = 0$ or $p - 4 = 0$ Solve each
$p = 0$ $p = 4$

78. $11t = 6t^2 + 3$ Write in standard form
$0 = 6t^2 - 11t + 3$ Factor
$0 = (2t - 3)(3t - 1)$ Set each factor equal to 0
$2t - 3 = 0$ or $3t - 1 = 0$ Solve each
$2t = 3$ $3t = 1$
$t = \dfrac{3}{2}$ $t = \dfrac{1}{3}$

79. $(x - 5)(x + 1) = -8$ FOIL
$x^2 - 4x - 5 = -8$ Write in standard form
$x^2 - 4x + 3 = 0$ Factor
$(x - 3)(x - 1) = 0$ Set each factor equal to 0
$x - 3 = 0$ or $x - 1 = 0$ Solve each
$x = 3$ $x = 1$

80. $2q(3q - 1) = 4$ Distribute $2q$
$6q^2 - 2q = 4$ Write in standard form
$6q^2 - 2q - 4 = 0$ Factor out 2
$2(3q^2 - q - 2) = 0$ Divide both sides by 2
$\dfrac{2(3q^2 - q - 2)}{2} = \dfrac{0}{2}$
$3q^2 - q - 2 = 0$ Factor
$(3q + 2)(q - 1) = 0$ Set each factor equal to 0
$3q + 2 = 0$ or $q - 1 = 0$ Solve each
$3q = -2$ $q = 1$
$q = \dfrac{-2}{3}$

144

81a. To find x-intercepts, set $y = 0$.

$y = x^2 + 6x$

$0 = x^2 + 6x$

$0 = x(x + 6)$

$x = 0$ or $x + 6 = 0$

$\qquad\qquad\qquad\qquad x = -6$

x-intercepts: $(0, 0)$ and $(-6, 0)$

To find y-intercept, set $x = 0$.

$y = x^2 + 6x$

$y = 0^2 + 6(0) = 0$

y-intercept: $(0, 0)$

b. To find vertex, find average of x-intercepts.

$x = \dfrac{0 + (-6)}{2} = \dfrac{-6}{2} = -3$

$y = x^2 + 6x$

$y = (-3)^2 + 6(-3) = 9 - 18 = -9$

vertex: $(-3, -9)$

c.

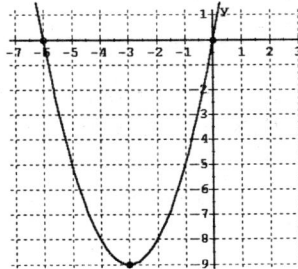

82a. To find x-intercepts, set $y = 0$.

$y = x^2 + 3x - 4$

$0 = x^2 + 3x - 4$

$0 = (x + 4)(x - 1)$

$x + 4 = 0$ or $x - 1 = 0$

$\quad x = -4 \qquad\qquad\qquad x = 1$

x-intercepts: $(-4, 0)$ and $(1, 0)$

To find y-intercept, set $x = 0$.

$y = x^2 + 3x - 4$

$y = 0^2 + 3(0) - 4 = -4$

y-intercept: $(0, -4)$

b. To find vertex, find average of x-intercepts.

$x = \dfrac{-4 + 1}{2} = \dfrac{-3}{2}$

$y = x^2 + 3x - 4$

$y = \left(\dfrac{-3}{2}\right)^2 + 3\left(\dfrac{-3}{2}\right) - 4$

$y = \dfrac{9}{4} - \dfrac{9}{2} - 4$

$y = \dfrac{9}{4} - \dfrac{18}{4} - \dfrac{16}{4}$

$y = \dfrac{-25}{4}$

vertex: $\left(\dfrac{-3}{2}, \dfrac{-25}{4}\right)$

c.

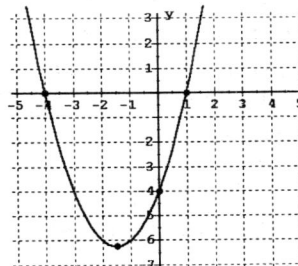

83. $2t^2 + 6t + 3 = 0 \qquad a = 2, \, b = 6, \, c = 3$

$t = \dfrac{-b \pm \sqrt{b^2 - 4ac}}{2a}$

$= \dfrac{-6 \pm \sqrt{6^2 - 4(2)(3)}}{2(2)}$

$= \dfrac{-6 \pm \sqrt{36 - 24}}{4}$

$= \dfrac{-6 \pm \sqrt{12}}{4}$

145

84.

$$\frac{x^2}{4} + 1 = \frac{13}{12}x \qquad \text{Multiply both sides by LCD} = 12$$

$$12 \cdot \left(\frac{x^2}{4} + 1\right) = 12 \cdot \frac{13}{12}x \qquad \text{Distribute 12 to each term}$$

$$12 \cdot \frac{x^2}{4} + 12 \cdot 1 = 12 \cdot \frac{13}{12}x$$

$$3x^2 + 12 = 13x \qquad \text{Write in standard form}$$

$$3x^2 - 13x + 12 = 0 \qquad a = 3, \ b = -13, \ c = 12$$

$$x = \frac{-b \pm \sqrt{b^2 - 4ac}}{2a}$$

$$= \frac{-(-13) \pm \sqrt{(-13)^2 - 4(3)(12)}}{2(3)}$$

$$= \frac{13 \pm \sqrt{169 - 144}}{6}$$

$$= \frac{13 \pm \sqrt{25}}{6}$$

$$= \frac{13 \pm 5}{6}$$

$$x = \frac{13 + 5}{6} = \frac{18}{6} = 3$$

$$x = \frac{13 - 5}{6} = \frac{8}{6} = \frac{4}{3}$$

85.

$$h = -16t^2 + 40 + 80$$

$$0 = -16t^2 + 40t + 80 \qquad \text{Factor out } -8$$

$$0 = -8(2t - 5t - 10) \qquad \text{Divide both sides by } -8$$

$$\frac{0}{-8} = \frac{-8(2t^2 - 5t - 10)}{-8}$$

$$0 = 2t^2 - 5t - 10 \qquad a = 2, \ b = -5, \ c = -10$$

$$t = \frac{-b \pm \sqrt{b^2 - 4ac}}{2a}$$

$$= \frac{-(-5) \pm \sqrt{(-5)^2 - 4(2)(-10)}}{2(2)}$$

$$= \frac{5 \pm \sqrt{25 + 80}}{4}$$

$$= \frac{5 \pm \sqrt{105}}{4}$$

If $t = \dfrac{5 - \sqrt{105}}{4} \approx -1.31$ (a negative number) which is not possible.

The flight is $\dfrac{5 + \sqrt{105}}{4} \approx 3.81$ sec.

86.

$$S = \frac{1}{2}n^2 + \frac{1}{2}n$$

$$325 = \frac{1}{2}n^2 + \frac{1}{2}n \qquad \text{Multiply both sides by LCD} = 2$$

$$2 \cdot 325 = 2 \cdot \left(\frac{1}{2}n^2 + \frac{1}{2}n\right) \qquad \text{Distribute 2 to each term}$$

$$2 \cdot 325 = 2 \cdot \frac{1}{2}n^2 + 2 \cdot \frac{1}{2}n$$

$$650 = n^2 + n \qquad \text{Write in standard form}$$

$$0 = n^2 + n - 650 \qquad \text{Factor}$$

$$0 = (n - 25)(n + 26) \qquad \text{Set each factor equal to 0}$$

$$n - 25 = 0 \quad \text{or} \quad n + 26 = 0$$

$$n = 25 \qquad\qquad n = -26 \leftarrow \text{Not possible}$$

There are 25 counting numbers.

87a. $R = p(160 - 2p)$

b. $\quad = 160p - 2p^2$

$$M = 160p - 2p^2 - (80 + 24p)$$
$$= 160p - 2p^2 - 80 - 24p$$
$$= -2p^2 + 136p - 80$$

c.

$$M = -2p^2 + 136p - 80$$

$$2200 = -2p^2 + 136p - 80$$

$$2p^2 - 136p + 2280 = 0 \qquad \text{Factor out 2}$$

$$2(p^2 - 68p + 1140) = 0 \qquad \text{Divide both sides by 2}$$

$$\frac{2(p^2 - 68p + 1140)}{2} = \frac{0}{2}$$

$$p^2 - 68p + 1140 = 0 \qquad \text{Factor}$$

$$(p - 30)(p - 38) = 0 \qquad \text{Set each factor equal to 0}$$

$$p - 30 = 0 \quad \text{or} \quad p - 38 = 0$$

$$p = 30 \qquad\qquad p = 38$$

They charged 30 cents/lb or 38 cents/lb.

88. $AC + BC = 17$

$\qquad\quad AC = 17 - BC$

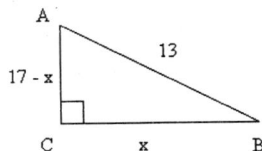

$BC: x$

$AC: 17 - x$

$$a^2 + b^2 = c^2$$

$$x^2 + (17 - x)^2 = 13^2$$

$$x^2 + (17 - x)(17 - x) = 13^2 \qquad \text{FOIL}$$

$$x^2 + 289 - 34x + x^2 = 169 \qquad \text{Write in standard form}$$

$$2x^2 - 34x + 120 = 0 \qquad \text{Factor out 2}$$

$$2(x^2 - 17x + 60) = 0 \qquad \text{Divide both sides by 2}$$

$$\frac{2(x^2 - 17x + 60)}{2} = \frac{0}{2}$$

$$x^2 - 17x + 60 = 0 \qquad \text{Factor}$$

$$(x - 5)(x - 12) = 0 \qquad \text{Set each factor equal to 0}$$

$$x - 5 = 0 \quad \text{or} \quad x - 12 = 0$$

$$x = 5 \qquad\qquad x = 12$$

$BC = x = 5$ mi or 12 mi.

89.
$$a^2 + b^2 = c^2$$
$$m^2 + 8^2 = (m + 5)^2$$
$m^2 + 8^2 = (m + 5)(m + 5)$ FOIL

$m^2 + 64 = m^2 + 10m + 25$ Subtract m^2 from both sides

$\qquad 64 = 10m + 25$ Subtract 25 from both sides

$\qquad 39 = 10m$ Divide both sides by 10

$\qquad 3.9 = m$

The sides are $m = 3.9$, $m + 5 = 8.9$, and 8.

90.
$$a^2 + b^2 = c^2$$
$$k^2 + (k - 1)^2 = (k + 1)^2$$
$k^2 + (k - 1)(k - 1) = (k + 1)(k + 1)$ FOIL

$k^2 + k^2 - 2k + 1 = k^2 + 2k + 1$ Collect like terms

$2k^2 - 2k + 1 = k^2 + 2k + 1$ Write in standard form

$\qquad k^2 - 4k = 0$ Factor out k

$\qquad k(k - 4) = 0$ Set each factor equal to 0

$k = 0 \quad$ or $\quad k - 4 = 0$

$\qquad\qquad k = 4$

The sides are $k = 4$, $k - 1 = 3$, and $k + 1 = 5$.

CHAPTER 5

Homework 5.1

1. More than three terms are allowed.   **3.** Division by a variable is not allowed.

In Problems **5-9**, other examples possible.

5. A polynomial of degree three has one term of degree three and no terms with a higher degree.
Examples: x^3 or $2x^3 + x^2 - x + 3$
A trinomial has exactly three terms and can have any degree.
Examples: $x^2 - x + 3$ or $2x^{11} + 8x^4 - 3x$

7. $2x^4$ **9.** $x^2 + 5x - 6$

11a. Is a polynomial   **b.** Is not a polynomial because the variable is in the denominator.
   c. Is not a polynomial because the variable is in the denominator.   **d.** Is a polynomial

13. 2nd degree   **15.** 7th degree   **17.** 4th degree

19. $-1.9x^3 + x + 6.4$ **21.** $-2x^2 + 6xy + 2y^3$

23. $2 - z^2 - 2z^3$
$2 - (-2)^2 - 2(-2)^3$
$= 2 - 4 - 2(-8)$
$= 2 - 4 + 16$
$= -2 + 16$
$= 14$

25. $2a^4 + 3a^2 - 3a$
$2(1.6)^4 + 3(1.6)^2 - 3(1.6)$
$= 2(6.5536) + 3(2.56) - 3(1.6)$
$= 13.1072 + 7.68 - 4.8$
$= 15.9872$

27. $-abc^2$
$-(-3)(2)(2)^2$
$= -(-3)(2)(4)$
$= 24$

29. $B = P(r^3 + 3r^2 + 3r + 1)$
$= 1000[(0.07)^3 + 3(0.07)^2 + 3(0.07) + 1]$
$= 1000[0.000343 + 3(0.0049) + 3(0.07) + 1]$
$= 1000[0.000343 + 0.0147 + 0.21 + 1]$
$= 1000[1.225043]$
$\approx \$1225.04$

31. $0.04s^2 + 0.6s$
$0.04(50)^2 + 0.6(50)$
$= 0.04(2500) + 0.6(50)$
$= 100 + 30$
$= 130$ ft; No

35. $n^3 + 1$
$(10)^3 + 1$
$= 1000 + 1$
$= 1001$

37. $x[x(x + 3) + 4] + 1$
$= x[x^2 + 3x + 4] + 1$
$= x^3 + 3x^2 + 4x + 1$

39. $x\{x[x(x - 7) - 5] + 8\} - 3$
$= x\{x[x^2 - 7x - 5] + 8\} - 3$
$= x\{x^3 - 7x^2 - 5x + 8\} - 3$
$= x^4 - 7x^3 - 5x^2 + 8x - 3$

41a. Before expanding:
$x[x(x + 3) + 4] + 1$
$2[2(2 + 3) + 4] + 1$
$= 2[2(5) + 4] + 1$
$= 2[10 + 4] + 1$
$= 2[14] + 1$
$= 28 + 1$
$= 29$

After expanding:
$x^3 + 3x^2 + 4x + 1$
$(2)^3 + 3(2)^2 + 4(2) + 1$
$= 8 + 3(4) + 4(2) + 1$
$= 8 + 12 + 8 + 1$
$= 29$
When evaluating mentally,
easier before expanding.

b. $x[x(x + 3) + 4] + 1$
$0.8[0.8(0.8 + 3) + 4] + 1$
$= 6.632$

43. $6b^3 - 2b^3 - (-8b^3) = 12b^3$

45. $6x - 3y + 5xy - 6y + xy = 6x + 6xy - 9y$

47. The exponent should not change when combining like terms. $6w^3 + 8w^3 = 14w^3$

49. Unlike terms cannot be combined. $6 + 3x^2$ cannot be simplified.

51. The sign of each term in parentheses must be changed.
$4t^2 + 7 - (3t^2 - 5)$
$= 4t^2 + 7 - 3t^2 + 5$
$= t^2 + 12$

53. $(2y^3 - 4y^2 - y) + (6y^2 + 2y + 1)$ Remove parentheses
$= 2y^3 - 4y^2 - y + 6y^2 + 2y + 1$ Combine like terms
$= 2y^3 + 2y^2 + y + 1$

55. $(5x^3 + 3x^2 - 4x + 8) - (2x^3 - 4x - 3)$ Change signs of each term of second polynomial
$= 5x^3 + 3x^2 - 4x + 8 - 2x^3 + 4x + 3$ Combine like terms
$= 3x^3 + 3x^2 + 11$

57.
$\begin{array}{r} 8x^2 - 3x + 4 \\ + \underline{-2x^2 + 5x - 7} \\ 6x^2 + 2x - 3 \end{array}$

59.
$\begin{array}{r} -3x^2 + 4x - 2 \\ - (\ 4x^2 - 3x - 1\ ) \end{array} \Rightarrow \begin{array}{r} -3x^2 + 4x - 2 \\ + \underline{-4x^2 + 3x + 1} \\ -7x^2 + 7x - 1 \end{array}$ Change signs of each term and add vertically

61a. $12x - 0.3x^2 - (50 + 3x)$
$= 12x - 0.3x^2 - 50 - 3x$
$= -0.3x^2 + 9x - 50$

b.
$-0.3x^2 + 9x - 50$
$-0.3(10)^2 + 9(10) - 50$
$= -0.3(100) + 9(10) - 50$
$= -30 + 90 - 50$
$= \$10$

$-0.3x^2 + 9x - 50$
$-0.3(15)^2 + 9(15) - 50$
$= -0.3(225) + 9(15) - 50$
$= -67.5 + 135 - 50$
$= \$17.50$

$-0.3x^2 + 9x - 50$
$-0.3(20)^2 + 9(20) - 50$
$= -0.3(400) + 9(20) - 50$
$= -120 + 180 - 50$
$= \$10$

63a. $12t^2 - 24t + 34 - (-16t^2 + 32t + 6)$
$= 12t^2 - 24t + 34 + 16t^2 - 32t - 6$
$= 28t^2 - 56t + 28$

b.
$28t^2 - 56t + 28$
$28(0)^2 - 56(0) + 28$
$= 0 - 0 + 28$
$= 28$ ft

$28t^2 - 56t + 28$
$28\left(\dfrac{1}{2}\right)^2 - 56\left(\dfrac{1}{2}\right) + 28$
$= 28\left(\dfrac{1}{4}\right) - 56\left(\dfrac{1}{2}\right) + 28$
$= 7 - 28 + 28$
$= 7$ ft

c.
$28t^2 - 56t + 28 = 0$
$28(t^2 - 2t + 1) = 0$
$\dfrac{28(t^2 - 2t + 1)}{28} = \dfrac{0}{28}$
$t^2 - 2t + 1 = 0$
$(t - 1)(t - 1) = 0$
$t - 1 = 0$
$t = 1$
They will be at the same height in 1 sec.

65a. $0.30h$
b. $50 - h$
c. $0.25(50 - h)$

d. $0.30h + 0.25(50 - h)$
$= 0.30h + 12.5 - 0.25h$
$= -0.05h + 12.5$

67a. $2w$
b. $w + 30$
c. $2(w + 30)$

d. $2w + 20 + 2(w + 30)$
$= 2w + 20 + 2w + 60$
$= 4w + 80$ mi

150

69.

n	S_n	$\frac{1}{2}n^2 + \frac{1}{2}n$
1	1	$\frac{1}{2}(1)^2 + \frac{1}{2}(1) = 1$
2	3	$\frac{1}{2}(2)^2 + \frac{1}{2}(2) = 3$
3	6	$\frac{1}{2}(3)^2 + \frac{1}{2}(3) = 6$
4	10	$\frac{1}{2}(4)^2 + \frac{1}{2}(4) = 10$
5	15	$\frac{1}{2}(5)^2 + \frac{1}{2}(5) = 15$
6	21	$\frac{1}{2}(6)^2 + \frac{1}{2}(6) = 21$
7	28	$\frac{1}{2}(7)^2 + \frac{1}{2}(7) = 28$
8	36	$\frac{1}{2}(8)^2 + \frac{1}{2}(8) = 36$
9	45	$\frac{1}{2}(9)^2 + \frac{1}{2}(9) = 45$
10	55	$\frac{1}{2}(10)^2 + \frac{1}{2}(10) = 55$

71. Remove parentheses and combine like terms.
$$(2x + y - z) + (3x - 4y + 6z) - (5x - 9y - 3z)$$
$$= 2x + y - z + 3x - 4y + 6z - 5x + 9y + 3z$$
$$= 6y + 8z \qquad \text{Substitute } y = -3.6 \text{ and } z = 1.8$$
$$= 6(-3.6) + 8(1.8)$$
$$= -21.6 + 14.4$$
$$= -7.2$$

Homework 5.2

1. $x^3 \cdot x^6 = x^{3+6} = x^9$

3. $5^6 \cdot 5^8 = 5^{6+8} = 5^{14}$

5. $b^3(b)(b^5) = b^{3+1+5} = b^9$

7. $y^3 \cdot y^n = y^8$
$3 + n = 8$
$n = 5$

9. $a^n \cdot a^4 = a^8$
$n + 4 = 8$
$n = 4$

11. $3 \cdot 3^n = 3^3$
$1 + n = 3$
$n = 2$

13a. $2x^4(-3x^4) = (2)(-3)x^{4+4} = -6x^8$
b. $-x^4(-2x^2) = (-1)(-2)x^{4+2} = 2x^6$
c. $-x^4 \cdot y^3 = -x^4 y^3$ cannot be simplified
d. $-3x^5(3y^5) = (-3)(3)x^5 y^5 = -9x^5 y^5$

15a. $2x^4 - 3x^4 = (2 - 3)x^4 = -x^4$
b. $-x^4 - 2x^2$ cannot be simplified
c. $-x^4 + y^3$ cannot be simplified
d. $-3x^5 + 3y^5$ cannot be simplified

17a. $3a(8a) = 24a^2$
b. $3a + 8a = 11a$

18a. $-2b^3(5b^3) = -10b^6$
b. $-2b^3 + (5b^3) = 3b^3$

21a. $4p^2 q(-7p^2 q) = -28p^4 q^2$
b. $4p^2 q + (-7p^2 q) = -3p^2 q$

23a. $P = 2l + 2w$
$= 2(16m^2) + 2(9m^2)$
$= 32m^2 + 18m^2$
$= 50m^2$

b. $A = lw$
$= (16m^2)(9m^2)$
$= 144m^4$

25a. $P = 2l + 2w$
$= 2(12w^5) + 2(12w^5)$
$= 24w^5 + 24w^5$
$= 48w^5$

b. $A = lw$
$= (12w^5)(12w^5)$
$= 144w^{10}$

27a. $x^2 + x^2 = 2x^2$
b. $x^2(x^2) = x^4$
c. $x^2 - x^2 = 0$
d. $x^2(-x^2) = -x^4$

29a. $x + x^2$ cannot be simplified
b. $x(x^2) = x^3$
c. $x^2 - x$ cannot be simplified
d. $x^2(-x) = -x^3$

31. $w^3(-8w^4) = -8w^7$

33. $-5s^2(2s^5)t = -10s^7 t$

35. $-6xy^2(-3xy^3)(-2xy) = -36x^3y^6$

37a. $2x + 2x = 4x$

   b. $2x(-2x) = -4x^2$

   c. $2(x - 2x) = 2(-x) = -2x$

   d. $-2x - 2 - x = -3x - 2$

39a. $3b + 4b = 7b$

   b. $3b(4b) = 12b^2$

   c. $b^3 + b^4$ cannot be simplified

   d. $b^3(b^4) = b^7$

41. $-xy(x^2 + xy + y^2)$
$$= -xy(x^2) - xy(xy) - xy(y^2)$$
$$= -x^3y - x^2y^2 - xy^3$$

43. $(6 - st + 3s^2t^2)(-3s^2t^2)$
$$= 6(-3s^2t^2) - st(-3s^2t^2) + 3s^2t^2(-3s^2t^2)$$
$$= -18s^2t^2 + 3s^3t^3 - 9s^4t^4$$
$$= -9s^4t^4 + 3s^3t^3 - 18s^2t^2$$

45. $ax(x^2 + 2x - 3) - a(x^3 + 2x^2)$
$$= ax(x^2) + ax(2x) + ax(-3) - a(x^3) - a(2x^2)$$
$$= ax^3 + 2ax^2 - 3ax - ax^3 - 2ax^2$$
$$= -3ax$$

47. $3ab^2(2 + 3a) - 2ab(3ab + 2b)$
$$= 3ab^2(2) + 3ab^2(3a) - 2ab(3ab) - 2ab(2b)$$
$$= 6ab^2 + 9a^2b^2 - 6a^2b^2 - 4ab^2$$
$$= 2ab^2 + 3a^2b^2$$
$$= 3a^2b^2 + 2ab^2$$

49. $4a(a - 1)(a + 5)$
$$= 4a(a^2 + 5a - a - 5)$$
$$= 4a(a^2 + 4a - 5)$$
$$= 4a^3 + 16a^2 - 20a$$

51. $-x(2x - 1)^2$
$$= -x(2x - 1)(2x - 1)$$
$$= -x(4x^2 - 2x - 2x + 1)$$
$$= -x(4x^2 - 4x + 1)$$
$$= -4x^3 + 4x^2 - x$$

53. $s^2t^2(2s + t)(3s - t)$
$$= s^2t^2(6s^2 - 2st + 3st - t^2)$$
$$= s^2t^2(6s^2 + st - t^2)$$
$$= 6s^4t^2 + s^3t^3 - s^2t^4$$

55. $(x - 2)(x^2 - 3x + 2)$
$$= x^3 - 3x^2 + 2x - 2x^2 + 6x - 4$$
$$= x^3 - 5x^2 + 8x - 4$$

57. $(3x - 1)(9x^2 - 3x + 1)$
$$= 27x^3 - 9x^2 + 3x - 9x^2 + 3x - 1$$
$$= 27x^3 - 18x^2 + 6x - 1$$

59a. $(x + 1)(x + 2)(x + 3)$
$$= (x + 1)(x^2 + 3x + 2x + 6)$$
$$= (x + 1)(x^2 + 5x + 6)$$
$$= x^3 + 5x^2 + 6x + x^2 + 5x + 6$$
$$= x^3 + 6x^2 + 11x + 6$$

$x^3 + 6x^2 + 11x + 6$
$(-2)^3 + 6(-2)^2 + 11(-2) + 6$
$= -8 + 6(4) + 11(-2) + 6$
$= -8 + 24 - 22 + 6$
$= 0$

Note: when $x = -2$, the factor $x + 2 = -2 + 2 = 0$

   b. $x^3 + 6x^2 + 11x + 6$
$(-1)^3 + 6(-1)^2 + 11(-1) + 6$
$= -1 + 6(1) + 11(-1) + 6$
$= -1 + 6 - 11 + 6$
$= 0$

Note: when $x = -1$, the factor $x + 1 = -1 + 1 = 0$

$x^3 + 6x^2 + 11x + 6$
$(-3)^3 + 6(-3)^2 + 11(-3) + 6$
$= -27 + 6(9) + 11(-3) + 6$
$= -27 + 54 - 33 + 6$
$= 0$

Note: when $x = -3$, the factor $x + 3 = -3 + 3 = 0$

61a. $(2x - 1)(2x - 1)(2x - 1)$
$= (2x - 1)(4x^2 - 2x - 2x + 1)$
$= (2x - 1)(4x^2 - 4x + 1)$
$= 8x^3 - 8x^2 + 2x - 4x^2 + 4x - 1$
$= 8x^3 - 12x^2 + 6x - 1$

b. $8x^3 - 12x^2 + 6x - 1 = 0$
$(2x - 1)(2x - 1)(2x - 1) = 0$
$2x - 1 = 0$ Note: The other factors
$2x = 1$ are the same and will
$x = \dfrac{1}{2}$ give the same solution

63a. One number: n
Other number: $16 - n$

b. $n(16 - n) = 16n - n^2$

65a. Width: w
Length: $w + 3$
Height: $w - 2$

b. $V = lwh$
$= (w + 3)w(w - 2)$
$= w(w + 3)(w - 2)$
$= w(w^2 + w - 6)$
$= w^3 + w^2 - 6w$

c. $S = 2lw + 2lh + 2wh$
$= 2(w + 3)w + 2(w + 3)(w - 2) + 2w(w - 2)$
$= 2w(w + 3) + 2(w^2 + w - 6) + 2w(w - 2)$
$= 2w^2 + 6w + 2w^2 + 2w - 12 + 2w^2 - 4w$
$= 6w^2 + 4w - 12$

67. $10^3(8 \cdot 10^4) = 80{,}000{,}000$

69. $(3 \cdot 10^2)(2 \cdot 10^2) = 60{,}000$

71. $(3.3 \cdot 10^2)(2 \cdot 10^2) = 66{,}000$

Homework 5.3

1. $\dfrac{a^6}{a^3} = a^{6-3} = a^3$

3. $\dfrac{3^9}{3^4} = 3^{9-4} = 3^5$

5. $\dfrac{z^6}{z^9} = \dfrac{1}{z^{9-6}} = \dfrac{1}{z^3}$

In **7-11**, other examples possible.

7. For $t = 2$: $t^2 \cdot t^3 = 2^2 \cdot 2^3 = 4 \cdot 8 = 32$
$t^6 = 2^6 = 64$

9. For $v = 2$: $\dfrac{v^8}{v^2} = \dfrac{2^8}{2^2} = \dfrac{256}{4} = 64$
$v^4 = 2^4 = 16$

11. For $n = 2$: $\dfrac{n^3}{n^5} = \dfrac{2^3}{2^5} = \dfrac{8}{32} = \dfrac{1}{4}$
$n^2 = 2^2 = 4$

13. $\dfrac{2x^3y}{8x^4y^5} = \dfrac{2}{8} \cdot \dfrac{1}{x^{4-3}} \cdot \dfrac{1}{y^{5-1}} = \dfrac{1}{4} \cdot \dfrac{1}{x} \cdot \dfrac{1}{y^4} = \dfrac{1}{4xy^4}$

15. $\dfrac{-12bx^4}{8bx^2} = \dfrac{-12}{8} \cdot \dfrac{b}{b} \cdot \dfrac{x^{4-2}}{1} = \dfrac{-3}{2} \cdot \dfrac{1}{1} \cdot \dfrac{x^2}{1} = \dfrac{-3x^2}{2}$

17. $\dfrac{-15x^3y^2}{-3x^3y^4} = \dfrac{-15}{-3} \cdot \dfrac{x^3}{x^3} \cdot \dfrac{1}{y^{4-2}} = \dfrac{5}{1} \cdot \dfrac{1}{1} \cdot \dfrac{1}{y^2} = \dfrac{5}{y^2}$

19. 2

21. $4ab$

23. $9x^2 - 12x^5 + 3x^3 = 3x^2(3 - 4x^3 + x)$

25. $14x^3y - 35x^2y^2 + 21xy^3 = 7xy(2x^2 - 5xy + 3y^2)$

27. $-b^2 - bc - ab = -b(b + c + a)$

29. $-4k^4 + 4k^2 - 2k = -2k(2k^3 - 2k + 1)$

31. $2x(x + 6) - 3(x + 6) = (x + 6)(2x - 3)$

33. $3x^2(2x + 3) - (2x + 3) = (2x + 3)(3x^2 - 1)$

35. $21 - 4x - x^2$
$= -x^2 - 4x + 21$
$= -1(x^2 + 4x - 21)$
$= -1(x + 7)(x - 3)$

37. $24a + 81 - a^2$
$= -a^2 + 24a + 81$
$= -1(a^2 - 24a - 81)$
$= -1(a - 27)(a + 3)$

39. $2x^2 + 10x + 12$
$= 2(x^2 + 5x + 6)$
$= 2(x + 2)(x + 3)$

41. $4a^2b + 12ab - 7b$
$= b(4a^2 + 12a - 7)$
$= b(2a + 7)(2a - 1)$

43. $4z^3 + 10z^2 + 6z$
$= 2z(2z^2 + 5z + 3)$
$= 2z(2z + 3)(z + 1)$

45. $18a^2b - 9ab - 27b$
$= 9b(2a^2 - a - 3)$
$= 9b(2a - 3)(a + 1)$

47. $x^2 - 5xy + 6y^2$
$= (x - 2y)(x - 3y)$

49. $x^2 + 4ax - 77a^2$
$= (x + 11a)(x - 7a)$

51. $4x^3 + 12x^2y + 8xy^2$
$= 4x(x^2 + 3xy + 2y^2)$
$= 4x(x + 2y)(x + y)$

53. $9a^3b + 9a^2b^2 - 18ab^3$
$= 9ab(a^2 + ab - 2b^2)$
$= 9ab(a + 2b)(a - b)$

55. $2t^2 - 5st - 3s^2$
$= (2t + s)(t - 3s)$

57. $4b^2y^2 + 5by + 1$
$= (4by + 1)(by + 1)$

59. $12ab^2 + 15a^2b + 3a^3$
$= 3a(4b^2 + 5ab + a^2)$
$= 3a(4b + a)(b + a)$

Homework 5.4

1. $(8t^4)^2 = 64t^8$

3. $(-12a^2)^2 = 144a^4$

5. $(10h^2k)^2 = 100h^4k^2$

7. $(a + b)^2 = a^2 + 2ab + b^2$
$(3a + b)^2 = (3a)^2 + 2(3a)(b) + (b)^2$
$= 9a^2 + 6ab + b^2$

9. $(a - b)^2 = a^2 - 2ab + b^2$
$(7b^3 - 6)^2 = (7b^3)^2 - 2(7b^3)(6) + (6)^2$
$= 49b^6 - 84b^3 + 36$

11. $(a + b)^2 = a^2 + 2ab + b^2$
$(2h + 5k^4)^2 = (2h)^2 + 2(2h)(5k^4) + (5k^4)^2$
$= 4h^2 + 20hk^4 + 25k^8$

13. $(a - b)(a + b) = a^2 - b^2$
$(3p - 4)(3p + 4) = (3p)^2 - (4)^2$
$= 9p^2 - 16$

15. $(a - b)(a + b) = a^2 - b^2$
$(2x^2 - 1)(2x^2 + 1) = (2x^2)^2 - (1)^2$
$= 4x^4 - 1$

17. $(a + b)(a - b) = a^2 - b^2$
$(h^2 + 7t)(h^2 - 7t) = (h^2)^2 - (7t)^2$
$= h^4 - 49t^2$

19. $y^2 + 6y + 9 = (y + 3)^2$

21. $m^2 - 30m + 225 = (m - 15)^2$

23. $x^2 + 4xy + 4y^2 = (x + 2y)^2$

25. $x^2 - 9 = (x + 3)(x - 3)$

27. $36 - a^2b^2 = (6 + ab)(6 - ab)$

29. $64y^2 - 49x^2 = (8y + 7x)(8y - 7x)$

154

31. $a^4 + 10a^2 + 25 = (a^2 + 5)^2$

33. $36y^8 - 49 = (6y^4 + 7)(6y^4 - 7)$

35. $16x^6 - 9y^4 = (4x^3 + 3y^2)(4x^3 - 3y^2)$

37. $3a^2 - 75$
$= 3(a^2 - 25)$
$= 3(a + 5)(a - 5)$

39. $2a^3 - 12a^2 + 18a$
$= 2a(a^2 - 6a + 9)$
$= 2a(a - 3)^2$

41. $9x^7 - 81x^3$
$= 9x^3(x^4 - 9)$
$= 9x^3(x^2 + 3)(x^2 - 3)$

43. $12h^2 + 3k^6$
$= 3(4h^2 + k^6)$

45. $81x^8 - y^4$
$= (9x^4 + y^2)(9x^4 - y^2)$
$= (9x^4 + y^2)(3x^2 + y)(3x^2 - y)$

47. $162a^4b^8 - 2a^8$
$= 2a^4(81b^8 - a^4)$
$= 2a^4(9b^4 + a^2)(9b^4 - a^2)$
$= 2a^4(9b^4 + a^2)(3b^2 + a)(3b^2 - a)$

49. The area of the large rectangle is $(a + b)^2$.
The sum of the areas of the four sub-rectangles is $a^2 + 2ab + b^2$.

51a. $(a - b)^3$
$= (a - b)(a - b)(a - b)$
$= (a - b)(a^2 - 2ab + b^2)$
$= a^3 - 2a^2b + ab^2 - a^2b + 2ab^2 - b^3$
$= a^3 - 3a^2b + 3ab^2 - b^3$

b. $(a - b)^3 = a^3 - 3a^2b + 3ab^2 - b^3$
$(2x - 3)^3 = (2x)^3 - 3(2x)^2(3) + 3(2x)(3)^2 - (3)^3$
$= 8x^3 - 36x^2 + 54x - 27$

c. $(a - b)^3 = (5 - 2)^3 = 3^3 = 27$
$a^3 - b^3 = 5^3 - 2^3 = 125 - 8 = 117$

53a. $(a + b)(a^2 - ab + b^2)$
$= a^3 - a^2b + ab^2 + a^2b - ab^2 + b^3$
$= a^3 + b^3$

b. $a^3 + b^3 = (a + b)(a^2 - ab + b^2)$

c. $x^3 + 8 =$
$x^3 + 2^3 = (x + 2)[x^2 - (x)(2) + 2^2]$
$= (x + 2)(x^2 - 2x + 4)$

Homework 5.5

1a.

n	s
5	240
6	200
10	120
12	100
20	60
25	48
30	40
40	30

c.

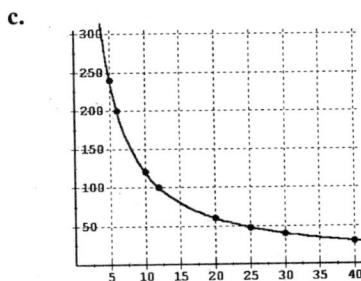

b. $s = \dfrac{1200}{n}$

d. $\dfrac{3}{2}$

3a. $k = 10(36) = 360;\quad c = \dfrac{360}{p}$

b.

p	c
6	60
9	40
10	36
12	30
15	24
18	20
20	18

c.

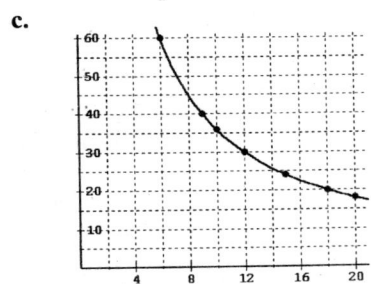

d. $\dfrac{2}{3}$

5. $k = 4.5(8) = 36$

$t = \dfrac{36}{w}$

w	t
4.5	8
7.2	5
7.5	4.8

7. $k = 20(200) = 4000$

$C = \dfrac{4000}{R}$

R	C
20	200
16	250
125	32

9. $k = 2(8) = 4(4) = 16; \quad y = \dfrac{16}{x}$

11. $k = 10(12) = 20(6) = 120; \quad y = \dfrac{120}{x}$

13. Yes; The product of the variables is constant. $50(286) = 55(260) = 71.5(200) = 14{,}300 = k$

15. No; The product of the variables is not constant. $20(74.6) \neq 50(67.3) \neq 200(40)$

17.

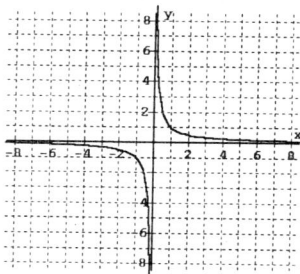

a.

x	2	4	5	8
y	$\frac{1}{2}$	$\frac{1}{4}$	$\frac{1}{5}$	$\frac{1}{8}$

As x gets large and is positive, y gets close to zero and is positive.

b.

x	1	$\frac{1}{2}$	$\frac{1}{4}$	$\frac{1}{8}$
y	1	2	4	8

As x gets close to zero and is positive, y gets large and is positive.

c.

x	-2	-3	-6	-8
y	$-\frac{1}{2}$	$-\frac{1}{3}$	$-\frac{1}{6}$	$-\frac{1}{8}$

As x gets large and is negative, y gets close to zero and is negative.

d.

x	-1	$-\frac{1}{2}$	$-\frac{1}{3}$	$-\frac{1}{5}$
y	-1	-2	-3	-5

As x gets close to zero and is negative, y gets large and is negative.

19.

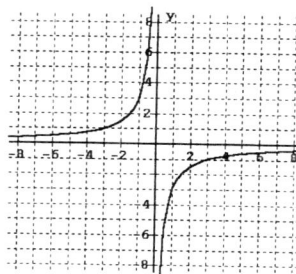

a.

x	2	4	5	8
y	$-\frac{3}{2}$	$-\frac{3}{4}$	$-\frac{3}{5}$	$-\frac{3}{8}$

As x gets large and is positive, y gets close to zero and is negative.

b.

x	1	$\frac{1}{2}$	$\frac{1}{4}$	$\frac{1}{8}$
y	-3	-6	-12	-24

As x gets close to zero and is positive, y gets large and is negative.

c.

x	-2	-3	-6	-8
y	$\frac{3}{2}$	1	$\frac{1}{2}$	$\frac{3}{8}$

As x gets large and is negative, y gets close to zero and is positive.

d.

x	-1	$-\frac{1}{2}$	$-\frac{1}{3}$	$-\frac{1}{5}$
y	3	6	9	15

As x gets close to zero and is negative, y gets large and is positive.

21. C

23. E

25. B

Midchapter 5 Review

1. $\dfrac{x}{2} + 3$ is a linear expression equivalent to $\frac{1}{2}x + 3$.

$\dfrac{2}{x} + 3$ has a variable in the denominator, therefore it can not be a polynomial.

2. $\sqrt{2}x^2 + 3\sqrt{2}x + 1$ is a polynomial because the variable is not under the radical, and the coefficients are real numbers.

In **3-8**, other examples possible.

3. 5th degree; $2x^5 + x^4 - 3x^2 + 1$

4. 3rd degree; $(x^3 + 1) + (x^2 + 2) = x^3 + x^2 + 3$

5. $(x + 1) + (x - 1) = 2x$ (one term)
$(x + 1) + (x + 1) = 2x + 2$ (two terms)
$(x^2 + 1) + (x + 1) = x^2 + x + 2$ (three terms)
$(x^3 + x^2) + (x + 1) = x^3 + x^2 + x + 1$ (four terms)

6. 5th degree; $x^3(x^2 + 1) = x^5 + x^3$

7. Square of monomial: $(3x)^2 = 9x^2$ (one term)
Square of binomial: $(x + 3)^2 = x^2 + 6x + 9$ (three terms)

8. The product of the sum and the difference of two terms is the difference of the two squares.
$(x + 3)(x - 3) = x^2 - 9$

9. Sum of two squares: $a^2 + b^2$ Square of a binomial: $(a + b)^2 = a^2 + 2ab + b^2$

10. If the variables x and y satisfy the equation $xy = k$, divide both sides by x.

$$\frac{xy}{x} = \frac{k}{x}$$

$$y = \frac{k}{x}$$ By definition, this means that y varies inversely with x.

11. $-\dfrac{x^6}{720} + \dfrac{x^4}{24} - \dfrac{x^2}{2} + 1$; 6th degree

12. $10n^4 + 10^8 n^2 + 10^6$: 4th degree

13. $-16t^2 + 50t + 5$

$-16\left(\dfrac{1}{2}\right)^2 + 50\left(\dfrac{1}{2}\right) + 5$

$= -16\left(\dfrac{1}{4}\right) + 50\left(\dfrac{1}{2}\right) + 5$

$= -4 + 25 + 5$

$= 26$

14. $\dfrac{1}{2}n^2 + \dfrac{1}{2}n$

$\dfrac{1}{2}(100)^2 + \dfrac{1}{2}(100)$

$= \dfrac{1}{2}(10{,}000) + \dfrac{1}{2}(100)$

$= 5000 + 50$

$= 5050$

15. $\dfrac{1}{6}z^3 + \dfrac{1}{2}z^2 + z + 1$

$\dfrac{1}{6}(-1)^3 + \dfrac{1}{2}(-1)^2 + (-1) + 1$

$= \dfrac{1}{6}(-1) + \dfrac{1}{2}(1) + (-1) + 1$

$= \dfrac{-1}{6} + \dfrac{1}{2} + (-1) + 1$

$= \dfrac{-1}{6} + \dfrac{3}{6} + \dfrac{-6}{6} + \dfrac{6}{6}$

$= \dfrac{2}{6}$

$= \dfrac{1}{3}$

16. $p^4 + 4p^3 + 6p^2 + 4p + 1$
$(-2)^4 + 4(-2)^3 + 6(-2)^2 + 4(-2) + 1$
$= 16 + 4(-8) + 6(4) + 4(-2) + 1$
$= 16 - 32 + 24 - 8 + 1$
$= -16 + 24 - 8 + 1$
$= 8 - 8 + 1$
$= 1$

17.

$4x^2 - 12xy + 9y^2$
$4(-3)^2 - 12(-3)(-2) + 9(-2)^2$
$= 4(9) - 12(-3)(-2) + 9(4)$
$= 36 - 72 + 36$
$= -36 + 36$
$= 0$

18.

$2R^4S$
$2(150)^4(0.01)$
$= 2(506{,}250{,}000)(0.01)$
$= 10{,}125{,}000$

19a.

$\dfrac{1}{24}n^4 - \dfrac{1}{4}n^3 + \dfrac{11}{24}n^2 - \dfrac{1}{4}n$

$\dfrac{1}{24}(20)^4 - \dfrac{1}{4}(20)^3 + \dfrac{11}{24}(20)^2 - \dfrac{1}{4}(20)$

$= \dfrac{1}{24}(160{,}000) - \dfrac{1}{4}(8000) + \dfrac{11}{24}(400) - \dfrac{1}{4}(20)$

$= \dfrac{20{,}000}{3} - 2000 + \dfrac{550}{3} - 5$

$= \dfrac{20{,}000}{3} - \dfrac{6000}{3} + \dfrac{550}{3} - \dfrac{15}{3}$

$= \dfrac{14{,}535}{3}$

$= 4845$

b.

$\dfrac{1}{24}n^4 - \dfrac{1}{4}n^3 + \dfrac{11}{24}n^2 - \dfrac{1}{4}n$

$\dfrac{1}{24}(3)^4 - \dfrac{1}{4}(3)^3 + \dfrac{11}{24}(3)^2 - \dfrac{1}{4}(3)$

$= \dfrac{1}{24}(81) - \dfrac{1}{4}(27) + \dfrac{11}{24}(9) - \dfrac{1}{4}(3)$

$= \dfrac{27}{8} - \dfrac{27}{4} + \dfrac{33}{8} - \dfrac{3}{4}$

$= \dfrac{27}{8} - \dfrac{54}{8} + \dfrac{33}{8} - \dfrac{6}{8}$

$= \dfrac{0}{8}$

$= 0$

Similarly, for $n = 2$ and $n = 1$, the polynomial $= 0$.

c.

$\dfrac{1}{24}n^4 - \dfrac{1}{4}n^3 + \dfrac{11}{24}n^2 - \dfrac{1}{4}n$

$\dfrac{1}{24}(4)^4 - \dfrac{1}{4}(4)^3 + \dfrac{11}{24}(4)^2 - \dfrac{1}{4}(4)$

$= \dfrac{1}{24}(256) - \dfrac{1}{4}(64) + \dfrac{11}{24}(16) - \dfrac{1}{4}(4)$

$= \dfrac{32}{3} - 16 + \dfrac{22}{3} - 1$

$= \dfrac{32}{3} - \dfrac{48}{3} + \dfrac{22}{3} - \dfrac{3}{3}$

$= \dfrac{3}{3}$

$= 1$

There is only one way to choose 4 items from a list of 4; choose all 4 items.

20a.

$125x^3 - 750x^2 + 4000$
$125(2)^3 - 750(2)^2 + 4000$
$= 125(8) - 750(4) + 4000$
$= 1000 - 3000 + 4000$
$= 2000$ ft

b.

$125x^3 - 750x^2 + 4000$
$125(0)^3 - 750(0)^2 + 4000$
$= 125(0) - 750(0) + 4000$
$= 0 - 0 + 4000$
$= 4000$ ft

When $x = 0$, her altitude is 4000 ft which confirms that her altitude before the descent was 4000 ft.

c.

$125x^3 - 750x^2 + 4000$
$125(4)^3 - 750(4)^2 + 4000$
$= 125(64) - 750(16) + 4000$
$= 8000 - 12{,}000 + 4000$
$= 0$ ft

When $x = 4$, her altitude is 0 ft means that the plane has landed after traveling 4 mi horizontally.

21. $(3a^2 - 4a - 7) - (a^2 - 5a + 2)$
$= 3a^2 - 4a - 7 - a^2 + 5a - 2$
$= 2a^2 + a - 9$

22. $(5ab^2 + 6a^2b) - (3ab^2 + 5ab)$
$= 5ab^2 + 6a^2b - 3ab^2 - 5ab$
$= 2ab^2 + 6a^2b - 5ab$

23. $\dfrac{13c^2d}{26c^3d} = \dfrac{13}{26} \cdot \dfrac{1}{c^{3-2}} \cdot \dfrac{d}{d} = \dfrac{1}{2} \cdot \dfrac{1}{c} \cdot \dfrac{1}{1} = \dfrac{1}{2c}$

24. $\dfrac{12m^4 + 4m}{4m} = \dfrac{12m^4}{4m} + \dfrac{4m}{4m} = \dfrac{12}{4}m^{4-1} + \dfrac{4m}{4m} = 3m^3 + 1$

25. $7q^2(8 - 7q^2 - q^4)$
$= 7q^2 \cdot 8 - 7q^2 \cdot 7q^2 - 7q^2 \cdot q^4$
$= 56q^2 - 49q^4 - 7q^6$

26. $2k^2(-3km)(m^3k)$
$= 2(-3) \cdot k^{2+1+1} \cdot m^{1+3}$
$= -6k^4m^4$

27. $(9v + 5w)(9v - 5w)$
$= (9v)^2 - (5w)^2$
$= 81v^2 - 25w^2$

28. $(x^2 + 1)^2$
$= (x^2)^2 + 2(x^2)(1) + (1)^2$
$= x^4 + 2x^2 + 1$

29. $-3p^2(p + 2)(p - 5)$
$= -3p^2(p^2 - 5p + 2p - 10)$
$= -3p^2(p^2 - 3p - 10)$
$= -3p^4 + 9p^3 + 30p^2$

30. $12rs^2(3r - s)(r + 4s)$
$= 12rs^2(3r^2 + 12rs - rs - 4s^2)$
$= 12rs^2(3r^2 + 11rs - 4s^2)$
$= 36r^3s^2 + 132r^2s^3 - 48rs^4$

31. $(2x - 3)(4x^2 + 6x + 9)$
$= 8x^3 + 12x^2 + 18x - 12x^2 - 18x - 27$
$= 8x^3 - 27$

32. $(3x + 2)(3x + 2)(3x + 2)$
$= (3x + 2)(9x^2 + 6x + 6x + 4)$
$= (3x + 2)(9x^2 + 12x + 4)$
$= 27x^3 + 36x^2 + 12x + 18x^2 + 24x + 8$
$= 27x^3 + 54x^2 + 36x + 8$

33a. $P = 2l + 2w$
$= 2(3a^4) + 2(2a^4)$
$= 6a^4 + 4a^4$
$= 10a^4$

b. $A = lw$
$= (3a^4)(2a^4)$
$= 6a^8$

34a. $P = 2l + 2w$
$= 2(7xy) + 2(5xy)$
$= 14xy + 10xy$
$= 24xy$

b. $A = lw$
$= (7xy)(5xy)$
$= 35x^2y^2$

35a. $P = 2l + 2w$
$= 2(m + 2n) + 2(3m - n)$
$= 2m + 4n + 6m - 2n$
$= 8m + 2n$

b. $A = lw$
$= (m + 2n)(3m - n)$
$= 3m^2 - mn + 6mn - 2n^2$
$= 3m^2 + 5mn - 2n^2$

36a. $P = 2l + 2w$
$= 2(4w - 9) + 2(4w + 9)$
$= 8w - 18 + 8w + 18$
$= 16w$

b. $A = lw$
$= (4w + 9)(4w - 9)$
$= (4w)^2 - (9)^2$
$= 16w^2 - 81$

37. $(4a - 3b - 2c) - (a + 6b - 5c) + (3a + 9b - 2c)$
$= 4a - 3b - 2c - a - 6b + 5c + 3a + 9b - 2c$
$= 6a + c$
$6a + c = 6(-6.3) + 5.2 = -37.8 + 5.2 = -32.6$

159

38a. $400 - x$

   b. $0.72x$

   c. $0.48(400 - x)$

   d. $0.72x + 0.48(400 - x)$
   $= 0.72x + 192 - 0.48x$
   $= 0.24x + 192$ people

39. Equation: $2x^2 + x - 3 = 0$
$$(2x + 3)(x - 1) = 0$$
$2x + 3 = 0 \quad$ or $\quad x - 1 = 0$
$$2x = -3 \qquad\qquad x = 1$$
$$x = \frac{-3}{2}$$

40. Polynomial: $a^2 - 9 = (a + 3)(a - 3)$

41. Polynomial: $2x^2 + x - 3 = (2x + 3)(x - 1)$

42. Equation: $a^2 = 9$
$$a^2 - 9 = 0$$
$$(a + 3)(a - 3) = 0$$
$a + 3 = 0 \quad$ or $\quad a - 3 = 0$
$$a = -3 \qquad\qquad a = 3$$

43. Equation: $2x^2 + x = 0$
$$x(2x + 1) = 0$$
$x = 0 \quad$ or $\quad 2x + 1 = 0$
$$2x = -1$$
$$x = \frac{-1}{2}$$

44. Equation: $a - 9 = 0$
$$a = 9$$

45. Equation: $2x + 3 = 0$
$$2x = -3$$
$$x = \frac{-3}{2}$$

46. Equation: $a^2 = 9a$
$$a^2 - 9a = 0$$
$$a(a - 9) = 0$$
$a = 0 \quad$ or $\quad a - 9 = 0$
$$a = 9$$

47. Polynomial: $2x^3 - 2x$
$$= 2x(x^2 - 1)$$
$$= 2x(x + 1)(x - 1)$$

48. Polynomial: $a^4 - 16$
$$= (a^2 + 4)(a^2 - 4)$$
$$= (a^2 + 4)(a + 2)(a - 2)$$

49. Equation: $p^3 - p = 0$
$$p(p^2 - 1) = 0$$
$$p(p + 1)(p - 1) = 0$$
$p = 0 \quad$ or $\quad p + 1 = 0 \quad$ or $\quad p - 1 = 0$
$$p = -1 \qquad\qquad p = 1$$

50. Equation: $n(n - 3)(n + 3) = 0$
$n = 0 \quad$ or $\quad n - 3 = 0 \quad$ or $\quad n + 3 = 0$
$$n = 3 \qquad\qquad n = -3$$

51. $12x^5 - 8x^4 + 20x^3 = 4x^3(3x^2 - 2x + 5)$

52. $9a^4b^2 + 6a^3b^3 - 3a^2b^4 = 3a^2b^2(3a^2 + 2ab - b^2)$

53. $-10d^4 + 20d^3 - 5d^2 = -5d^2(2d^2 - 4d + 1)$

54. $-6m^3n - 18m^2n + 6mn = -6mn(m^2 + 3m - 1)$

55. $7q(q - 3) - (q - 3) = (q - 3)(7q - 1)$

56. $-r^3(3r + 2) + 4(3r + 2) = (3r + 2)(-r^3 + 4)$

160

57. $3z^3 - 12z$
$= 3z(z^2 - 4)$
$= 3z(z + 2)(z - 2)$

58. $-4x^3y + 8x^2y - 4xy$
$= -4xy(x^2 - 2x + 1)$
$= -4xy(x - 1)^2$

59. $a^4 + 10a^2 + 25$
$= (a^2 + 5)^2$

60. $4x^8 - 64$
$= 4(x^8 - 16)$
$= 4(x^4 + 4)(x^4 - 4)$
$= 4(x^4 + 4)(x^2 + 2)(x^2 - 2)$

61. $2a^2b^6 + 32a^6b^2$
$= 2a^2b^2(b^4 + 16a^4)$

62. $4p^2q^4 + 32p^3q^3 + 64p^4q^2$
$= 4p^2q^2(q^2 + 8pq + 16p^2)$
$= 4p^2q^2(q + 4p)^2$

63. $-2a^4b - 4a^3b^2 + 30a^2b^3$
$= -2a^2b(a^2 + 2ab - 15b^2)$
$= -2a^2b(a + 5b)(a - 3b)$

64. $15r^3s^2 + 39r^2s^3 - 18rs^4$
$= 3rs^2(5r^2 + 13rs - 6s^2)$
$= 3rs^2(5r - 2s)(r + 3s)$

65. $k = 2.4(7.5) = 18$;
$t = \dfrac{18}{v}$

v	t
2.4	7.5
3.6	5
15	1.2

66. $k = 48(75) = 3600$;
$V = \dfrac{3600}{P}$

P	V
48	75
60	60
240	15

67. $3(5) = 15 = k$; $\quad y = \dfrac{15}{x}$

68. $2(12) = 4(6) = 24 = k$; $\quad y = \dfrac{24}{x}$

69. No; The product of the variables is not constant. $25(263) \neq 56(232) \neq 72(216) \ldots$

70. No; The product of the variables is not constant. $1(2) \neq 2(2.5) \neq 2.5(2.4) \ldots$

Homework 5.6

1a. $\dfrac{x+1}{x-3} = \dfrac{4+1}{4-3} = \dfrac{5}{1} = 5$
$\dfrac{x+1}{x-3} = \dfrac{-4+1}{-4-3} = \dfrac{-3}{-7} = \dfrac{3}{7}$

b. Fraction is undefined if the denominator is zero.
Denominator $\rightarrow x - 3 = 0$
$x = 3$

3a. $\dfrac{2a - a^2}{a^2 + 1} = \dfrac{2(3) - 3^2}{3^2 + 1} = \dfrac{6 - 9}{9 + 1} = \dfrac{-3}{10}$
$\dfrac{2a - a^2}{a^2 + 1} = \dfrac{2(-1) - (-1)^2}{(-1)^2 + 1} = \dfrac{-2 - 1}{1 + 1} = \dfrac{-3}{2}$

b. Fraction is undefined if the denominator is zero.
$a^2 + 1$ is always positive for any real number a.
Therefore, fraction is defined for all real numbers.

5. (a) For $x = 1$: $\dfrac{1-x}{x+1} = \dfrac{1-1}{1+1} = \dfrac{0}{2} = 0$.

(b) For $x = 1$: $\dfrac{1+x}{1-x} = \dfrac{1+1}{1-1} = \dfrac{2}{0}$ which is undefined.

(c) For $x = 1$: $\dfrac{2x}{x^2-1} = \dfrac{2(1)}{1^2-1} = \dfrac{2}{0}$ which is undefined.

(d) For $x = 1$: $\dfrac{x-2}{x^2-2x+1} = \dfrac{1-2}{1^2-2(1)+1} = \dfrac{-1}{1-2+1} = \dfrac{-1}{0}$ which is undefined.

7a. $\dfrac{18}{x-4}$ dollars per gallon

b. $\dfrac{18}{14-4} = \dfrac{18}{10} = 1.8$
If the tank holds 14 gal, then gas costs \$1.80/gal.

9a. $\dfrac{2800}{h-10}$ miles per hour

b. $\dfrac{2800}{50-10} = \dfrac{2800}{40} = 70$
If he drove 50 hr, then his average speed was 70 mph.

11a. $\dfrac{200}{2x-1}$ square centimeters

b. $\dfrac{200}{2(13)-1} = \dfrac{200}{26-1} = \dfrac{200}{25} = 8$
If $x = 13$, then the area of cross-section is 8 cm^2.

13. $\dfrac{9x^5 - 6x^2 - 2}{3x^2}$
$= \dfrac{9x^5}{3x^2} - \dfrac{6x^2}{3x^2} - \dfrac{2}{3x^2}$
$= 3x^3 - 2 - \dfrac{2}{3x^2}$

15. $\dfrac{3n^3 - 3n^2 + 2n - 3}{3n^2}$
$= \dfrac{3n^3}{3n^2} - \dfrac{3n^2}{3n^2} + \dfrac{2n}{3n^2} - \dfrac{3}{3n^2}$
$= n - 1 + \dfrac{2n-3}{3n^2}$

17. $\dfrac{2x^2y^2 - 4xy^2 + 6xy}{2xy^2}$
$= \dfrac{2x^2y^2}{2xy^2} - \dfrac{4xy^2}{2xy^2} + \dfrac{6xy}{2xy^2}$
$= x - 2 + \dfrac{3}{y}$

19. $\dfrac{6x}{8x} = \dfrac{\cancel{2} \cdot 3 \cdot \cancel{x}}{\cancel{2} \cdot 2 \cdot 2 \cdot \cancel{x}} = \dfrac{3}{4}$

21. $\dfrac{-3a^3b^3}{15a^4b} = \dfrac{-\cancel{3} \cdot \cancel{a} \cdot \cancel{a} \cdot \cancel{a} \cdot \cancel{b} \cdot b \cdot b}{\cancel{3} \cdot 5 \cdot \cancel{a} \cdot \cancel{a} \cdot \cancel{a} \cdot a \cdot \cancel{b}} = \dfrac{-b^2}{5a}$

23. $\dfrac{3x+12}{6} = \dfrac{\cancel{3}(x+4)}{2 \cdot \cancel{3}} = \dfrac{x+4}{2}$

25. $\dfrac{2a^2}{2a^2 - 6a} = \dfrac{\cancel{2} \cdot \cancel{a} \cdot a}{\cancel{2} \cdot \cancel{a}(a-3)} = \dfrac{a}{a-3}$

27. $\dfrac{3(a+b)}{4(a+b)} = \dfrac{3}{4}$

29. $\dfrac{x-4}{x^2-3x-4} = \dfrac{x-4}{(x-4)(x+1)} = \dfrac{1}{x+1}$

31. (b) $\dfrac{x+2}{y+2}$ Cannot be reduced

33. (a) $\dfrac{2x+4}{4} = \dfrac{\cancel{2}(x+2)}{\cancel{2} \cdot 2} = \dfrac{x+2}{2}$

35. (a) $\dfrac{y^2-1}{y-1} = \dfrac{(y+1)(y-1)}{y-1} = y+1$

37. $\dfrac{x+4}{x-4}$ Cannot be reduced

39. $\dfrac{x+3z}{z+3x}$ Cannot be reduced

41a. 30 min   **b.**   50 min   **c.** $t = \dfrac{300}{10-v} = \dfrac{300}{10-4} = \dfrac{300}{6} = 50$ min

  d. $t = 2\dfrac{1}{2}$ hr $= \dfrac{5}{2}(60) = 150$ min, then $v = 8$ mph.

  e.   As the current increases, the time increases.

        If the current is 10 mph, the team will not be able to row upstream. $\frac{300}{10-v} = \frac{300}{10-10} = \frac{300}{0}$ is undefined.

43. $\dfrac{b-2}{4-2b} = \dfrac{-(2-b)}{2(2-b)} = \dfrac{-1}{2}$ **45.** $\dfrac{a-b}{a^2-b^2} = \dfrac{a-b}{(a+b)(a-b)} = \dfrac{1}{a+b}$

47. $\dfrac{(3x+2y)^2}{4y^2-9x^2} = \dfrac{(3x+2y)(3x+2y)}{(2y+3x)(2y-3x)} = \dfrac{3x+2y}{2y-3x}$

49. $\dfrac{3a-a^2}{a^2-2a-3} = \dfrac{-a(a-3)}{(a-3)(a+1)} = \dfrac{-a}{a+1}$ **51.**   (b), (c), (d)   **53.**   (a), (d)   **55.**   (b), (c)

57.   The 3 in the numerator is not a factor; no cancellation is possible.

       If you divide 3 into each term in the numerator, $\dfrac{3x+4}{3} = \dfrac{3x}{3} + \dfrac{4}{3} = x + \dfrac{4}{3}$.

59.   The $5z$ in the numerator is not a factor; no cancellation is possible.

Homework 5.7

1. $\dfrac{2}{3x^3} \cdot \dfrac{9x^2}{4} = \dfrac{\not{2}}{\not{3} \cdot \not{x} \cdot \not{x} \cdot x} \cdot \dfrac{\not{3} \cdot 3 \cdot \not{x} \cdot \not{x}}{\not{2} \cdot 2} = \dfrac{3}{2x}$ **3.** $\dfrac{2b}{3} \cdot \dfrac{4}{b+1} = \dfrac{8b}{3(b+1)}$

5. $\dfrac{3x-9}{5x-15} \cdot \dfrac{10x-5}{8x-4} = \dfrac{3(x-3)}{\not{5}(x-3)} \cdot \dfrac{\not{5}(2x-1)}{4(2x-1)} = \dfrac{3}{4}$

7. $\dfrac{5a+25}{5a} \cdot \dfrac{10a}{2a+10} = \dfrac{\not{5}(a+5)}{\not{5} \cdot \not{a}} \cdot \dfrac{\not{2} \cdot 5 \cdot \not{a}}{\not{2}(a+5)} = 5$

9. $\dfrac{2}{3}x = \dfrac{2}{3} \cdot \dfrac{x}{1} = \dfrac{2x}{3}$ **11.** $\dfrac{3}{4}(a-b) = \dfrac{3}{4} \cdot \dfrac{a-b}{1} = \dfrac{3(a-b)}{4}$ or $\dfrac{3a-3b}{4}$

13. $\dfrac{-2}{t^2}\left(4t^3 - \dfrac{t^2}{8} + \dfrac{3t}{2}\right)$ **15.** $\dfrac{4}{3}v\left(\dfrac{2}{3}v - \dfrac{6}{v^2} - \dfrac{3}{4v}\right)$

       $= \dfrac{-2}{t^2}\left(\dfrac{4t^3}{1}\right) - \dfrac{-2}{t^2}\left(\dfrac{t^2}{8}\right) + \dfrac{-2}{t^2}\left(\dfrac{3t}{2}\right)$ $= \dfrac{4v}{3}\left(\dfrac{2v}{3}\right) - \dfrac{4v}{3}\left(\dfrac{6}{v^2}\right) - \dfrac{4v}{3}\left(\dfrac{3}{4v}\right)$

       $= -8t + \dfrac{1}{4} - \dfrac{3}{t}$ $= \dfrac{8v^2}{9} - \dfrac{8}{v} - 1$

17. $\dfrac{4V}{D} \cdot \dfrac{LR}{DV} = \dfrac{4 \cdot \not{V}}{D} \cdot \dfrac{L \cdot R}{D \cdot \not{V}} = \dfrac{4LR}{D^2}$

19.
$$\frac{2L}{c}\left(1 + \frac{V^2}{c^2}\right)$$
$$= \frac{2L}{c}(1) + \frac{2L}{c}\left(\frac{V^2}{c^2}\right)$$
$$= \frac{2L}{c} + \frac{2LV^2}{c^3}$$

21.
$$\frac{q}{8\pi}\left(\frac{3}{R} - \frac{a^2}{R^3}\right)$$
$$= \frac{q}{8\pi}\left(\frac{3}{R}\right) - \frac{q}{8\pi}\left(\frac{a^2}{R^3}\right)$$
$$= \frac{3q}{8\pi R} - \frac{a^2 q}{8\pi R^3}$$

23.
$$\frac{12c}{21d} \div \frac{24c}{27d}$$
$$= \frac{2 \cdot 2 \cdot 3 \cdot \cancel{c}}{3 \cdot 7 \cdot \cancel{d}} \cdot \frac{\cancel{3} \cdot 3 \cdot 3 \cdot \cancel{d}}{2 \cdot 2 \cdot 2 \cdot 3 \cdot \cancel{c}}$$
$$= \frac{9}{14}$$

25.
$$\frac{2ab^3}{3} \div 4a^2b$$
$$= \frac{\cancel{2} \cdot \cancel{a} \cdot \cancel{b} \cdot b \cdot b}{3} \cdot \frac{1}{\cancel{2} \cdot 2 \cdot \cancel{a} \cdot a \cdot \cancel{b}}$$
$$= \frac{b^2}{6a}$$

27.
$$1 \div \frac{x}{2y}$$
$$= 1 \cdot \frac{2y}{x}$$
$$= \frac{2y}{x}$$

29.
$$\frac{a^2 - ab}{ab} \div \frac{2a - 2b}{3ab}$$
$$= \frac{a(a - b)}{\cancel{a} \cdot \cancel{b}} \cdot \frac{3 \cdot \cancel{a} \cdot \cancel{b}}{2(a - b)}$$
$$= \frac{3a}{2}$$

31.
$$\frac{3xy + x}{y^2 - y} \div \frac{3y + 1}{xy}$$
$$= \frac{x(3y + 1)}{\cancel{y}(y - 1)} \cdot \frac{x \cdot \cancel{y}}{3y + 1}$$
$$= \frac{x^2}{y - 1}$$

33.
$$\frac{6a^2 - 12a}{3a + 9} \div \frac{8a^2 - 4a^3}{15 + 5a}$$
$$= \frac{6a(a - 2)}{3(a + 3)} \cdot \frac{5(3 + a)}{4a^2(2 - a)}$$
$$= \frac{\cancel{2} \cdot \cancel{3} \cdot \cancel{a} \cdot (-1)(2 - a)}{\cancel{3}(a + 3)} \cdot \frac{5(3 + a)}{\cancel{2} \cdot 2 \cdot \cancel{a} \cdot a(2 - a)}$$
$$= \frac{-5}{2a}$$

35.
$$\frac{c^2 - 6c + 5}{c^2 + 2c - 15} \div \frac{c^2 - 3c - 10}{c^2 + 3c + 2}$$
$$= \frac{(c - 5)(c - 1)}{(c + 5)(c - 3)} \cdot \frac{(c + 2)(c + 1)}{(c - 5)(c + 2)}$$
$$= \frac{(c - 1)(c + 1)}{(c + 5)(c - 3)} \text{ or } \frac{c^2 - 1}{c^2 + 2c - 15}$$

37.
$$\frac{5}{2a} + \frac{3}{2a}$$
$$= \frac{5 + 3}{2a}$$
$$= \frac{8}{2a}$$
$$= \frac{4}{a}$$

39.
$$\frac{3}{x - 1} + \frac{5}{x - 1}$$
$$= \frac{3 + 5}{x - 1}$$
$$= \frac{8}{x - 1}$$

41.
$$\frac{x - 2y}{3x} + \frac{x + 3y}{3x}$$
$$= \frac{(x - 2y) + (x + 3y)}{3x}$$
$$= \frac{2x + y}{3x}$$

43.
$$\frac{m^2+1}{m-1}-\frac{2m}{m-1}$$
$$=\frac{(m^2+1)-(2m)}{m-1}$$
$$=\frac{m^2-2m+1}{m-1}$$
$$=\frac{(m-1)(m-1)}{m-1}$$
$$=m-1$$

45.
$$\frac{z^2-2}{z+2}-\frac{z+4}{2+z}$$
$$=\frac{(z^2-2)-(z+4)}{z+2}$$
$$=\frac{z^2-2-z-4}{z+2}$$
$$=\frac{z^2-z-6}{z+2}$$
$$=\frac{(z-3)(z+2)}{z+2}$$
$$=z-3$$

47.
$$\frac{b+1}{b^2-2b+1}-\frac{5-3b}{b^2-2b+1}$$
$$=\frac{(b+1)-(5-3b)}{b^2-2b+1}$$
$$=\frac{b+1-5+3b}{b^2-2b+1}$$
$$=\frac{4b-4}{b^2-2b+1}$$
$$=\frac{4(b-1)}{(b-1)(b-1)}$$
$$=\frac{4}{b-1}$$

49a.
$$\frac{2m^2-8}{3m^2-3}\cdot\frac{6-6m}{2m^2+4m}$$
$$=\frac{2(m^2-4)}{3(m^2-1)}\cdot\frac{6(1-m)}{2m(m+2)}$$
$$=\frac{2(m+2)(m-2)}{\not3(m+1)(m-1)}\cdot\frac{\not2\cdot\not3\cdot(-1)(m-1)}{\not2\cdot m(m+2)}$$
$$=\frac{-2(m-2)}{m(m+1)}\quad\text{or}\quad\frac{4-2m}{m(m+1)}$$

b.
$$\frac{-2(m-2)}{m(m+1)}$$
$$\frac{-2(3-2)}{3(3+1)}$$
$$=\frac{-2(1)}{3(4)}$$
$$=\frac{-2}{12}$$
$$=\frac{-1}{6}$$

51.
$$2x(x+2)\left(\frac{1}{2x}+\frac{x}{x+2}-1\right)$$
$$=\frac{2x(x+2)}{1}\left(\frac{1}{2x}\right)+\frac{2x(x+2)}{1}\left(\frac{x}{x+2}\right)-2x(x+2)(1)$$
$$=x+2+2x^2-2x^2-4x$$
$$=-3x+2$$

53.
$$\frac{y(y+1)}{(y-1)(y+1)}+\frac{3(y-1)}{(y+1)(y-1)}$$
$$=\frac{y(y+1)+3(y-1)}{(y-1)(y+1)}$$
$$=\frac{y^2+y+3y-3}{(y-1)(y+1)}$$
$$=\frac{y^2+4y-3}{(y-1)(y+1)}$$

55.

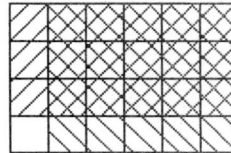

$$\frac{5}{6}\cdot\frac{3}{4}=\frac{15}{24}\quad\text{or}\quad\frac{5}{8}$$

57a. $\dfrac{1}{2}x=\dfrac{1}{2}\cdot\dfrac{x}{1}=\dfrac{x}{2}$

b. $x\div\dfrac{1}{2}=\dfrac{x}{1}\cdot\dfrac{2}{1}=2x$

c. $\dfrac{1}{2}\div x=\dfrac{1}{2}\cdot\dfrac{1}{x}=\dfrac{1}{2x}$

59a. $\dfrac{1}{a+b}$

b. $\dfrac{3}{4}\cdot\dfrac{1}{a+b}=\dfrac{3}{4(a+b)}$

c. $\dfrac{1}{a+b}\div\dfrac{3}{4}=\dfrac{1}{a+b}\cdot\dfrac{4}{3}=\dfrac{4}{3(a+b)}$

165

61. $\dfrac{x+3}{\cancel{4}} \cdot \cancel{4} = x + 3$

63. $\cancel{6}^{3}\,\cancel{n} \cdot \dfrac{2n+3}{\cancel{2}\,\cancel{n}} = 6n + 9$

65. $\dfrac{6w-1}{\cancel{-3}\,\cancel{\psi}} \cdot \cancel{-3}^{\,-3}\,\cancel{\psi} = -18w + 3 \ $ or $\ 3 - 18w$

Homework 5.8

1. $\dfrac{3}{x} - \dfrac{4}{y}$

$= \dfrac{3 \cdot y}{x \cdot y} - \dfrac{4 \cdot x}{y \cdot x}$

$= \dfrac{3y - 4x}{xy}$

3. $1 - \dfrac{3}{a}$

$= \dfrac{1 \cdot a}{1 \cdot a} - \dfrac{3}{a}$

$= \dfrac{a - 3}{a}$

5. $\dfrac{1}{x} + \dfrac{x}{y} + \dfrac{x}{z}$

$= \dfrac{1 \cdot yz}{x \cdot yz} + \dfrac{x \cdot xz}{y \cdot xz} + \dfrac{x \cdot xy}{z \cdot xy}$

$= \dfrac{yz + x^2 z + x^2 y}{xyz}$

7. $\dfrac{u-4}{3} + \dfrac{6}{v}$

$= \dfrac{(u-4) \cdot v}{3 \cdot v} + \dfrac{6 \cdot 3}{v \cdot 3}$

$= \dfrac{uv - 4v}{3v} + \dfrac{18}{3v}$

$= \dfrac{uv - 4v + 18}{3v}$

9. $\dfrac{3}{x} - \dfrac{2}{x+1}$

$= \dfrac{3 \cdot (x+1)}{x \cdot (x+1)} - \dfrac{2 \cdot x}{(x+1) \cdot x}$

$= \dfrac{3x + 3}{x(x+1)} - \dfrac{2x}{x(x+1)}$

$= \dfrac{x + 3}{x(x+1)}$

11. $\dfrac{3}{n+3} + \dfrac{4n}{n-3}$

$= \dfrac{3 \cdot (n-3)}{(n+3)(n-3)} + \dfrac{4n \cdot (n+3)}{(n-3)(n+3)}$

$= \dfrac{3n - 9}{(n+3)(n-3)} + \dfrac{4n^2 + 12n}{(n-3)(n+3)}$

$= \dfrac{4n^2 + 15n - 9}{(n+3)(n-3)}$

13a. $A = lw$

$= \dfrac{12}{x} \cdot \dfrac{12}{x-2}$

$= \dfrac{144}{x(x-2)}$ sq ft

b. $P = 2l + 2w$

$= 2 \cdot \dfrac{12}{x} + 2 \cdot \dfrac{12}{x-2}$

$= \dfrac{24}{x} + \dfrac{24}{x-2}$

$= \dfrac{24 \cdot (x-2)}{x \cdot (x-2)} + \dfrac{24 \cdot x}{(x-2) \cdot x}$

$= \dfrac{24x - 48}{x(x-2)} + \dfrac{24x}{x(x-2)}$

$= \dfrac{48x - 48}{x(x-2)}$ ft

15a. $\dfrac{100}{x-10}$

b. $\dfrac{100}{x+10}$

c.
$$\frac{100}{x-10}+\frac{100}{x+10}$$
$$=\frac{100\cdot(x+10)}{(x-10)(x+10)}+\frac{100\cdot(x-10)}{(x+10)(x-10)}$$
$$=\frac{100x+1000}{(x-10)(x+10)}+\frac{100x-1000}{(x+10)(x-10)}$$
$$=\frac{200x}{(x-10)(x+10)}$$

d. (a)
$$\frac{100}{x-10}=\frac{100}{150-10}=\frac{100}{140}=\frac{5}{7}$$

(b)
$$\frac{100}{x+10}=\frac{100}{150+10}=\frac{100}{160}=\frac{5}{8}$$

(c)
$$\frac{200x}{(x-10)(x+10)}$$
$$\frac{200(150)}{(150-10)(150+10)}$$
$$=\frac{30{,}000}{(140)(160)}$$
$$=\frac{30{,}000}{22{,}400}$$
$$=\frac{75}{56}\quad\text{(Note: }\frac{5}{7}+\frac{5}{8}=\frac{75}{56}\text{)}$$

17a. $\dfrac{1}{d}$

b. $\dfrac{1}{d-5}$

c.
$$\frac{1}{d}+\frac{1}{d-5}$$
$$=\frac{1\cdot(d-5)}{d\cdot(d-5)}+\frac{1\cdot d}{(d-5)\cdot d}$$
$$=\frac{d-5}{d(d-5)}+\frac{d}{d(d-5)}$$
$$=\frac{2d-5}{d(d-5)}\ \text{ of bag per day}$$

d.
$$\frac{d(d-5)}{2d-5}\ \text{ days per bag}$$
$$\frac{25(25-5)}{2(25)-5}$$
$$=\frac{25(20)}{50-5}$$
$$=\frac{500}{45}$$
$$=\frac{100}{9}\ \text{ or }11\frac{1}{9}\ \text{ days}$$

19.
$$h-\frac{3}{h+2}$$
$$=\frac{h\cdot(h+2)}{1\cdot(h+2)}-\frac{3}{h+2}$$
$$=\frac{h^2+2h}{h+2}-\frac{3}{h+2}$$
$$=\frac{h^2+2h-3}{h+2}$$

21.
$$\frac{v+1}{v}+\frac{1}{v-1}$$
$$=\frac{(v+1)(v-1)}{v\cdot(v-1)}+\frac{1\cdot v}{(v-1)\cdot v}$$
$$=\frac{v^2-1}{v(v-1)}+\frac{v}{v(v-1)}$$
$$=\frac{v^2+v-1}{v(v-1)}$$

23.
$$\frac{2}{x}+\frac{x}{x-2}-2$$
$$=\frac{2\cdot(x-2)}{x\cdot(x-2)}+\frac{x\cdot x}{(x-2)\cdot x}-\frac{2\cdot x(x-2)}{1\cdot x(x-2)}$$
$$=\frac{2x-4}{x(x-2)}+\frac{x^2}{x(x-2)}-\frac{2x^2-4x}{x(x-2)}$$
$$=\frac{(2x-4)+(x^2)-(2x^2-4x)}{x(x-2)}$$
$$=\frac{2x-4+x^2-2x^2+4x}{x(x-2)}$$
$$=\frac{-x^2+6x-4}{x(x-2)}$$

25a.
$$2x=2\cdot x$$
$$4x^2=2\cdot2\cdot x\cdot x$$
$$\text{LCD}=2\cdot2\cdot x\cdot x=4x^2$$

b.
$$\frac{5}{2x}+\frac{3}{4x^2}$$
$$=\frac{5\cdot 2x}{2x\cdot 2x}+\frac{3}{4x^2}$$
$$=\frac{10x}{4x^2}+\frac{3}{4x^2}$$
$$=\frac{10x+3}{4x^2}$$

27a.
$$8z = 2 \cdot 2 \cdot 2 \cdot z$$
$$6z = 2 \cdot 3 \cdot z$$
$$\text{LCD} = 2 \cdot 2 \cdot 2 \cdot 3 \cdot z = 24z$$

b.
$$\frac{2z-3}{8z} + \frac{z-2}{6z}$$
$$= \frac{(2z-3)\cdot 3}{8z \cdot 3} + \frac{(z-2)\cdot 4}{6z \cdot 4}$$
$$= \frac{6z-9}{24z} + \frac{4z-8}{24z}$$
$$= \frac{10z-17}{24z}$$

29a.
$$2a - b = (2a - b)$$
$$8a - 4b = 4 \cdot (2a - b)$$
$$\text{LCD} = 4 \cdot (2a - b)$$

b.
$$\frac{3}{2a-b} + \frac{1}{8a-4b}$$
$$= \frac{3 \cdot 4}{(2a-b)\cdot 4} + \frac{1}{4(2a-b)}$$
$$= \frac{12}{4(2a-b)} + \frac{1}{4(2a-b)}$$
$$= \frac{13}{4(2a-b)}$$

31a. $\text{LCD} = x(d-x)$

b.
$$\frac{1}{x} + \frac{1}{d-x}$$
$$= \frac{1 \cdot (d-x)}{x \cdot (d-x)} + \frac{1 \cdot x}{(d-x)\cdot x}$$
$$= \frac{d-x}{x(d-x)} + \frac{x}{x(d-x)}$$
$$= \frac{d}{x(d-x)}$$

33a. $\text{LCD} = 2a$

b.
$$a + \frac{N-a^2}{2a}$$
$$= \frac{a \cdot 2a}{1 \cdot 2a} + \frac{N-a^2}{2a}$$
$$= \frac{2a^2}{2a} + \frac{N-a^2}{2a}$$
$$= \frac{a^2 + N}{2a}$$

35a. $\text{LCD} = 2at$

b.
$$\frac{1}{2} \cdot \frac{a}{t} - \frac{m}{a}$$
$$= \frac{a \cdot a}{2t \cdot a} - \frac{m \cdot 2t}{a \cdot 2t}$$
$$= \frac{a^2}{2at} - \frac{2mt}{2at}$$
$$= \frac{a^2 - 2mt}{2at}$$

37a.
$$4\pi r = 2 \cdot 2 \cdot \pi \cdot r$$
$$2\pi r^2 = 2 \cdot \pi \cdot r \cdot r$$
$$\text{LCD} = 2 \cdot 2 \cdot \pi \cdot r \cdot r = 4\pi r^2$$

b.
$$\frac{q}{4\pi r} + \frac{qa}{2\pi r^2}$$
$$= \frac{q \cdot r}{4\pi r \cdot r} + \frac{qa \cdot 2}{2\pi r^2 \cdot 2}$$
$$= \frac{qr}{4\pi r^2} + \frac{2aq}{4\pi r^2}$$
$$= \frac{qr + 2aq}{4\pi r^2}$$

39a. $\text{LCD} = (c-V)(c+V)$

b.
$$\frac{L}{c-V} + \frac{L}{c+V}$$
$$= \frac{L \cdot (c+V)}{(c-V)(c+V)} + \frac{L \cdot (c-V)}{(c+V)(c-V)}$$
$$= \frac{Lc + LV}{(c-V)(c+V)} + \frac{Lc - LV}{(c+V)(c-V)}$$
$$= \frac{2Lc}{(c-V)(c+V)}$$

41a. $\text{LCD} = r(r-a)(r+a)$

b.
$$\frac{q}{r-a} - \frac{2q}{r} + \frac{q}{r+a}$$
$$= \frac{q \cdot r(r+a)}{(r-a)r(r+a)} - \frac{2q(r-a)(r+a)}{r(r-a)(r+a)} + \frac{q \cdot r(r-a)}{(r+a)r(r-a)}$$
$$= \frac{qr(r+a)}{r(r-a)(r+a)} - \frac{2q(r^2-a^2)}{r(r-a)(r+a)} + \frac{qr(r-a)}{r(r+a)(r-a)}$$
$$= \frac{qr^2 + aqr}{r(r-a)(r+a)} - \frac{2qr^2 - 2a^2q}{r(r-a)(r+a)} + \frac{qr^2 - aqr}{r(r+a)(r-a)}$$
$$= \frac{(qr^2 + aqr) - (2qr^2 - 2a^2q) + (qr^2 - aqr)}{r(r-a)(r+a)}$$
$$= \frac{qr^2 + aqr - 2qr^2 + 2a^2q + qr^2 - aqr}{r(r-a)(r+a)}$$
$$= \frac{2a^2q}{r(r-a)(r+a)}$$

43. $$\dfrac{\dfrac{3x}{y}}{\dfrac{x}{2y^2}} = \dfrac{2y^2\left(\dfrac{3x}{y}\right)}{2y^2\left(\dfrac{x}{2y^2}\right)} = \dfrac{\dfrac{2y^2}{1}\left(\dfrac{3x}{y}\right)}{\dfrac{2y^2}{1}\left(\dfrac{x}{2y^2}\right)} = \dfrac{6xy}{x} = 6y$$

45. $$\dfrac{1-\dfrac{1}{6}}{2+\dfrac{2}{3}} = \dfrac{6\left(1-\dfrac{1}{6}\right)}{6\left(2+\dfrac{2}{3}\right)} = \dfrac{6(1)-\dfrac{6}{1}\left(\dfrac{1}{6}\right)}{6(2)+\dfrac{6}{1}\left(\dfrac{2}{3}\right)} = \dfrac{6-1}{12+4} = \dfrac{5}{16}$$

47. $$\dfrac{n}{\dfrac{p}{q}+1} = \dfrac{q(n)}{q\left(\dfrac{p}{q}+1\right)} = \dfrac{q(n)}{\dfrac{q}{1}\left(\dfrac{p}{q}\right)+q(1)} = \dfrac{nq}{p+q}$$

49. $$\dfrac{4-\dfrac{1}{x^2}}{2-\dfrac{1}{x}} = \dfrac{x^2\left(4-\dfrac{1}{x^2}\right)}{x^2\left(2-\dfrac{1}{x}\right)} = \dfrac{x^2(4)-\dfrac{x^2}{1}\left(\dfrac{1}{x^2}\right)}{x^2(2)-\dfrac{x^2}{1}\left(\dfrac{1}{x}\right)} = \dfrac{4x^2-1}{2x^2-x} = \dfrac{(2x+1)(2x-1)}{x(2x-1)} = \dfrac{2x+1}{x}$$

51. $$\dfrac{\dfrac{b^2}{d}+d}{2} = \dfrac{d\left(\dfrac{b^2}{d}+d\right)}{d(2)} = \dfrac{\dfrac{d}{1}\left(\dfrac{b^2}{d}\right)+d(d)}{d(2)} = \dfrac{b^2+d^2}{2d}$$

53. $$\dfrac{1+\dfrac{x^2}{y^2}}{\dfrac{x}{y}} = \dfrac{y^2\left(1+\dfrac{x^2}{y^2}\right)}{y^2\left(\dfrac{x}{y}\right)} = \dfrac{y^2(1)+\dfrac{y^2}{1}\left(\dfrac{x^2}{y^2}\right)}{\dfrac{y^2}{1}\left(\dfrac{x}{y}\right)} = \dfrac{y^2+x^2}{xy}$$

55. $$m = \dfrac{y_2-y_1}{x_2-x_1} = \dfrac{\dfrac{-5}{3}-2}{\dfrac{3}{2}-\dfrac{-4}{3}} = \dfrac{6\left(\dfrac{-5}{3}-2\right)}{6\left(\dfrac{3}{2}-\dfrac{-4}{3}\right)} = \dfrac{\dfrac{6}{1}\left(\dfrac{-5}{3}\right)-6(2)}{\dfrac{6}{1}\left(\dfrac{3}{2}\right)-\dfrac{6}{1}\left(\dfrac{-4}{3}\right)} = \dfrac{-10-12}{9-(-8)} = \dfrac{-22}{17}$$

57a.

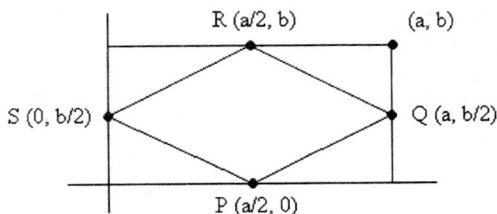

169

b.
$$m_{\overline{PQ}} = \frac{y_2 - y_1}{x_2 - x_1} = \frac{\frac{b}{2} - 0}{a - \frac{a}{2}} = \frac{2\left(\frac{b}{2}\right)}{2\left(a - \frac{a}{2}\right)} = \frac{\frac{2}{1}\left(\frac{b}{2}\right)}{2(a) - \frac{2}{1}\left(\frac{a}{2}\right)} = \frac{b}{2a - a} = \frac{b}{a}$$

$$m_{\overline{QR}} = \frac{y_2 - y_1}{x_2 - x_1} = \frac{b - \frac{b}{2}}{\frac{a}{2} - a} = \frac{2\left(b - \frac{b}{2}\right)}{2\left(\frac{a}{2} - a\right)} = \frac{2(b) - \frac{2}{1}\left(\frac{b}{2}\right)}{2\left(\frac{a}{2}\right) - 2(a)} = \frac{2b - b}{a - 2a} = \frac{b}{-a} = \frac{-b}{a}$$

$$m_{\overline{RS}} = \frac{y_2 - y_1}{x_2 - x_1} = \frac{\frac{b}{2} - b}{0 - \frac{a}{2}} = \frac{2\left(\frac{b}{2} - b\right)}{2\left(\frac{-a}{2}\right)} = \frac{\frac{2}{1}\left(\frac{b}{2}\right) - 2(b)}{\frac{2}{1}\left(\frac{-a}{2}\right)} = \frac{b - 2b}{-a} = \frac{-b}{-a} = \frac{b}{a}$$

$$m_{\overline{SP}} = \frac{y_2 - y_1}{x_2 - x_1} = \frac{0 - \frac{b}{2}}{\frac{a}{2} - 0} = \frac{2\left(\frac{-b}{2}\right)}{2\left(\frac{a}{2}\right)} = \frac{-b}{a}$$

59.
$$\left(1 - \frac{k}{n}\right)\left(1 + \frac{k}{n}\right) = (1)^2 - \left(\frac{k}{n}\right)^2 = 1 - \frac{k^2}{n^2} \quad \text{or} \quad \frac{n^2 - k^2}{n^2}$$

61.
$$\frac{2d}{c} \cdot \frac{1}{1 - \left(\frac{u}{c}\right)^2} = \frac{2d}{c\left(1 - \frac{u^2}{c^2}\right)} = \frac{2d}{c(1) - \frac{c}{1}\left(\frac{u^2}{c^2}\right)} = \frac{2d}{c - \frac{u2}{c}} = \frac{c(2d)}{c\left(c - \frac{u^2}{c}\right)} = \frac{c(2d)}{c(c) - \frac{c}{1}\left(\frac{u^2}{c}\right)}$$
$$= \frac{2cd}{c^2 - u^2}$$

63.
$$\frac{\frac{L}{F}}{\frac{L}{F} - 1} \cdot \frac{K}{N} = \frac{\left(\frac{L}{F}\right)\frac{K}{1}}{\left(\frac{L}{F} - 1\right)\frac{N}{1}} = \frac{\frac{LK}{F}}{\frac{LN}{F} - N} = \frac{F\left(\frac{LK}{F}\right)}{F\left(\frac{LN}{F} - N\right)} = \frac{F\left(\frac{LK}{F}\right)}{\frac{F}{1}\left(\frac{LN}{F}\right) - F(N)} = \frac{LK}{LN - FN}$$

65. The denominators, $x - 1$ and $1 - x$, are opposites. If we multiply the numerator and the denominator of the second fraction by -1. the fractions will have the same denominator.

$$\frac{2x + 3}{x - 1} + \frac{2x - 5}{1 - x}$$
$$= \frac{2x + 3}{x - 1} + \frac{(-1)(2x - 5)}{(-1)(1 - x)}$$
$$= \frac{2x + 3}{x - 1} + \frac{5 - 2x}{x - 1}$$
$$= \frac{8}{x - 1}$$

170

67. First, multiply numerator and denominator of the second fraction by -1, then factor the denominators to find the LCD.

$$\frac{5}{2p-4}-\frac{2}{6-3p}$$

$$=\frac{5}{2p-4}-\frac{(-1)(2)}{(-1)(6-3p)}$$

$$=\frac{5}{2p-4}-\frac{-2}{3p-6}$$

$$=\frac{5}{2(p-2)}-\frac{-2}{3(p-2)}$$

$$=\frac{5\cdot 3}{2(p-2)\cdot 3}-\frac{-2\cdot 2}{3(p-2)\cdot 2}$$

$$=\frac{15}{6(p-2)}-\frac{-4}{6(p-2)}$$

$$=\frac{15-(-4)}{6(p-2)}$$

$$=\frac{19}{6(p-2)}$$

69. Factor the denominators to find the LCD.

$$\frac{-2}{m^2+3m}+\frac{1}{m^2-9}$$

$$=\frac{-2}{m(m+3)}+\frac{1}{(m+3)(m-3)}$$

$$=\frac{-2\cdot(m-3)}{m(m+3)(m-3)}+\frac{1\cdot m}{(m+3)(m-3)m}$$

$$=\frac{-2m+6}{m(m+3)(m-3)}+\frac{1m}{m(m+3)(m-3)}$$

$$=\frac{-m+6}{m(m+3)(m-3)}$$

71. Factor the denominators to find the LCD.

$$\frac{4}{k^2-3k}+\frac{1}{k^2+k}$$

$$=\frac{4}{k(k-3)}+\frac{1}{k(k+1)}$$

$$=\frac{4\cdot(k+1)}{k(k-3)(k+1)}+\frac{1\cdot(k-3)}{k(k+1)(k-3)}$$

$$=\frac{4k+4}{k(k-3)(k+1)}+\frac{k-3}{k(k+1)(k-3)}$$

$$=\frac{5k+1}{k(k-3)(k+1)}$$

73. Factor the denominators to find the LCD.

$$\frac{x-1}{x^2+3x}+\frac{x}{x^2+6x+9}$$

$$=\frac{x-1}{x(x+3)}+\frac{x}{(x+3)(x+3)}$$

$$=\frac{(x-1)(x+3)}{x(x+3)(x+3)}+\frac{x\cdot x}{(x+3)(x+3)x}$$

$$=\frac{x^2+2x-3}{x(x+3)(x+3)}+\frac{x^2}{x(x+3)(x+3)}$$

$$=\frac{2x^2+2x-3}{x(x+3)(x+3)}\quad\text{or}\quad\frac{2x^2+2x-3}{x(x+3)^2}$$

75. $\dfrac{A(T_2-T_1)}{\dfrac{L_1}{K_1}+\dfrac{L_2}{K_2}}=\dfrac{K_1K_2\cdot A(T_2-T_1)}{K_1K_2\cdot\left(\dfrac{L_1}{K_1}+\dfrac{L_2}{K_2}\right)}=\dfrac{K_1K_2\cdot A(T_2-T_1)}{\dfrac{K_1K_2}{1}\left(\dfrac{L_1}{K_1}\right)+\dfrac{K_1K_2}{1}\left(\dfrac{L_2}{K_2}\right)}=\dfrac{AK_1K_2(T_2-T_1)}{K_2L_1+K_1L_2}$

77. $\dfrac{1-\dfrac{2h}{m}}{m-\dfrac{4h^2}{m}}=\dfrac{m\left(1-\dfrac{2h}{m}\right)}{m\left(m-\dfrac{4h^2}{m}\right)}=\dfrac{m(1)-\dfrac{m}{1}\left(\dfrac{2h}{m}\right)}{m(m)-\dfrac{m}{1}\left(\dfrac{4h^2}{m}\right)}=\dfrac{m-2h}{m^2-4h^2}=\dfrac{m-2h}{(m+2h)(m-2h)}=\dfrac{1}{m+2h}$

Homework 5.9

1.
$$\frac{5x}{2} - 1 = x + \frac{1}{2}$$ Multiply both sides by LCD $= 2$

$$2\left(\frac{5x}{2} - 1\right) = \left(x + \frac{1}{2}\right)2$$ Distribute 2 to each term

$$2\left(\frac{5x}{2}\right) - 2(1) = 2(x) + 2\left(\frac{1}{2}\right)$$

$$5x - 2 = 2x + 1$$ Subtract $2x$ from both sides
$$3x - 2 = 1$$ Add 2 to both sides
$$3x = 3$$ Divide both sides by 3
$$x = 1$$

3.
$$\frac{t}{6} - \frac{7}{3} = \frac{2t}{9} - \frac{t}{4}$$ Multiply both sides by LCD $= 36$

$$36\left(\frac{t}{6} - \frac{7}{3}\right) = \left(\frac{2t}{9} - \frac{t}{4}\right)36$$ Distribute 36 to each term

$$36\left(\frac{t}{6}\right) - 36\left(\frac{7}{3}\right) = 36\left(\frac{2t}{9}\right) - 36\left(\frac{t}{4}\right)$$

$$6t - 84 = 8t - 9t$$ Combine like terms on right side
$$6t - 84 = -t$$ Add t to both sides
$$7t - 84 = 0$$ Add 84 to both sides
$$7t = 84$$ Divide both sides by 7
$$t = 12$$

5.
$$\frac{2}{3}(x - 1) + x = 6$$ Multiply both sides by LCD $= 3$

$$3\left[\frac{2}{3}(x - 1) + x\right] = [6]3$$ Distribute 3 to each term

$$3\left(\frac{2}{3}\right)(x - 1) + 3(x) = 18$$

$$2(x - 1) + 3x = 18$$ Distribute 2
$$2x - 2 + 3x = 18$$ Combine like terms on left side
$$5x - 2 = 18$$ Add 2 to both sides
$$5x = 20$$ Divide both sides by 5
$$x = 4$$

7.
$$\frac{3x^2}{2} - \frac{x}{4} = \frac{1}{2}$$ Multiply both sides by LCD $= 4$

$$4\left(\frac{3x^2}{2} - \frac{x}{4}\right) = \left(\frac{1}{2}\right)4$$ Distribute 4 to each term

$$4\left(\frac{3x^2}{2}\right) - 4\left(\frac{x}{4}\right) = \left(\frac{1}{2}\right)4$$

$$6x^2 - x = 2$$ Write in standard form
$$6x^2 - x - 2 = 0$$ Factor
$$(3x - 2)(2x + 1) = 0$$ Set each factor equal to 0

(Problem continued on next page)

172

$$3x - 2 = 0 \quad \text{or} \quad 2x + 1 = 0 \qquad \text{Solve each}$$
$$3x = 2 \qquad\qquad 2x = -1$$
$$x = \frac{2}{3} \qquad\qquad x = \frac{-1}{2}$$

9.
$$2 + \frac{5}{2x} = \frac{3}{x} + \frac{3}{2} \qquad\qquad \text{Multiply both sides by LCD} = 2x$$

$$2x\left(2 + \frac{5}{2x}\right) = \left(\frac{3}{x} + \frac{3}{2}\right)2x \qquad \text{Distribute } 2x \text{ to each term}$$

$$2x(2) + 2x\left(\frac{5}{2x}\right) = 2x\left(\frac{3}{x}\right) + 2x\left(\frac{3}{2}\right)$$

$$4x + 5 = 6 + 3x \qquad\qquad \text{Subtract } 3x \text{ from both sides}$$
$$x + 5 = 6 \qquad\qquad \text{Subtract 5 from both sides}$$
$$x = 1$$

11.
$$1 + \frac{1}{x(x-1)} = \frac{3}{x} \qquad\qquad \text{Multiply both sides by LCD} = x(x-1)$$

$$x(x-1)\left[1 + \frac{1}{x(x-1)}\right] = \left[\frac{3}{x}\right]x(x-1) \quad \text{Distribute } x(x-1) \text{ to each term}$$

$$x(x-1)[1] + x(x-1)\left[\frac{1}{x(x-1)}\right] = \left[\frac{3}{x}\right]x(x-1)$$
$$x(x-1) + 1 = 3(x-1)$$
$$x^2 - x + 1 = 3x - 3 \qquad\qquad \text{Write in standard form}$$
$$x^2 - 4x + 4 = 0 \qquad\qquad \text{Factor}$$
$$(x-2)(x-2) = 0 \qquad\qquad \text{Set the factor equal to 0}$$
$$x - 2 = 0 \qquad\qquad \text{Note: The other factor is the same}$$
$$x = 2 \qquad\qquad\qquad \text{and will give the same solution}$$

13.
$$\frac{4}{x-1} - \frac{4}{x+2} = \frac{3}{7} \qquad \begin{array}{l}\text{Multiply both sides by}\\ \text{LCD} = 7(x-1)(x+2)\end{array}$$

$$7(x-1)(x+2)\left(\frac{4}{x-1} - \frac{4}{x+2}\right) = \left(\frac{3}{7}\right)7(x-1)(x+2) \qquad \begin{array}{l}\text{Distribute LCD}\\ \text{to each term}\end{array}$$

$$7(x-1)(x+2)\left(\frac{4}{x-1}\right) - 7(x-1)(x+2)\left(\frac{4}{x+2}\right) = \left(\frac{3}{7}\right)7(x-1)(x+2)$$

$$7(x+2)4 - 7(x-1)4 = 3(x-1)(x+2) \qquad \text{FOIL on right side}$$
$$28(x+2) - 28(x-1) = 3(x^2 + x - 2) \qquad \text{Distribute}$$
$$28x + 56 - 28x + 28 = 3x^2 + 3x - 6 \qquad \text{Combine like terms}$$
$$84 = 3x^2 + 3x - 6 \qquad \text{Write in standard form}$$
$$0 = 3x^2 + 3x - 90 \qquad \text{Factor out 3}$$
$$0 = 3(x^2 + x - 30) \qquad \text{Divide both sides by 3}$$
$$\frac{0}{3} = \frac{3(x^2 + x - 30)}{3}$$
$$0 = x^2 + x - 30 \qquad \text{Factor}$$
$$0 = (x+6)(x-5) \qquad \text{Set each factor equal to 0}$$
$$x + 6 = 0 \quad \text{or} \quad x - 5 = 0 \quad \text{Solve each}$$
$$x = -6 \qquad\qquad x = 5$$

15.

$$\frac{1}{x-2} + \frac{1}{x+2} = \frac{4}{x^2-4}$$

Factor the denominator on right to find the LCD

$$\frac{1}{x-2} + \frac{1}{x+2} = \frac{4}{(x+2)(x-2)}$$

Multiply both sides by LCD $= (x-2)(x+2)$

$$(x-2)(x+2)\left[\frac{1}{x-2} + \frac{1}{x+2}\right] = \left[\frac{4}{(x+2)(x-2)}\right](x+2)(x-2)$$

$$(x-2)(x+2)\left[\frac{1}{x-2}\right] + (x-2)(x+2)\left[\frac{1}{x+2}\right] = \left[\frac{4}{(x-2)(x+2)}\right](x-2)(x+2)$$

$$x+2+x-2 = 4$$

Combine like terms

$$2x = 4$$

Divide both sides by 2

$$x = 2$$

But the original fractions are undefined if $x = 2$ or -2, therefore, $x \neq 2$ and there is NO solution.

17.

$$\frac{15x}{1+x^2} = 6$$

Multiply both sides by LCD $= 1 + x^2$

$$(1+x^2)\left(\frac{15x}{1+x^2}\right) = (6)(1+x^2)$$

$$15x = 6 + 6x^2$$

Write in standard form

$$0 = 6x^2 - 15x + 6$$

Factor out 3

$$0 = 3(2x^2 - 5x + 2)$$

Divide both sides by 3

$$\frac{0}{3} = \frac{3(2x^2 - 5x + 2)}{3}$$

$$0 = 2x^2 - 5x + 2$$

Factor

$$0 = (2x-1)(x-2)$$

Set each factor equal to 0

$$2x - 1 = 0 \quad \text{or} \quad x - 2 = 0 \quad \text{Solve each}$$

$$2x = 1 \qquad\qquad x = 2$$

$$x = \frac{1}{2}$$

19.

$$C = \frac{25p}{100-p}$$

Substitute 25 for C

$$25 = \frac{25p}{100-p}$$

Multiply both sides by LCD $= 100 - p$

$$(100-p)(25) = \left(\frac{25p}{100-p}\right)(100-p)$$

$$2500 - 25p = 25p$$

Subtract $25p$ from both sides

$$2500 - 50p = 0$$

Subtract 2500 from both sides

$$-50p = -2500$$

Divide both sides by -50

$$p = 50\%$$

21a. $\dfrac{34}{164} \approx 0.207$

b. $\dfrac{x+34}{x+164}$

174

c.
$$\frac{x + 34}{x + 164} = 0.350 \qquad \text{Multiply both sides by LCD} = x + 164$$

$$(x + 164)\left(\frac{x + 34}{x + 164}\right) = (0.350)(x + 164)$$

$$x + 34 = 0.35x + 57.4 \qquad \text{Subtract } 0.35x \text{ from both sides}$$

$$0.65x + 34 = 57.4 \qquad \text{Subtract } 34 \text{ from both sides}$$

$$0.65x = 23.4 \qquad \text{Divide both sides by } 0.65$$

$$x = 36 \text{ hits}$$

23a. $AE = 1; \quad ED = x - 1; \quad CD = 1$

b. $\dfrac{AD}{AB} = \dfrac{CD}{ED}$

$$\frac{x}{1} = \frac{1}{x - 1} \qquad \text{(other answers possible)}$$

c.
$$\frac{x}{1} = \frac{1}{x - 1} \qquad \text{Multiply both sides by } x - 1$$

$$(x - 1)\left(\frac{x}{1}\right) = \left(\frac{1}{x - 1}\right)(x - 1)$$

$$x^2 - x = 1 \qquad \text{Write in standard form}$$

$$x^2 - x - 1 = 0 \qquad a = 1, b = -1, c = -1$$

$$x = \frac{-b \pm \sqrt{b^2 - 4ac}}{2a}$$

$$= \frac{-(-1) \pm \sqrt{(-1)^2 - 4(1)(-1)}}{2(1)}$$

$$= \frac{1 \pm \sqrt{1 + 4}}{2}$$

$$= \frac{1 \pm \sqrt{5}}{2}$$

If $x = \dfrac{1 - \sqrt{5}}{2} \approx -0.6$ (a negative number) which is not possible.

$$\frac{AD}{AB} = \frac{x}{1} = x = \frac{1 + \sqrt{5}}{2} \quad \leftarrow \text{ The golden ratio}$$

25.
$$V = \frac{hT}{P} \qquad \text{Multiply both sides by } P$$

$$P(V) = \left(\frac{hT}{P}\right)P$$

$$PV = hT \qquad \text{Divide both sides by } h$$

$$\frac{PV}{h} = \frac{hT}{h}$$

$$\frac{PV}{h} = T$$

27.
$$a = \frac{F}{m + M}$$
Multiply both sides by $m + M$

$$(m + M)(a) = \left(\frac{F}{m + M}\right)(m + M)$$

$$am + aM = F$$
Subtract aM from both sides

$$am = F - aM$$
Divide both sides by a

$$\frac{am}{a} = \frac{F - aM}{a}$$

$$m = \frac{F - aM}{a}$$

29.
$$m = \frac{y - k}{x - h}$$
Multiply both sides by $x - h$

$$(x - h)(m) = \left(\frac{y - k}{x - h}\right)(x - h)$$

$$mx - hm = y - k$$
Add hm to both sides

$$mx = y - k + hm$$
Divide both sides by m

$$\frac{mx}{m} = \frac{y - k + hm}{m}$$

$$x = \frac{y - k + hm}{m}$$

31.
$$\frac{1}{R} = \frac{1}{A} + \frac{1}{B}$$
Multiply both sides by RAB

$$RAB\left(\frac{1}{R}\right) = \left(\frac{1}{A} + \frac{1}{B}\right)RAB$$
Distribute RAB to each term

$$RAB\left(\frac{1}{R}\right) = \left(\frac{1}{A}\right)RAB + \left(\frac{1}{B}\right)RAB$$

$$AB = BR + AR$$
Subtract AR from both sides

$$AB - AR = BR$$
Factor out A

$$A(B - R) = BR$$
Divide both sides by $B - R$

$$\frac{A(B - R)}{B - R} = \frac{BR}{B - R}$$

$$A = \frac{BR}{B - R}$$

33.
$$w = 0.622\frac{e}{P - e}$$
Multiply both sides by $P - e$

$$(P - e)(w) = \left(0.622\frac{e}{P - e}\right)(P - e)$$

$$Pw - ew = 0.622e$$
Add ew to both sides

$$Pw = ew + 0.622e$$
Factor out e

$$Pw = e(w + 0.622)$$
Divide both sides by $w + 0.622$

$$\frac{Pw}{w + 0.622} = \frac{e(w + 0.622)}{w + 0.622}$$

$$\frac{Pw}{w + 0.622} = e$$

35.

$$I = \dfrac{E}{R + \dfrac{r}{n}}$$ Multiply both sides by $R + \frac{r}{n}$

$$\left(R + \dfrac{r}{n}\right)(I) = \left(\dfrac{E}{R + \dfrac{r}{n}}\right)\left(R + \dfrac{r}{n}\right)$$

$$IR + \dfrac{Ir}{n} = E$$ Multiply both sides by n

$$n\left(IR + \dfrac{Ir}{n}\right) = (E)n$$ Distribute n to each term

$$n(IR) + n\left(\dfrac{Ir}{n}\right) = (E)n$$

$$InR + Ir = En$$ Subtract Ir from both sides

$$InR = En - Ir$$ Divide both sides by In

$$\dfrac{InR}{In} = \dfrac{En - Ir}{In}$$

$$R = \dfrac{En - Ir}{In}$$

37.

$$r = \dfrac{dc}{1 - ec}$$ Multiply both sides by $1 - ec$

$$(1 - ec)(r) = \left(\dfrac{dc}{1 - ec}\right)(1 - ec)$$

$$r - ecr = dc$$ Subtract r from both sides

$$-ecr = dc - r$$ Divide both sides by $-cr$

$$\dfrac{-ecr}{-cr} = \dfrac{dc - r}{-cr}$$ Multiply numerator and denominator
      of the fraction on right by -1

$$e = \dfrac{(-1)(dc - r)}{(-1)(-cr)}$$

$$e = \dfrac{r - dc}{cr}$$

39.

	Distance	Rate	Time
Express	180	$r + 20$	$\frac{180}{r+20}$
Freight	120	r	$\frac{120}{r}$

$$\dfrac{180}{r + 20} = \dfrac{120}{r}$$ Multiply both sides by LCD $= r(r + 20)$

$$r(r + 20)\left(\dfrac{180}{r + 20}\right) = \left(\dfrac{120}{r}\right)r(r + 20)$$

$$180r = 120(r + 20)$$ Distribute 120

$$180r = 120r + 2400$$ Subtract $120r$ from both sides

$$60r = 2400$$ Divide both sides by 60

$$r = 40$$

The speed of the freight train is $r = 40$ mph, and the speed of the express train is $r + 20 = 60$ mph.

41.

	Distance	Rate	Time
Sam	360	x	$\frac{360}{x}$
Reg	360	$x + 20$	$\frac{360}{x+20}$

$$\frac{360}{x} - \frac{360}{x + 20} = 3 \qquad \text{Multiply both sides by}$$
$$\text{LCD} = x(x + 20)$$

$$x(x + 20)\left(\frac{360}{x} - \frac{360}{x + 20}\right) = (3)x(x + 20) \qquad \text{Distribute } x(x + 20) \text{ to each term}$$

$$x(x + 20)\left(\frac{360}{x}\right) - x(x + 20)\left(\frac{360}{x + 20}\right) = (3)x(x + 20)$$

$$360(x + 20) - 360x = 3x(x + 20)$$

$$360x + 7200 - 360x = 3x^2 + 60x \qquad \text{Combine like terms}$$

$$7200 = 3x^2 + 60x \qquad \text{Write in standard form}$$

$$0 = 3x^2 + 60x - 7200 \qquad \text{Factor out 3}$$

$$0 = 3(x^2 + 20x - 2400) \qquad \text{Divide both sides by 3}$$

$$\frac{0}{3} = \frac{3(x^2 + 20x - 2400)}{3}$$

$$0 = x^2 + 20x - 2400 \qquad \text{Factor}$$

$$0 = (x - 40)(x + 60) \qquad \text{Set each factor equal to 0}$$

$$x - 40 = 0 \quad \text{or} \quad x + 60 = 0$$

$$x = 40 \qquad\qquad x = -60 \leftarrow \text{Not possible}$$

Sam's speed is $x = 40$ mph, and Reginald's speed is $x + 20 = 60$ mph.

43.

	Rate	Time	Fraction of Book
New Press	$\frac{1}{10}$	$x - 4$	$\frac{x-4}{10}$
Old Press	$\frac{1}{18}$	x	$\frac{x}{18}$

$$\frac{x - 4}{10} + \frac{x}{18} = 1 \qquad \text{Multiply both sides by LCD} = 90$$

$$90\left(\frac{x - 4}{10} + \frac{x}{18}\right) = (1)90 \qquad \text{Distribute 90 to each term}$$

$$90\left(\frac{x - 4}{10}\right) + 90\left(\frac{x}{18}\right) = (1)90$$

$$9(x - 4) + 5x = 90$$

$$9x - 36 + 5x = 90 \qquad \text{Combine like terms}$$

$$14x - 36 = 90 \qquad \text{Add 36 to both sides}$$

$$14x = 126 \qquad \text{Divide both sides by 14}$$

$$x = 9$$

The complete printing will take 9 hr.

45. Note: 4 hr $= 4 \times 60 = 240$ min

	Rate	Time	Fraction of Tank
Fill	$\frac{1}{30}$	x	$\frac{x}{30}$
Drain	$\frac{1}{240}$	x	$\frac{x}{240}$

(Problem continued on next page)

178

$$\frac{x}{30} - \frac{x}{240} = 1 \qquad \text{Multiply both sides by LCD} = 240$$

$$240\left(\frac{x}{30} - \frac{x}{240}\right) = (1)240 \qquad \text{Distribute 240 to each term}$$

$$240\left(\frac{x}{30}\right) - 240\left(\frac{x}{240}\right) = (1)240$$

$$8x - x = 240 \qquad \text{Combine like terms}$$

$$7x = 240 \qquad \text{Divide both sides by 7}$$

$$x = \frac{240}{7}$$

It will take $x = \frac{240}{7} = 34\frac{2}{7}$ min to fill.

47a. When adding fractions, we must build each fraction to an equivalent fraction with the LCD as its denominator.

$$\frac{x-1}{4} + \frac{3x}{5} \qquad \begin{array}{l}\text{Since the LCD} = 20, \text{ multiply numerator and denominator by} \\ \text{5 for the first fraction and by 4 for the second fraction}\end{array}$$

$$= \frac{(x-1)\cdot 5}{4\cdot 5} + \frac{3x\cdot 4}{5\cdot 4} \qquad \text{Simplify the numerators}$$

$$= \frac{5x-5}{20} + \frac{12x}{20} \qquad \text{Add the numerators and keep the same denominator}$$

$$= \frac{17x-5}{20}$$

b. When solving an equation, we can multiply both sides of the equation by the LCD to clear fractions.

$$\frac{x-1}{4} + \frac{3x}{5} = 1 \qquad \text{Multiply both sides by LCD} = 20$$

$$20\left(\frac{x-1}{4} + \frac{3x}{5}\right) = (1)20 \qquad \text{Distribute 20 to each term}$$

$$20\left(\frac{x-1}{4}\right) + 20\left(\frac{3x}{5}\right) = (1)20$$

$$5(x-1) + 4(3x) = 20$$

$$5x - 5 + 12x = 20 \qquad \text{Combine like terms}$$

$$17x - 5 = 20 \qquad \text{Add 5 to both sides}$$

$$17x = 25 \qquad \text{Divide both sides by 17}$$

$$x = \frac{25}{17}$$

49. Since $x = 1$, dividing by $x - 1$, in the fourth step, is dividing by zero.

51. The left side of the equation has been multiplied by 6, but the right side has not.

$$\frac{a}{2} - \frac{a}{3} = 1 \qquad \begin{array}{l}\text{Multiply both sides} \\ \text{by LCD} = 6\end{array}$$

$$6\left(\frac{a}{2} - \frac{a}{3}\right) = (1)6 \qquad \begin{array}{l}\text{Distribute 6} \\ \text{to each term}\end{array}$$

$$6\left(\frac{a}{2}\right) - 6\left(\frac{a}{3}\right) = (1)6$$

$$3a - 2a = 6 \qquad \text{Combine like terms}$$

$$a = 6$$

53. In the third equation, dividing by m is division by zero since $m = 0$ is one of the solutions. Never divide both sides by an expression containing a variable because you may lose a solution.

$$m^2 - 6m = 0 \qquad \text{Factor}$$

$$m(m-6) = 0 \qquad \text{Set each factor equal to 0}$$

$$m = 0 \quad \text{or} \quad m - 6 = 0 \qquad \text{Solve each}$$

$$m = 6$$

55. 2 days

Chapter 5 Summary and Review

1. <u>First law of exponents:</u> To multiply two powers with the same base, add the exponents and leave the base unchanged.
$$a^m \cdot a^n = a^{m+n} \qquad x^5 \cdot x^2 = x^{5+2} = x^7$$

2. <u>Second law of exponents:</u> To divide two powers with the same base, subtract the smaller exponent from the larger one and leave the base unchanged.
  1. If the larger exponent occurs in the numerator, put the power in the numerator.
  2. If the larger exponent occurs in the denominator, put the power in the denominator.

$$\frac{a^m}{a^n} = a^{m-n} \text{ if } m > n \qquad \frac{x^5}{x^2} = x^{5-2} = x^3$$
$$\frac{a^m}{a^n} = \frac{1}{a^{n-m}} \text{ if } n > m \qquad \frac{x^2}{x^5} = \frac{1}{x^{5-2}} = \frac{1}{x^3}$$

3. <u>Squares of binomials:</u>
$$(a+b)^2 = a^2 + 2ab + b^2$$
$$(a-b)^2 = a^2 - 2ab + b^2$$

4. <u>Factoring the difference of two squares:</u>
$$a^2 - b^2 = (a+b)(a-b)$$

5. <u>Inverse variation:</u>
$$y = \frac{k}{x} \quad \text{where } k \text{ is the constant of variation.}$$

6. <u>Fundamental principal of fractions:</u> We can multiply or divide the numerator and denominator by the same nonzero factor, and the resulting fraction will be equivalent to the original fraction.
$$\frac{a \cdot c}{b \cdot c} = \frac{a}{b} \quad \text{if } b, c \neq 0$$

7. To add two polynomials, remove the parentheses and combine like terms. To subtract one polynomial from another, change the sign of each term in the polynomial being subtracted and combine like terms.

8. To multiply two monomials, multiply the coefficients, and use the first law of exponents to find the products of the variables.

9. To divide one monomial by another, divide the coefficients, and use the second law of exponents to find the quotients of the variables.

10. When factoring a polynomial, begin by looking for a common factor.

11. $a^2 - b^2 = (a+b)(a-b)$
$a^2 + b^2$ cannot be factored

12. Two variables vary inversely if their product is constant.

13. <u>To reduce an algebraic fraction:</u>
  1. Factor the numerator and denominator completely.
  2. Divide numerator and denominator by any common factors.

14. <u>To multiply algebraic fractions:</u>
  1. Factor each numerator and denominator completely.
  2. Divide out any common factors that appear in a numerator and a denominator.
  3. Multiply the remaining factors of the numerator and the denominator.

15. To divide algebraic fractions, multiply the first fraction by the reciprocal of the second fraction.

16. To add or subtract like fractions:
1. Add or subtract the numerators.
2. Keep the same denominator.
3. Reduce if necessary.

17. To add or subtract unlike fractions:
1. Find the least common denominator (LCD) for the fractions.
2. Build each fraction to an equivalent one with the LCD as the denominator.
3. Add or subtract the numerators and keep the same denominator.
4. Reduce if necessary.

18. To simplify a complex fraction:
1. Find the LCD for all simple fractions inside the complex fraction.
2. Multiply numerator and denominator of the complex fraction by the LCD.
3. Reduce if necessary.

19. To solve an equation that contains fractions, first multiply both sides of the equation by the LCD to clear fractions.

20. Work rate $\times$ time $=$ work completed

21.
$$\frac{5}{4}x^2 + \frac{1}{2}xy - 2y^2$$
$$\frac{5}{4}(-4)^2 + \frac{1}{2}(-4)(2) - 2(2)^2$$
$$= \frac{5}{4}(16) + \frac{1}{2}(-8) - 2(4)$$
$$= 20 - 4 - 8$$
$$= 16 - 8$$
$$= 8$$

22.
$$\frac{n^4}{4} + \frac{n^3}{2} + \frac{n^2}{4}$$
$$\frac{5^4}{4} + \frac{5^3}{2} + \frac{5^2}{4}$$
$$= \frac{625}{4} + \frac{125}{2} + \frac{25}{4}$$
$$= \frac{625}{4} + \frac{250}{4} + \frac{25}{4}$$
$$= \frac{900}{4}$$
$$= 225$$

23a. Is a polynomial.
b. Is a polynomial.
c. Is not a polynomial because the variable is in the denominator.
d. Is not a polynomial because the . variable is under the radical

24.
$$x(3x[2x(x-2)+1]-4)$$
$$= x(3x[2x^2 - 4x + 1] - 4)$$
$$= x(6x^3 - 12x^2 + 3x - 4)$$
$$= 6x^4 - 12x^3 + 3x^2 - 4x$$

25. $8p^2q - 3pq^2 - 2pq^2 + p^2q = 9p^2q - 5pq^2$

26. $1.7m^3 + 2.6 - 0.3m - 1.4m^2 - 1.2m^3 + 4.5m^2 + 1.1$
$= 0.5m^3 + 3.1m^2 - 0.3m + 3.7$

27. $(4b^3 + 2b^2 - 3b + 7) - (-2b + 3 - b^3 - 7b)$
$= 4b^3 + 2b^2 - 3b + 7 + 2b - 3 + b^3 + 7b$
$= 5b^3 + 2b^2 + 6b + 4$

28. $(8w^6 - 5w^4 - 3w^2) + (2w^4 - 8w^2 + 4)$
$= 8w^6 - 5w^4 - 3w^2 + 2w^4 - 8w^2 + 4$
$= 8w^6 - 3w^4 - 11w^2 + 4$

29a. $r + 1$

   b. $2(r + 1) = 2r + 2$

   c. $3r$

   d. $3r - (2r + 2)$
$$= 3r - 2r - 2$$
$$= r - 2$$

30. $(2x^3 + x^2 + 5x - 7) + (2x^2 + 3x - 2) - (x^3 + 5x^2 - 2x + 9)$
$$= 2x^3 + x^2 + 5x - 7 + 2x^2 + 3x - 2 - x^3 - 5x^2 + 2x - 9$$
$$= x^3 - 2x^2 + 10x - 18$$
$$= (-3)^3 - 2(-3)^2 + 10(-3) - 18$$
$$= -27 - 2(9) + 10(-3) - 18$$
$$= -27 - 18 - 30 - 18$$
$$= -93$$

31a. $3x^2 + x^2 = 4x^2$

   b. $3x^2 \cdot x^2 = 3x^4$

32a. $5b^3 - b^3 = 4b^3$

   b. $5b^3(-b^3) = -5b^6$

33a. Cannot be simplified

   b. $7a^4 \cdot a^6 = 7a^{10}$

34a. Cannot be simplified

   b. $2r^4(-r^3) = -2r^7$

35a. $\dfrac{3b^9}{9b^3} = \dfrac{b^6}{3}$

   b. $\dfrac{9b^3}{3b^9} = \dfrac{3}{b^6}$

36a. $\dfrac{4m^8}{m^8} = 4$

   b. $\dfrac{m^4}{8m^4} = \dfrac{1}{8}$

37. $(5m^2n)(-6m^3n^2)$
$$= (5)(-6)m^{2+3}n^{1+2}$$
$$= -30m^5n^3$$

38. $-7qr^2(2q^4 - 1)$
$$= -7qr^2(2q^4) - 7qr^2(-1)$$
$$= -14q^5r^2 + 7qr^2$$

39. $\dfrac{-21x^4y^3}{3xy^3} = \dfrac{-21}{3} \cdot \dfrac{x^{4-1}}{1} \cdot \dfrac{y^3}{y^3} = \dfrac{-7}{1} \cdot \dfrac{x^3}{1} \cdot \dfrac{1}{1} = -7x^3$

40. $\dfrac{3a^2b}{6a^4b^3} = \dfrac{3}{6} \cdot \dfrac{1}{a^{4-2}} \cdot \dfrac{1}{b^{3-1}} = \dfrac{1}{2} \cdot \dfrac{1}{a^2} \cdot \dfrac{1}{b^2} = \dfrac{1}{2a^2b^2}$

41. $-2y(y - 4)(y + 3)$
$$= -2y(y^2 + 3y - 4y - 12)$$
$$= -2y(y^2 - y - 12)$$
$$= -2y^3 + 2y^2 + 24y$$

42. $6p^4(2p + 1)(p - 4)$
$$= 6p^4(2p^2 - 8p + p - 4)$$
$$= 6p^4(2p^2 - 7p - 4)$$
$$= 12p^6 - 42p^5 - 24p^4$$

43. $(d - 2)(d^2 - 4d + 4)$
$$= d^3 - 4d^2 + 4d - 2d^2 + 8d - 8$$
$$= d^3 - 6d^2 + 12d - 8$$

44. $(2k + 1)(k - 2)(k + 3)$
$$= (2k + 1)(k^2 + 3k - 2k - 6)$$
$$= (2k + 1)(k^2 + k - 6)$$
$$= 2k^3 + 2k^2 - 12k + k^2 + k - 6$$
$$= 2k^3 + 3k^2 - 11k - 6$$

45. $9x^3y(4x - 3y)^2$
$$= 9x^3y[(4x)^2 - 2(4x)(3y) + (3y)^2]$$
$$= 9x^3y[16x^2 - 24xy + 9y^2]$$
$$= 144x^5y - 216x^4y^2 + 81x^3y^3$$

46. $(a - 3)^3$
$$= (a - 3)(a - 3)^2$$
$$= (a - 3)[(a)^2 - 2(a)(3) + (3)^2]$$
$$= (a - 3)[a^2 - 6a + 9]$$
$$= a^3 - 6a^2 + 9a - 3a^2 + 18a - 27$$
$$= a^3 - 9a^2 + 27a - 27$$

47. Width: w
Length: $w + 10$
Height: $w - 2$
$V = lwh$
$= (w + 10)w(w - 2)$
$= w(w + 10)(w - 2)$
$= w(w^2 - 2w + 10w - 20)$
$= w(w^2 + 8w - 20)$
$= w^3 + 8w^2 - 20w$

48a. $p(140 - 2p) = 140p - 2p^2$

b. $140p - 2p^2$
$140(4) - 2(4)^2$
$= 140(4) - 2(16)$
$= 560 - 32$
$= \$528$

49. $30w^9 - 42w^4 + 54w^8$
$= 6w^4(5w^5 - 7 + 9w^4)$

50. $45x^2y^2 + 18x^2y^3 - 27x^3y^3$
$= 9x^2y^2(5 + 2y - 3xy)$

51. $-vw^5 - vw^4 + vw^2$
$= -vw^2(w^3 + w^2 - 1)$

52. $5(x - 2) - x^2(x - 2)$
$= (x - 2)(5 - x^2)$

53. $2q^4 + 6q^3 - 80q^2$
$= 2q^2(q^2 + 3q - 40)$
$= 2q^2(q + 8)(q - 5)$

54. $32 - b^4 - 14b^2$
$= -b^4 - 14b^2 + 32$
$= -1(b^4 + 14b^2 - 32)$
$= -1(b^2 + 16)(b^2 - 2)$

55. $x^2 - 3xy + 2y^2$
$= (x - 2y)(x - y)$

56. $y^2 - 3by - 28b^2$
$= (y - 7b)(y + 4b)$

57. $3a^2b + 12ab^2 + 9b^3$
$= 3b(a^2 + 4ab + 3b^2)$
$= 3b(a + 3b)(a + b)$

58. $80t^4 - 28t^3 - 24t^2$
$= 4t^2(20t^2 - 7t - 6)$
$= 4t^2(5t + 2)(4t - 3)$

59. $9x^2y^2 + 3xy - 2$
$= (3xy + 2)(3xy - 1)$

60. $4a^3x - 2a^2x^2 - 12ax^3$
$= 2ax(2a^2 - ax - 6x^2)$
$= 2ax(2a + 3x)(a - 2x)$

61. $h^2 - 24h + 144$
$= (h - 12)(h - 12)$
or $(h - 12)^2$

62. $9t^2 - 30tv + 25v^2$
$= (3t - 5v)(3t - 5v)$
or $(3t - 5v)^2$

63. $x^2 + 144 \rightarrow$ The <u>sum</u> of squares cannot be factored.

64. $98n^4 - 8n^2$
$= 2n^2(49n^2 - 4)$
$= 2n^2(7n + 2)(7n - 2)$

65. $w^8 + 12w^4 + 36$
$= (w^4 + 6)(w^4 + 6)$
or $(w^4 + 6)^2$

66. $q^6 - 14q^3 + 49$
$= (q^3 - 7)(q^3 - 7)$
or $(q^3 - 7)^2$

67. $3s^4 - 48$
$= 3(s^4 - 16)$
$= 3(s^2 + 4)(s^2 - 4)$
$= 3(s^2 + 4)(s + 2)(s - 2)$

68. $2y^4 - 2$
$= 2(y^4 - 1)$
$= 2(y^2 + 1)(y^2 - 1)$
$= 2(y^2 + 1)(y + 1)(y - 1)$

69. (c); $k = 2(24) = 4(12) = 6(8) = 48$; $y = \dfrac{48}{x}$

70. (b); $k = 0.5(12) = 3(2) = 8(0.75) = 6$

71a. $k = 6(20) = 120$; $I = \dfrac{120}{R}$

b. $I = \dfrac{120}{R} = \dfrac{120}{4} = 30$ amps

72. Two variables vary <u>inversely</u> means that when one variable increases, the other variable decreases by the same factor.

Two variables vary <u>directly</u> means that when one variable increases, the other variable also increases by the same factor.

73a. $\dfrac{s^2 - s}{s^2 + 3s - 10}$

$\dfrac{(-2)^2 - (-2)}{(-2)^2 + 3(-2) - 10}$

$= \dfrac{4 + 2}{4 - 6 - 10}$

$= \dfrac{6}{-12}$

$= \dfrac{-1}{2}$

b. $s^2 + 3s - 10 = 0$

$(s + 5)(s - 2) = 0$

$s + 5 = 0$ or $s - 2 = 0$

$s = -5 \qquad\quad s = 2$

74a. $\dfrac{1}{x + 5}$

b. $7 \cdot \dfrac{1}{x + 5} = \dfrac{7}{x + 5}$

75. $\dfrac{a + 3}{b + 3}$ cannot be reduced

76. $\dfrac{5x + 7}{5x}$ cannot be reduced

77. $\dfrac{10 + 2y}{2y} = \dfrac{\cancel{2}(5 + y)}{\cancel{2} \cdot y} = \dfrac{5 + y}{y}$

78. $\dfrac{3x^2 - 1}{1 - 3x^2} = \dfrac{-(1 - 3x^2)}{1 - 3x^2} = -1$

79. $\dfrac{v - 2}{v^2 - 4} = \dfrac{v - 2}{(v + 2)(v - 2)} = \dfrac{1}{v + 2}$

80. $\dfrac{q^5 - q^4}{q^4} = \dfrac{q^4(q - 1)}{q^4} = q - 1$

81. $\dfrac{-3x}{6x^2 + 9x} = \dfrac{-\cancel{3} \cdot \cancel{x}}{\cancel{3} \cdot \cancel{x}(2x + 3)} = \dfrac{-1}{2x + 3}$

82. $\dfrac{x^2 + 5x + 6}{x^2 - 4} = \dfrac{(x + 2)(x + 3)}{(x + 2)(x - 2)} = \dfrac{x + 3}{x - 2}$

83a. $\dfrac{3}{8} + \dfrac{5}{12} = \dfrac{3 \cdot 3}{8 \cdot 3} + \dfrac{5 \cdot 2}{12 \cdot 2} = \dfrac{9}{24} + \dfrac{10}{24} = \dfrac{19}{24}$

b. $\dfrac{3}{8} \cdot \dfrac{5}{12} = \dfrac{\cancel{3}}{2 \cdot 2 \cdot 2} \cdot \dfrac{5}{2 \cdot 2 \cdot \cancel{3}} = \dfrac{5}{32}$

84a. $\dfrac{2}{x} + \dfrac{1}{x + 2}$

$= \dfrac{2 \cdot (x + 2)}{x \cdot (x + 2)} + \dfrac{1 \cdot x}{(x + 2) \cdot x}$

$= \dfrac{2x + 4}{x(x + 2)} + \dfrac{x}{x(x + 2)}$

$= \dfrac{3x + 4}{x(x + 2)}$

b. $\dfrac{2}{x} \cdot \dfrac{1}{x + 2} = \dfrac{2}{x(x + 2)}$

85a. $\dfrac{3x}{2x + 2} + \dfrac{x + 1}{6x}$

$= \dfrac{3x}{2(x + 1)} + \dfrac{x + 1}{6x}$

$= \dfrac{3x \cdot 3x}{2(x + 1) \cdot 3x} + \dfrac{(x + 1)(x + 1)}{6x \cdot (x + 1)}$

$= \dfrac{9x^2}{6x(x + 1)} + \dfrac{x^2 + 2x + 1}{6x(x + 1)}$

$= \dfrac{10x^2 + 2x + 1}{6x(x + 1)}$

b. $\dfrac{3x}{2x + 2} \cdot \dfrac{x + 1}{6x} = \dfrac{\cancel{3} \cdot \cancel{x}}{2(x + 1)} \cdot \dfrac{x + 1}{2 \cdot \cancel{3} \cdot \cancel{x}} = \dfrac{1}{4}$

86a. $\dfrac{x+1}{x-1} + \dfrac{1}{x^2-1}$

$= \dfrac{x+1}{x-1} + \dfrac{1}{(x+1)(x-1)}$

$= \dfrac{(x+1)(x+1)}{(x-1)(x+1)} + \dfrac{1}{(x+1)(x-1)}$

$= \dfrac{x^2+2x+1}{(x-1)(x+1)} + \dfrac{1}{(x+1)(x-1)}$

$= \dfrac{x^2+2x+2}{(x+1)(x-1)}$

b. $\dfrac{x+1}{x-1} \cdot \dfrac{1}{x^2-1}$

$= \dfrac{x+1}{x-1} \cdot \dfrac{1}{(x+1)(x-1)}$

$= \dfrac{1}{(x-1)^2}$

87a. $2 + \dfrac{1}{x} = \dfrac{2 \cdot x}{1 \cdot x} + \dfrac{1}{x} = \dfrac{2x}{x} + \dfrac{1}{x} = \dfrac{2x+1}{x}$

b. $2 \cdot \dfrac{1}{x} = \dfrac{2}{1} \cdot \dfrac{1}{x} = \dfrac{2}{x}$

88a. $x + \dfrac{1}{x+2}$

$= \dfrac{x \cdot (x+2)}{1 \cdot (x+2)} + \dfrac{1}{x+2}$

$= \dfrac{x^2+2x}{x+2} + \dfrac{1}{x+2}$

$= \dfrac{x^2+2x+1}{x+2}$

b. $x \cdot \dfrac{1}{x+2} = \dfrac{x}{1} \cdot \dfrac{1}{x+2} = \dfrac{x}{x+2}$

89. $\dfrac{4c^2d}{3} \div 6cd^2$

$= \dfrac{\cancel{2} \cdot 2 \cdot \cancel{c} \cdot c \cdot \cancel{d}}{3} \cdot \dfrac{1}{\cancel{2} \cdot 3 \cdot \cancel{c} \cdot \cancel{d} \cdot d}$

$= \dfrac{2c}{9d}$

90. $\dfrac{u^2-2uv}{uv} \div \dfrac{3u-6v}{2uv}$

$= \dfrac{u(u-2v)}{\cancel{u} \cdot \cancel{v}} \cdot \dfrac{2 \cdot \cancel{u} \cdot \cancel{v}}{3(u-2v)}$

$= \dfrac{2u}{3}$

91. $\dfrac{2m^2-m-1}{m+1} - \dfrac{m^2-m}{m+1}$

$= \dfrac{2m^2-m-1-(m^2-m)}{m+1}$

$= \dfrac{2m^2-m-1-m^2+m}{m+1}$

$= \dfrac{m^2-1}{m+1}$

$= \dfrac{(m+1)(m-1)}{m+1}$

$= m-1$

92. $\dfrac{3}{2p} + \dfrac{7}{6p^2}$

$= \dfrac{3 \cdot 3p}{2p \cdot 3p} + \dfrac{7}{6p^2}$

$= \dfrac{9p}{6p^2} + \dfrac{7}{6p^2}$

$= \dfrac{9p+7}{6p^2}$

93. $\dfrac{5q}{q-3} - \dfrac{7}{q} + 3$

$= \dfrac{5q \cdot q}{(q-3)\cdot q} - \dfrac{7\cdot(q-3)}{q\cdot(q-3)} + \dfrac{3\cdot q(q-3)}{1\cdot q(q-3)}$

$= \dfrac{5q^2}{q(q-3)} - \dfrac{7q-21}{q(q-3)} + \dfrac{3q^2-9q}{q(q-3)}$

$= \dfrac{5q^2 - (7q-21) + (3q^2-9q)}{q(q-3)}$

$= \dfrac{5q^2 - 7q + 21 + 3q^2 - 9q}{q(q-3)}$

$= \dfrac{8q^2 - 16q + 21}{q(q-3)}$

94. $\dfrac{2w}{w^2-4} + \dfrac{4}{w^2+4w+4}$ Factor denominators to find LCD

$= \dfrac{2w}{(w+2)(w-2)} + \dfrac{4}{(w+2)(w+2)}$ $\text{LCD} = (w+2)(w-2)(w+2)$

$= \dfrac{2w(w+2)}{(w+2)(w-2)(w+2)} + \dfrac{4(w-2)}{(w+2)(w+2)(w-2)}$

$= \dfrac{2w^2+4w}{(w+2)(w-2)(w+2)} + \dfrac{4w-8}{(w+2)(w+2)(w-2)}$

$= \dfrac{2w^2+8w-8}{(w+2)(w-2)(w+2)}$ or $\dfrac{2w^2+8w-8}{(w+2)^2(w-2)}$

95a. $\dfrac{5}{x-2}$ **b.** $\dfrac{5}{x+2}$ **c.** $\dfrac{5}{x-2} + \dfrac{5}{x+2}$

$= \dfrac{5\cdot(x+2)}{(x-2)(x+2)} + \dfrac{5\cdot(x-2)}{(x+2)(x-2)}$

$= \dfrac{5x+10}{(x-2)(x+2)} + \dfrac{5x-10}{(x+2)(x-2)}$

$= \dfrac{10x}{(x+2)(x-2)}$

96. $\dfrac{2-\dfrac{a}{b}}{2-\dfrac{b}{a}} = \dfrac{ab\left(2-\dfrac{a}{b}\right)}{ab\left(2-\dfrac{b}{a}\right)} = \dfrac{ab(2) - \dfrac{ab}{1}\left(\dfrac{a}{b}\right)}{ab(2) - \dfrac{ab}{1}\left(\dfrac{b}{a}\right)} = \dfrac{2ab-a^2}{2ab-b^2}$

97. $\dfrac{3p-\dfrac{q^2}{3p}}{1-\dfrac{q}{3p}} = \dfrac{3p\left(3p-\dfrac{q^2}{3p}\right)}{3p\left(1-\dfrac{q}{3p}\right)} = \dfrac{3p(3p) - \dfrac{3p}{1}\left(\dfrac{q^2}{3p}\right)}{3p(1) - \dfrac{3p}{1}\left(\dfrac{q}{3p}\right)} = \dfrac{9p^2-q^2}{3p-q} = \dfrac{(3p+q)(3p-q)}{3p-q} = 3p+q$

98.

$$q - \frac{16}{q} = 6 \qquad \text{Multiply both sides by LCD} = q$$

$$q\left(q - \frac{16}{q}\right) = (6)q \qquad \text{Distribute } q \text{ to each term}$$

$$q(q) - q\left(\frac{16}{q}\right) = (6)q$$

$$q^2 - 16 = 6q \qquad \text{Write in standard form}$$

$$q^2 - 6q - 16 = 0 \qquad \text{Factor}$$

$$(q - 8)(q + 2) = 0 \qquad \text{Set each factor equal to 0}$$

$$q - 8 = 0 \quad \text{or} \quad q + 2 = 0 \qquad \text{Solve each}$$

$$q = 8 \qquad\qquad q = -2$$

99.

$$\frac{2 - x}{5x} = \frac{4}{15x} - \frac{1}{6} \qquad \text{Multiply both sides by LCD} = 30x$$

$$30x\left(\frac{2 - x}{5x}\right) = \left(\frac{4}{15x} - \frac{1}{6}\right)30x \qquad \text{Distribute } 30x \text{ to each term}$$

$$30x\left(\frac{2 - x}{5x}\right) = 30x\left(\frac{4}{15x}\right) - 30x\left(\frac{1}{6}\right)$$

$$6(2 - x) = 8 - 5x$$

$$12 - 6x = 8 - 5x \qquad \text{Add } 5x \text{ to both sides}$$

$$12 - x = 8 \qquad \text{Subtract 12 from both sides}$$

$$-x = -4 \qquad \text{Divide both sides by } -1$$

$$x = 4$$

100.

$$\frac{9}{m + 2} + \frac{2}{m} = 2 \qquad \begin{array}{l}\text{Multiply both sides by}\\ \text{LCD} = m(m + 2)\end{array}$$

$$m(m + 2)\left(\frac{9}{m + 2} + \frac{2}{m}\right) = (2)m(m + 2) \qquad \text{Distribute } m(m + 2) \text{ to each term}$$

$$m(m + 2)\left(\frac{9}{m + 2}\right) + m(m + 2)\left(\frac{2}{m}\right) = (2)m(m + 2)$$

$$9m + 2(m + 2) = 2m(m + 2)$$

$$9m + 2m + 4 = 2m^2 + 4m \qquad \text{Combine like terms}$$

$$11m + 4 = 2m^2 + 4m \qquad \text{Write in standard form}$$

$$0 = 2m^2 - 7m - 4 \qquad \text{Factor}$$

$$0 = (2m + 1)(m - 4) \qquad \text{Set each factor equal to 0}$$

$$2m + 1 = 0 \quad \text{or} \quad m - 4 = 0 \qquad \text{Solve each}$$

$$2m = -1 \qquad\qquad m = 4$$

$$m = \frac{-1}{2}$$

101.

$$\frac{15}{x^2 - 3x} + \frac{4}{x} = \frac{5}{x - 3}$$ Factor denominators to find LCD

$$\frac{15}{x(x - 3)} + \frac{4}{x} = \frac{5}{x - 3}$$ Multiply both sides by LCD $= x(x - 3)$

$$x(x - 3)\left[\frac{15}{x(x - 3)} + \frac{4}{x}\right] = \left[\frac{5}{x - 3}\right]x(x - 3)$$ Distribute $x(x - 3)$ to each term

$$x(x - 3)\left[\frac{15}{x(x - 3)}\right] + x(x - 3)\left[\frac{4}{x}\right] = \left[\frac{5}{x - 3}\right]x(x - 3)$$

$$15 + 4(x - 3) = 5x$$

$$15 + 4x - 12 = 5x$$ Combine like terms

$$4x + 3 = 5x$$ Subtract $5x$ from both sides

$$-x + 3 = 0$$ Subtract 3 from both sides

$$-x = -3$$ Divide both sides by -1

$$x = 3$$

But the original fractions are undefined if $x = 0$ or 3, therefore, $x \neq 3$ and there is NO solution.

102.

$$\frac{1}{x} + \frac{1}{y} = \frac{1}{z}$$ Multiply both sides by LCD $= xyz$

$$xyz\left(\frac{1}{x} + \frac{1}{y}\right) = \left(\frac{1}{z}\right)xyz$$ Distribute xyz to each term

$$xyz\left(\frac{1}{x}\right) + xyz\left(\frac{1}{y}\right) = \left(\frac{1}{z}\right)xyz$$

$$yz + xz = xy$$ Subtract xz from both sides

$$yz = xy - xz$$ Factor out x

$$yz = x(y - z)$$ Divide both sides by $y - z$

$$\frac{yz}{y - z} = \frac{x(y - z)}{(y - z)}$$

$$\frac{yz}{y - z} = x$$

103.

$$y = \frac{2x + 3}{1 - x}$$ Multiply both sides by LCD $= 1 - x$

$$(1 - x)(y) = \left(\frac{2x + 3}{1 - x}\right)(1 - x)$$

$$y - xy = 2x + 3$$ Add xy to both sides

$$y = xy + 2x + 3$$ Subtract 3 from both sides

$$y - 3 = xy + 2x$$ Factor out x

$$y - 3 = x(y + 2)$$ Divide both sides by $y + 2$

$$\frac{y - 3}{y + 2} = \frac{x(y + 2)}{y + 2}$$

$$\frac{y - 3}{y + 2} = x$$

104a. adding fractions
subtracting fractions
simplifying complex fractions
solving equations with fractions

b. adding fractions
subtracting fractions

c. simplifying complex fractions
solving equations with fractions

188

105.

	Distance	Rate	Time
Before lunch	5	r	$\frac{5}{r}$
After lunch	8	$r+2$	$\frac{8}{r+2}$

$$\frac{5}{r} - \frac{8}{r+2} = 1 \qquad \text{Multiply both sides by LCD} = r(r+2)$$

$$r(r+2)\left(\frac{5}{r} - \frac{8}{r+2}\right) = (1)r(r+2) \qquad \text{Distribute } r(r+2) \text{ to each term}$$

$$r(r+2)\left(\frac{5}{r}\right) - r(r+2)\left(\frac{8}{r+2}\right) = (1)r(r+2)$$

$$5(r+2) - 8r = r(r+2)$$

$$5r + 10 - 8r = r^2 + 2r \qquad \text{Combine like terms}$$

$$-3r + 10 = r^2 + 2r \qquad \text{Write in standard form}$$

$$0 = r^2 + 5r - 10 \qquad \text{Use Quadratic Formula to solve}$$

$$r = \frac{-b \pm \sqrt{b^2 - 4ac}}{2a} \qquad a = 1, b = 5, c = -10$$

$$= \frac{-5 \pm \sqrt{5^2 - 4(1)(-10)}}{2(1)}$$

$$= \frac{-5 \pm \sqrt{25 + 40}}{2}$$

$$= \frac{-5 \pm \sqrt{65}}{2}$$

If $r = \dfrac{-5 - \sqrt{65}}{2} \approx -6.53$ (a negative number) which is not possible.

Her speed before lunch is $r = \dfrac{-5 + \sqrt{65}}{2} \approx 1.53$ mph

106.

	Rate	Time	Fraction of Book
Pipe	$\frac{1}{30}$	x	$\frac{x}{30}$
Hose	$\frac{1}{45}$	x	$\frac{x}{45}$

$$\frac{x}{30} + \frac{x}{45} = 1 \qquad \text{Multiply both sides by LCD} = 90$$

$$90\left(\frac{x}{30} + \frac{x}{45}\right) = (1)90 \qquad \text{Distribute 90 to each term}$$

$$90\left(\frac{x}{30}\right) + 90\left(\frac{x}{45}\right) = (1)90$$

$$3x + 2x = 90 \qquad \text{Combine like terms}$$

$$5x = 90 \qquad \text{Divide both sides by 5}$$

$$x = 18$$

It will take 18 hr to fill the pool.

CHAPTER 6

Homework 6.1

1a. $(t^3)^5 = t^{3 \cdot 5} = t^{15}$

b. $(b^4)^2 = b^{4 \cdot 2} = b^8$

c. $(w^{12})^{12} = w^{12 \cdot 12} = w^{144}$

5a. $\left(\dfrac{w}{2}\right)^6 = \dfrac{w^6}{2^6} = \dfrac{w^6}{64}$

b. $\left(\dfrac{5}{v}\right)^4 = \dfrac{5^4}{v^4} = \dfrac{625}{v^4}$

c. $\left(\dfrac{-m}{p}\right)^3 = \dfrac{(-m)^3}{p^3} = \dfrac{-m^3}{p^3}$

3a. $(5x)^3 = 5^3 x^3 = 125x^3$

b. $(-3wz)^4 = (-3)^4 w^4 z^4 = 81w^4 z^4$

c. $(-ab)^5 = (-a)^5 b^5 = -a^5 b^5$

7a. $x^3 \cdot x^6 = x^{3+6} = x^9$

b. $(x^3)^6 = x^{3 \cdot 6} = x^{18}$

c. $\dfrac{x^3}{x^6} = \dfrac{1}{x^{6-3}} = \dfrac{1}{x^3}$

d. $\dfrac{x^6}{x^3} = x^{6-3} = x^3$

9. The first law of exponents simplifies a product of powers with the same base by adding the exponents. Example: $x^5 \cdot x^2 = x^{5+2} = x^7$.
The third law of exponents simplifies a power raised to a power by multiplying the exponents. Example: $(x^5)^2 = x^{5 \cdot 2} = x^{10}$.
(other examples possible)

11. The 3 should be raised to the second power before multiplying. $2 \cdot 3^2 = 2 \cdot 9 = 18$

13. The exponent does not apply to the negative sign. $-10^2 = -10 \cdot 10 = -100$

15. When like bases are multiplied, the exponents should be *added*, not multiplied. $a^4 \cdot a^3 = a^{4+3} = a^7$

17. $(2p^3)^5 = 2^5(p^3)^5 = 32p^{15}$

19. $\left(\dfrac{-3}{q^4}\right)^5 = \dfrac{(-3)^5}{(q^4)^5} = \dfrac{-243}{q^{20}}$

21. $\left(\dfrac{-2h^2}{m^3}\right)^4 = \dfrac{(-2)^4(h^2)^4}{(m^3)^4} = \dfrac{16h^8}{m^{12}}$

23. $x^3(x^2)^5 = x^3 \cdot x^{10} = x^{3+10} = x^{13}$

25. $(2x^3 y)^2(xy^3)^4$
$= 2^2 \cdot (x^3)^2 \cdot y^2 \cdot x^4 \cdot (y^3)^4$
$= 4 \cdot x^6 \cdot y^2 \cdot x^4 \cdot y^{12}$
$= 4x^{10}y^{14}$

27. $[ab^2(a^2 b)^3]^3$
$= [a \cdot b^2 \cdot (a^2)^3 \cdot b^3]^3$
$= [a \cdot b^2 \cdot a^6 \cdot b^3]^3$
$= [a^7 \cdot b^5]^3$
$= (a^7)^3(b^5)^3$
$= a^{21}b^{15}$

29. $-a^2(-a)^2$
$= -a^2 \cdot a^2$
$= -a^4$

31. $-(-xy)^2(xy^2)$
$= -(-x)^2 \cdot y^2 \cdot x \cdot y^2$
$= -x^2 \cdot y^2 \cdot x \cdot y^2$
$= -x^3 y^4$

33. $-4p(-p^2q^2)^2(-q^3)^2$
$= -4 \cdot p \cdot (-p^2)^2 \cdot (q^2)^2 \cdot (-q^3)^2$
$= -4 \cdot p \cdot p^4 \cdot q^4 \cdot q^6$
$= -4p^5q^{10}$

35. $2y(y^3)^2 - 2y^4(3y)^3$
$= 2 \cdot y \cdot y^6 - 2 \cdot y^4 \cdot 3^3 \cdot y^3$
$= 2 \cdot y \cdot y^6 - 2 \cdot y^4 \cdot 27 \cdot y^3$
$= 2y^7 - 54y^7$
$= -52y^7$

37. $2a(a^2)^4 + 3a^2(a^6) - a^2(a^2)^3$
$= 2 \cdot a \cdot a^8 + 3 \cdot a^2 \cdot a^6 - a^2 \cdot a^6$
$= 2a^9 + 3a^8 - a^8$
$= 2a^9 + 2a^8$

39. $-3v^2(2v^3 - v^2) + v(-4v)^2$
$= -3v^2(2v^3) - 3v^2(-v^2) + v(-4)^2(v)^2$
$= -3 \cdot v^2 \cdot 2 \cdot v^3 - 3 \cdot v^2 \cdot (-v^2) + v \cdot 16 \cdot v^2$
$= -6v^5 + 3v^4 + 16v^3$

41a. $4x^2 + 2x^4$

b. $4x^2(2x^4) = 4 \cdot x^2 \cdot 2 \cdot x^4 = 8x^6$

43a. $(-x)^3x^4 = -x^3 \cdot x^4 = -x^7$

b. $[(-x^3)(-x)]^4 = [x^4]^4 = x^{16}$

45a. $(3x^2)^4(2x^4)^2$
$= 3^4 \cdot (x^2)^4 \cdot 2^2 \cdot (x^4)^2$
$= 81 \cdot x^8 \cdot 4 \cdot x^8$
$= 324x^{16}$

b. $(3x^2)^4 - (2x^4)^2$
$= 3^4 \cdot (x^2)^4 - 2^2 \cdot (x^4)^2$
$= 81x^8 - 4x^8$
$= 77x^8$

47a. $6x^3 - 3x^6$

b. $6x^3(-3x^6) = 6 \cdot x^3 \cdot (-3) \cdot x^6 = -18x^9$

49a. $6x^3 - 3x^3(x^3) = 6x^3 - 3 \cdot x^3 \cdot x^3 = 6x^3 - 3x^6$

b. $(6x^3 - 3x^3)x^3 = (3x^3)x^3 = 3 \cdot x^3 \cdot x^3 = 3x^6$

51. $b^3 \cdot b^n = b^9$
$b^{3+n} = b^9$
$3 + n = 9$
$n = 6$

53. $\dfrac{c^8}{c^n} = c^2$
$c^{8-n} = c^2$
$8 - n = 2$
$-n = -6$
$n = 6$

55. $\dfrac{n^3}{3^3} = 8$
$\left(\dfrac{n}{3}\right)^3 = 2^3$
$\dfrac{n}{3} = 2$
$n = 6$

57. $x^4 + x^6 = x^2(x^2 + x^4)$

59. $4m^4 - 4m^8 + 8m^{16} = 4m^4(1 - m^4 + 2m^{12})$

61. $-8^2 < 64$

63. $(-3)^5 = -3^5$

65. $\left(\dfrac{-7}{4}\right)^{11} < 0$

67. $6^{10} \cdot 4^{10} = 24^{10}$

69. $(8 - 2)^3 > 8 - 2^3$

71. $(17^4)^5 = (17^5)^4$

Homework 6.2

1. $5^{-2} = \dfrac{1}{5^2} = \dfrac{1}{25}$

3. $x^{-6} = \dfrac{1}{x^6}$

5. $(8x)^0 = 1$

7. $\left(\dfrac{3}{4}\right)^{-3} = \left(\dfrac{4}{3}\right)^3 = \dfrac{4^3}{3^3} = \dfrac{64}{27}$

191

9. $\left(\dfrac{b}{3}\right)^{-4} = \left(\dfrac{3}{b}\right)^4 = \dfrac{3^4}{b^4} = \dfrac{81}{b^4}$

11. $(2q)^{-5} = \dfrac{1}{(2q)^5} = \dfrac{1}{2^5 q^5} = \dfrac{1}{32q^5}$

13. $3 \cdot 4^{-3} = 3 \cdot \dfrac{1}{4^3} = \dfrac{3}{64}$

15. $4x^{-2} = 4 \cdot \dfrac{1}{x^2} = \dfrac{4}{x^2}$

17. $\dfrac{1}{2^3} = 2^{-3}$

19. $\dfrac{3}{5^2} = 3 \cdot 5^{-2}$

21. $\dfrac{1}{27} = \dfrac{1}{3^3} = 3^{-3}$

23. $\dfrac{x}{625} = \dfrac{x}{5^4} = 5^{-4}x$

25. $\dfrac{2}{z^2} = 2z^{-2}$

27. $\left(\dfrac{z}{10}\right)^5 = \dfrac{z^5}{10^5} = 10^{-5}z^5$

29. Any nonzero number raised to the zero power is 1.
$x^0 = 1$

31. A negative exponent is the reciprocal of the corresponding positive power (not a negative number).
$w^{-3} = \dfrac{1}{w^3}$

33. The exponent applies only to the base x.
$2x^{-4} = 2 \cdot x^{-4} = 2 \cdot \dfrac{1}{x^4} = \dfrac{2}{x^4}$

35. $x^{-3} \cdot x^8 = x^{-3+8} = x^5$

37. $5^{-4} \cdot 5^{-3} = 5^{-4+(-3)} = 5^{-7}$

39. $(3b^{-5})(5b^2) = (3 \cdot 5)(b^{-5+2}) = 15b^{-3}$

41. $\dfrac{c^{-7}}{c^{-4}} = c^{-7-(-4)} = c^{-7+4} = c^{-3}$

43. $\dfrac{8b^{-4}}{4b^{-8}} = \dfrac{8}{4} \cdot b^{-4-(-8)} = 2 \cdot b^{-4+8} = 2b^4$

45. $\dfrac{6^6}{6^{-2}} = 6^{6-(-2)} = 6^{6+2} = 6^8$

47. $\dfrac{1}{6^{-3}} = 1 \cdot 6^3 = 216$

49. $\dfrac{3}{2^{-6}} = 3 \cdot 2^6 = 3 \cdot 64 = 192$

51. $\dfrac{8x^3}{y^{-5}} = 8x^3 \cdot y^5 = 8x^3y^5$

53. $(8^{-2})^5 = 8^{-2 \cdot 5} = 8^{-10}$

55. $(w^{-6})^{-3} = w^{-6(-3)} = w^{18}$

57. $(d^6)^{-4} = d^{6(-4)} = d^{-24}$

59. $(pq)^{-5} = p^{-5}q^{-5}$

61. $(3x)^{-2} = 3^{-2}x^{-2}$

63. $5(2r)^{-3} = 5 \cdot 2^{-3}r^{-3}$

65. $(a^{-4}c^2)^{-3}$
$= (a^{-4})^{-3}(c^2)^{-3}$
$= a^{-4(-3)}c^{2(-3)}$
$= a^{12}c^{-6}$
$= \dfrac{a^{12}}{c^6}$

67. $(2u^2)^{-3}(u^{-4})^2$
$= 2^{-3}(u^2)^{-3}(u^{-4})^2$
$= 2^{-3}u^{2(-3)}u^{-4(2)}$
$= 2^{-3}u^{-6}u^{-8}$
$= 2^{-3}u^{-6+(-8)}$
$= 2^{-3}u^{-14}$
$= \dfrac{1}{2^3 u^{14}}$
$= \dfrac{1}{8u^{14}}$

69. $\dfrac{5k^{-3}(k^4)^{-3}}{6k^{-5}}$

$= \dfrac{5k^{-3}k^{4(-3)}}{6k^{-5}}$

$= \dfrac{5k^{-3}k^{-12}}{6k^{-5}}$

$= \dfrac{5k^{-3+(-12)}}{6k^{-5}}$

$= \dfrac{5k^{-15}}{6k^{-5}}$

$= \dfrac{5k^{-15-(-5)}}{6}$

$= \dfrac{5k^{-10}}{6}$

$= \dfrac{5}{6k^{10}}$

71. $\left(\dfrac{2p^{-3}}{p^2}\right)^{-2}$

$= \dfrac{2^{-2}(p^{-3})^{-2}}{(p^2)^{-2}}$

$= \dfrac{2^{-2}p^{-3(-2)}}{p^{2(-2)}}$

$= \dfrac{2^{-2}p^6}{p^{-4}}$

$= 2^{-2}p^{6-(-4)}$

$= 2^{-2}p^{10}$

$= \dfrac{p^{10}}{2^2}$

$= \dfrac{p^{10}}{4}$

73. $\dfrac{a^{-1}+2}{a^{-1}-2} = \dfrac{\frac{1}{a}+2}{\frac{1}{a}-2} = \dfrac{a\left(\frac{1}{a}+2\right)}{a\left(\frac{1}{a}-2\right)} = \dfrac{\frac{a}{1}\left(\frac{1}{a}\right)+a(2)}{\frac{a}{1}\left(\frac{1}{a}\right)-a(2)} = \dfrac{1+2a}{1-2a}$

75. $\dfrac{c^{-2}-1}{c^{-1}+1} = \dfrac{\frac{1}{c^2}-1}{\frac{1}{c}+1} = \dfrac{c^2\left(\frac{1}{c^2}-1\right)}{c^2\left(\frac{1}{c}+1\right)} = \dfrac{\frac{c^2}{1}\left(\frac{1}{c^2}\right)-c^2(1)}{\frac{c^2}{1}\left(\frac{1}{c}\right)+c^2(1)} = \dfrac{1-c^2}{c+c^2} = \dfrac{(1+c)(1-c)}{c(1+c)} = \dfrac{1-c}{c}$

77. $\dfrac{x^{-2}-y^{-2}}{x^{-1}-y^{-1}} = \dfrac{\frac{1}{x^2}-\frac{1}{y^2}}{\frac{1}{x}-\frac{1}{y}} = \dfrac{x^2y^2\left(\frac{1}{x^2}-\frac{1}{y^2}\right)}{x^2y^2\left(\frac{1}{x}-\frac{1}{y}\right)} = \dfrac{\frac{x^2y^2}{1}\left(\frac{1}{x^2}\right)-\frac{x^2y^2}{1}\left(\frac{1}{y^2}\right)}{\frac{x^2y^2}{1}\left(\frac{1}{x}\right)-\frac{x^2y^2}{1}\left(\frac{1}{y}\right)}$

$= \dfrac{y^2-x^2}{xy^2-x^2y} = \dfrac{(y+x)(y-x)}{xy(y-x)} = \dfrac{y+x}{xy}$

79. $\dfrac{v^{-2}+v^{-1}}{w^{-1}v^{-1}+w^{-1}} = \dfrac{\frac{1}{v^2}+\frac{1}{v}}{\frac{1}{wv}+\frac{1}{w}} = \dfrac{wv^2\left(\frac{1}{v^2}+\frac{1}{v}\right)}{wv^2\left(\frac{1}{wv}+\frac{1}{w}\right)} = \dfrac{\frac{wv^2}{1}\left(\frac{1}{v^2}\right)+\frac{wv^2}{1}\left(\frac{1}{v}\right)}{\frac{wv^2}{1}\left(\frac{1}{wv}\right)+\frac{wv^2}{1}\left(\frac{1}{w}\right)}$

$= \dfrac{w+wv}{v+v^2} = \dfrac{w(1+v)}{v(1+v)} = \dfrac{w}{v}$

81. $\dfrac{10^3}{10^{-2}} = 10^5$

83. $\dfrac{10^{-5}}{10^{-3}} = 10^{-2}$

85. $\dfrac{10}{10^{-1}} = 10^2$

87. $\dfrac{10^{-3}\times 10^{-2}}{10^{-6}} = 10^1$

89. $\dfrac{10^{-5}}{10^{-4}\times 10^7} = 10^{-8}$

Homework 6.3

1. $4.3 \times 10^4 = 43,000$

3. $8 \times 10^{-6} = 0.000\,008$

5. $0.002 \times 10^{-2} = 0.000\,02$

7. $234 = 2.34 \times 10^2$

9. $0.92 = 9.2 \times 10^{-1}$

11. $1,720,000 = 1.72 \times 10^6$

13. $4834 = 4.834 \times 10^3$

15. $0.072 = 7.2 \times 10^{-2}$

17. $0.000\,007 = 7 \times 10^{-6}$

19. $685,000,000 = 6.85 \times 10^8$

21. 56.74×10^4
$= (5.674 \times 10^1) \times 10^4$
$= 5.674 \times (10^1 \times 10^4)$
$= 5.674 \times 10^5$

23. 385×10^{-3}
$= (3.85 \times 10^2) \times 10^{-3}$
$= 3.85 \times (10^2 \times 10^{-3})$
$= 3.85 \times 10^{-1}$

25. Positive

27. Negative

29. Positive

31. Negative

33. $29,141 = 2.9141 \times 10^4$ ft

35. $0.000\,076 = 7.6 \times 10^{-5}$ cm

37. $5 \times 10^{-6} = 0.000\,005$ g

39. $5 \times 10^9 = 5,000,000,000$ yr

41. $(2,000,000)(0.000\,07)$
$= (2 \times 10^6) \times (7 \times 10^{-5})$
$= (2 \times 7) \times (10^6 \times 10^{-5})$
$= 14 \times 10^1$
$= 140$

43. $0.000\,036 \div 0.0009$
$= \dfrac{0.000\,036}{0.0009}$
$= \dfrac{3.6 \times 10^{-5}}{9 \times 10^{-4}}$
$= \dfrac{3.6}{9} \times \dfrac{10^{-5}}{10^{-4}}$
$= 0.4 \times 10^{-1}$
$= 0.04$

45. $\dfrac{(80,000,000,000)(0.0006)}{20,000}$
$= \dfrac{(8 \times 10^{10}) \times (6 \times 10^{-4})}{2 \times 10^4}$
$= \dfrac{(8 \times 6) \times (10^{10} \times 10^{-4})}{2 \times 10^4}$
$= \dfrac{48 \times 10^6}{2 \times 10^4}$
$= \dfrac{48}{2} \times \dfrac{10^6}{10^4}$
$= 24 \times 10^2$
$= 2400$

47. $(1.86 \times 10^5) \times (3.16 \times 10^7)$
$= 1.86 \boxed{\text{EE}} \ 5 \ \boxed{\times} \ 3.16 \ \boxed{\text{EE}} \ 7 \ \boxed{=}$
$= \boxed{5.8776 \quad 12}$
$\approx 5.88 \times 10^{12}$ mi

49a. $(4.676 \times 10^{12}) \div (3.6 \times 10^3)$
$= 4.676 \ \boxed{\text{EE}} \ 12 \ \boxed{\div} \ 3.6 \ \boxed{\text{EE}} \ 3 \ \boxed{=}$
$= \boxed{1.298888889 \quad 09}$
$\approx 1.30 \times 10^9$ ft

b. $1.30 \times 10^9 \div 5280$
$= 1.30 \ \boxed{\text{EE}} \ 9 \ \boxed{\div} \ 5280 \ \boxed{=}$
$= \boxed{2.462121212 \quad 05}$
$\approx 2.46 \times 10^5$ mi

51a. $200,000,000 \times 5,600,000,000$
$= (2 \times 10^8) \times (5.6 \times 10^9)$
$= 2\ \boxed{\text{EE}}\ 8\ \boxed{\times}\ 5.6\ \boxed{\text{EE}}\ 9\ \boxed{=}$
$= \boxed{1.12 \qquad 18}$
$= 1.12 \times 10^{18}$ insects

b. $(1.12 \times 10^{18}) \times (3 \times 10^{-4})$
$= 1.12\ \boxed{\text{EE}}\ 18\ \boxed{\times}\ 3\ \boxed{\text{EE}}\ 4\ \boxed{+/-}\ \boxed{=}$
$= \boxed{3.36 \qquad 14}$
$= 3.36 \times 10^{14}$ g

53. $5,800,000 \times 110,000,000,000 \div 5,600,000,000$
$(5.8 \times 10^6) \times (1.1 \times 10^{11}) \div (5.6 \times 10^9)$
$= 5.8\ \boxed{\text{EE}}\ 6\ \boxed{\times}\ 1.1\ \boxed{\text{EE}}\ 11\ \boxed{\div}\ 5.6\ \boxed{\text{EE}}\ 9\ \boxed{=}$
$= \boxed{113,928,571}$
$\approx 1.14 \times 10^8$ gal

Midchapter 6 Review

1. $b^2 \cdot b^4$ is a product of powers with the same base, therefore, the exponents are added.
$b^2 \cdot b^4 = b^{2+4} = b^6$

$(b^2)^4$ is a power raised to a power, therefore the exponents are multiplied.
$(b^2)^4 = b^{2 \cdot 4} = b^8$

2. In $2b^3$ only the b is cubed.

In $(2b)^3$ both the 2 and the b are cubed.

3. A negative exponent is the reciprocal of the corresponding positive power. Example: $2^{-3} = \dfrac{1}{2^3} = \dfrac{1}{8}$.

4a. $\dfrac{n^3}{n^6} = n^{3-6} = n^{-3}$

b. $\dfrac{n^3}{n^6} = \dfrac{1}{n^{6-3}} = \dfrac{1}{n^3}$

5. There should be exactly one nonzero digit to the left of the decimal point.

6. If the number is larger, the exponent on 10 will be positive.
If the number is a decimal fraction less than 1, the exponent on 10 will be negative.

7a. The power of a sum is not equal to the sum of the powers.

b. In a quotient of powers with the same base, subtract the exponents. $\dfrac{m^{-8}}{m^{-4}} = m^{-8-(-4)} = m^{-4}$

8a. $(a^{-1})^{-1} = a^{-1(-1)} = a^1 = a$

b. $\dfrac{1}{a^{-1}} = a^1 = a$

c. $\dfrac{1}{\frac{1}{a}} = \dfrac{1}{a^{-1}} = a^1 = a$

d. $\left(\dfrac{1}{a}\right)^{-1} = \left(\dfrac{a}{1}\right)^1 = a$

9. $(-2x^2)^3 = (-2)^3(x^2)^3 = -8x^6$

10. $\left(\dfrac{-v^3}{3w}\right)^4 = \dfrac{(-v^3)^4}{3^4 w^4} = \dfrac{v^{12}}{81w^4}$

11. $(3b^4)^2(4b)^3$
$= 3^2 \cdot (b^4)^2 \cdot 4^3 \cdot b^3$
$= 9 \cdot b^8 \cdot 64 \cdot b^3$
$= 576b^{11}$

12. $-x^4(-x^2)^4$
$= -x^4 \cdot x^8$
$= -x^{12}$

13. $[-2a(-3a)^2]^2$
$= [-2a \cdot (-3)^2 \cdot a^2]^2$
$= [-2 \cdot a \cdot 9 \cdot a^2]^2$
$= [-18a^3]^2$
$= (-18)^2(a^3)^2$
$= 324a^6$

14. $-2a^3 - 3a(-a^2) = -2a^3 + 3a^3 = a^3$

15. $2t^6 - 12t^{12} = 2t^6(1 - 6t^6)$

16. $6g^3 + 12g^9 - 9g^6 = 3g^3(2 + 4g^6 - 3g^3)$

17. $(2x)^{-4} = 2^{-4}x^{-4} = \dfrac{1}{2^4 x^4} = \dfrac{1}{16x^4}$

18. $3x^{-3}(-3x^4) = 3(-3) \cdot x^{-3+4} = -9x$

19. $\dfrac{8x^{-6}}{6x^{-2}} = \dfrac{8}{6} \cdot x^{-6-(-2)} = \dfrac{4}{3} \cdot x^{-4} = \dfrac{4}{3x^4}$

20. $5^{-2} \cdot 5^4 = 5^{-2+4} = 5^2 = 25$

21. $(3b^{-4})^{-2}(b^3 \cdot 3b^5)$
$= 3^{-2}(b^{-4})^{-2} \cdot (b^3 \cdot 3b^5)$
$= 3^{-2}b^{-4(-2)} \cdot 3b^{3+5}$
$= 3^{-2}b^8 \cdot 3b^8$
$= 3^{-2+1}b^{8+8}$
$= 3^{-1}b^{16}$
$= \dfrac{b^{16}}{3}$

22. $\dfrac{(-2c^{-2})^3}{4c^{-3}(c^9)}$
$= \dfrac{(-2)^3(c^{-2})^3}{4c^{-3}(c^9)}$
$= \dfrac{-8c^{-2\cdot3}}{4c^{-3+9}}$
$= \dfrac{-8c^{-6}}{4c^6}$
$= \dfrac{-8}{4} \cdot c^{-6-6}$
$= -2c^{-12}$
$= \dfrac{-2}{c^{12}}$

23. $\dfrac{2t^{-2} + t^{-1}}{2t^{-1} + t^0} = \dfrac{\dfrac{2}{t^2} + \dfrac{1}{t}}{\dfrac{2}{t} + 1} = \dfrac{t^2\left(\dfrac{2}{t^2} + \dfrac{1}{t}\right)}{t^2\left(\dfrac{2}{t} + 1\right)} = \dfrac{\dfrac{t^2}{1}\left(\dfrac{2}{t^2}\right) + \dfrac{t^2}{1}\left(\dfrac{1}{t}\right)}{\dfrac{t^2}{1}\left(\dfrac{2}{t}\right) + t^2(1)} = \dfrac{2 + t}{2t + t^2} = \dfrac{2 + t}{t(2 + t)} = \dfrac{1}{t}$

24. $\dfrac{4km^{-1} - (km)^{-1}}{k^{-1} - 2} = \dfrac{\dfrac{4k}{m} - \dfrac{1}{km}}{\dfrac{1}{k} - 2} = \dfrac{km\left(\dfrac{4k}{m} - \dfrac{1}{km}\right)}{km\left(\dfrac{1}{k} - 2\right)} = \dfrac{\dfrac{km}{1}\left(\dfrac{4k}{m}\right) - \dfrac{km}{1}\left(\dfrac{1}{km}\right)}{\dfrac{km}{1}\left(\dfrac{1}{k}\right) - km(2)}$
$= \dfrac{4k^2 - 1}{m - 2km} = \dfrac{(2k + 1)(2k - 1)}{-m(2k - 1)} = \dfrac{-(2k + 1)}{m}$

25. $23.4 \times 10^{-5} = 0.000\,234$

26. $0.086 \times 10^4 = 860$

27. $4800 = 4.8 \times 10^3$

28. $0.004\,27 = 4.27 \times 10^{-3}$

29. $0.0063 = 6.3 \times 10^{-3}$

30. $520,000,000 = 5.2 \times 10^8$

31.
$$\frac{0.000\,000\,8\,(4.8 \times 10^{24})}{32,000,000\,(1.2 \times 10^{-15})}$$
$$= \frac{(8 \times 10^{-7}) \times (4.8 \times 10^{24})}{(3.2 \times 10^7) \times (1.2 \times 10^{-15})}$$
$$= \frac{(8 \times 4.8) \times (10^{-7} \times 10^{24})}{(3.2 \times 1.2) \times (10^7 \times 10^{-15})}$$
$$= \frac{38.4 \times 10^{17}}{3.84 \times 10^{-8}}$$
$$= \frac{38.4}{3.84} \times \frac{10^{17}}{10^{-8}}$$
$$= 10 \times 10^{25}$$
$$= 10^{26}$$

32.
$$\frac{3.6 \times 10^{20}}{(0.000\,000\,000\,25)\,(1.2 \times 10^{32})}$$
$$= \frac{3.6 \times 10^{20}}{(2.5 \times 10^{-10}) \times (1.2 \times 10^{32})}$$
$$= \frac{3.6 \times 10^{20}}{(2.5 \times 1.2) \times (10^{-10} \times 10^{32})}$$
$$= \frac{3.6 \times 10^{20}}{3 \times 10^{22}}$$
$$= \frac{3.6}{3} \times \frac{10^{20}}{10^{22}}$$
$$= 1.2 \times 10^{-2}$$
$$= 0.012$$

33. $\$16,000,000 \times 4$ quarters per dollar
$$= 1.6 \times 10^7 \times 4$$
$$= (1.6 \times 4) \times 10^7$$
$$= 6.4 \times 10^7 \text{ quarters}$$

$$\frac{6.4 \times 10^7 \text{ quarters}}{10 \text{ quarters per oz}}$$
$$= 6.4 \times \frac{10^7}{10^1}$$
$$= 6.4 \times 10^6$$
$$= 6,400,000 \text{ oz}$$
or $\frac{6,400,000}{16} = 400,000$ lb

34. $\dfrac{2100 \text{ kernels}}{14 \text{ oz}}$
$$= \frac{2.1 \times 10^3}{1.4 \times 10}$$
$$= \frac{2.1}{1.4} \times \frac{10^3}{10^1}$$
$$= 1.5 \times 10^2 \text{ kernels per oz}$$

$$\frac{1,000,000 \text{ kernels}}{1.5 \times 10^2 \text{ kernels per oz}}$$
$$= \frac{1 \times 10^6}{1.5 \times 10^2}$$
$$= \frac{1}{1.5} \times \frac{10^6}{10^2}$$
$$= 0.66666\ldots \times 10^4$$
$$\approx 6667 \text{ oz}$$

Homework 6.4

1. True **3.** False **5.** False **7.** $\sqrt{y^8} = y^4$ **9.** $\sqrt{n^{36}} = n^{18}$

11. $\pm\sqrt{16x^4} = \pm 4x^2$

13. $-\sqrt{121a^2b^6} = -11ab^3$

15. $\sqrt{9(x+y)^2} = 3(x+y)$

17. $-\sqrt{\dfrac{64}{b^6}} = \dfrac{-\sqrt{64}}{\sqrt{b^6}} = \dfrac{-8}{b^3}$

19. $\sqrt{8} = \sqrt{4 \cdot 2} = \sqrt{4} \cdot \sqrt{2} = 2\sqrt{2}$

21. $-\sqrt{20} = -\sqrt{4 \cdot 5} = -\sqrt{4} \cdot \sqrt{5} = -2\sqrt{5}$

23. $\sqrt{125} = \sqrt{25 \cdot 5} = \sqrt{25} \cdot \sqrt{5} = 5\sqrt{5}$

25. $\sqrt{x^3} = \sqrt{x^2 \cdot x} = \sqrt{x^2} \cdot \sqrt{x} = x\sqrt{x}$

27. $-\sqrt{b^{11}} = -\sqrt{b^{10} \cdot b} = -\sqrt{b^{10}} \cdot \sqrt{b} = -b^5\sqrt{b}$

29. $\sqrt{p^{25}} = \sqrt{p^{24} \cdot p} = \sqrt{p^{24}} \cdot \sqrt{p} = p^{12}\sqrt{p}$

31. $\sqrt{8a^3}$
$= \sqrt{4a^2 \cdot 2a}$
$= \sqrt{4a^2}\sqrt{2a}$
$= 2a\sqrt{2a}$

33. $\pm\sqrt{72m^9}$
$= \pm\sqrt{36m^8 \cdot 2m}$
$= \pm\sqrt{36m^8}\sqrt{2m}$
$= \pm 6m^4\sqrt{2m}$

35. $\sqrt{\dfrac{x^8}{27}}$
$= \dfrac{\sqrt{x^8}}{\sqrt{27}}$
$= \dfrac{\sqrt{x^8}}{\sqrt{9 \cdot 3}}$
$= \dfrac{\sqrt{x^8}}{\sqrt{9}\sqrt{3}}$
$= \dfrac{x^4}{3\sqrt{3}}$

37. $\sqrt{48c^6d}$
$= \sqrt{16c^6 \cdot 3d}$
$= \sqrt{16c^6}\sqrt{3d}$
$= 4c^3\sqrt{3d}$

39. $-\sqrt{\dfrac{45}{4}b^2d^3}$
$= \dfrac{-\sqrt{45b^2d^3}}{\sqrt{4}}$
$= \dfrac{-\sqrt{9b^2d^2 \cdot 5d}}{\sqrt{4}}$
$= \dfrac{-\sqrt{9b^2d^2}\sqrt{5d}}{\sqrt{4}}$
$= \dfrac{-3bd\sqrt{5d}}{2}$

41. $\sqrt{\dfrac{9w^3}{28z}}$
$= \dfrac{\sqrt{9w^3}}{\sqrt{28z}}$
$= \dfrac{\sqrt{9w^2 \cdot w}}{\sqrt{4 \cdot 7z}}$
$= \dfrac{\sqrt{9w^2}\sqrt{w}}{\sqrt{4}\sqrt{7z}}$
$= \dfrac{3w\sqrt{w}}{2\sqrt{7z}}$

43. $3\sqrt{4x^3}$
$= 3\sqrt{4x^2 \cdot x}$
$= 3\sqrt{4x^2}\sqrt{x}$
$= 3 \cdot 2x\sqrt{x}$
$= 6x\sqrt{x}$

45. $-2a\sqrt{50a^3b^2}$
$= -2a\sqrt{25a^2b^2 \cdot 2a}$
$= -2a\sqrt{25a^2b^2}\sqrt{2a}$
$= -2a \cdot 5ab\sqrt{2a}$
$= -10a^2b\sqrt{2a}$

47. $-\dfrac{2}{3k}\sqrt{9b^3k^5}$
$= \dfrac{-2}{3k} \cdot \dfrac{\sqrt{9b^2k^4 \cdot bk}}{1}$
$= \dfrac{-2}{3k} \cdot \dfrac{\sqrt{9b^2k^4}\sqrt{bk}}{1}$
$= \dfrac{-2}{3k} \cdot \dfrac{3bk^2\sqrt{bk}}{1}$
$= \dfrac{-2}{3\!\!\!/k} \cdot \dfrac{3\!\!\!/\,b\,\cancel{k}\,k\sqrt{bk}}{1}$
$= -2bk\sqrt{bk}$

49. $\sqrt{3} + 2\sqrt{3} = 3\sqrt{3}$

51. Cannot be simplified

53. $2\sqrt{6} - 9\sqrt{6} = -7\sqrt{6}$

55. $\sqrt{20} + \sqrt{45} - 2\sqrt{80}$
$= \sqrt{4 \cdot 5} + \sqrt{9 \cdot 5} - 2\sqrt{16 \cdot 5}$
$= \sqrt{4}\sqrt{5} + \sqrt{9}\sqrt{5} - 2\sqrt{16}\sqrt{5}$
$= 2\sqrt{5} + 3\sqrt{5} - 2 \cdot 4\sqrt{5}$
$= 2\sqrt{5} + 3\sqrt{5} - 8\sqrt{5}$
$= -3\sqrt{5}$

57. $\sqrt{3} - 2\sqrt{12} - \sqrt{18}$
$= \sqrt{3} - 2\sqrt{4 \cdot 3} - \sqrt{9 \cdot 2}$
$= \sqrt{3} - 2\sqrt{4}\sqrt{3} - \sqrt{9}\sqrt{2}$
$= \sqrt{3} - 2 \cdot 2\sqrt{3} - 3\sqrt{2}$
$= \sqrt{3} - 4\sqrt{3} - 3\sqrt{2}$
$= -3\sqrt{3} - 3\sqrt{2}$

59.
$$\sqrt{8a} + \sqrt{18a} - 7\sqrt{2a}$$
$$= \sqrt{4\cdot 2a} + \sqrt{9\cdot 2a} - 7\sqrt{2a}$$
$$= \sqrt{4}\sqrt{2a} + \sqrt{9}\sqrt{2a} - 7\sqrt{2a}$$
$$= 2\sqrt{2a} + 3\sqrt{2a} - 7\sqrt{2a}$$
$$= -2\sqrt{2a}$$

61.
$$2\sqrt{5x^3} - x\sqrt{125x} - 3\sqrt{20x^2}$$
$$= 2\sqrt{x^2\cdot 5x} - x\sqrt{25\cdot 5x} - 3\sqrt{4x^2\cdot 5}$$
$$= 2\sqrt{x^2}\sqrt{5x} - x\sqrt{25}\sqrt{5x} - 3\sqrt{4x^2}\sqrt{5}$$
$$= 2x\sqrt{5x} - x\cdot 5\sqrt{5x} - 3\cdot 2x\sqrt{5}$$
$$= 2x\sqrt{5x} - 5x\sqrt{5x} - 6x\sqrt{5}$$
$$= -3x\sqrt{5x} - 6x\sqrt{5}$$

63. The square root of a sum does not equal the sum of the square roots. Addition under the radical should be done first. $\sqrt{36+64} = \sqrt{100} = 10$

65. When adding like radicals, add their coefficients and leave the radicand unchanged.
$\sqrt{3} + \sqrt{3} = 2\sqrt{3}$

67. The square root of a sum does not equal the sum of the square roots. Because 9 and x^2 are not like terms, the addition cannot be done, therefore, $\sqrt{9+x^2}$ cannot be simplified.

69. (c) **71.** (b) **73.** (a) **75.** $(2\sqrt{5})^2 = 2^2(\sqrt{5})^2 = 4\cdot 5 = 20$
Therefore, $2\sqrt{5} = \sqrt{20}$

77. $(2\sqrt{3})^2 = 2^2(\sqrt{3})^2 = 4\cdot 3 = 12$
Therefore, $2\sqrt{3} = \sqrt{12}$

79. $(3\sqrt{6})^2 = 3^2(\sqrt{6})^2 = 9\cdot 6 = 54$
Therefore, $3\sqrt{6} = \sqrt{54}$

81. Valid

83. Invalid; $r^2 - s^2 \neq (r-s)^2$

Homework 6.5

1. $\sqrt{8}\sqrt{2} = \sqrt{16} = 4$

3.
$$\sqrt{2x}\sqrt{3x}$$
$$= \sqrt{6x^2}$$
$$= \sqrt{x^2\cdot 6}$$
$$= \sqrt{x^2}\sqrt{6}$$
$$= x\sqrt{6}$$

5.
$$(3\sqrt{8a})(a\sqrt{18a})$$
$$= 3a\sqrt{144a^2}$$
$$= 3a\cdot 12a$$
$$= 36a^2$$

7.
$$\sqrt{2}(3+\sqrt{3})$$
$$= \sqrt{2}\cdot 3 + \sqrt{2}\cdot\sqrt{3}$$
$$= 3\sqrt{2} + \sqrt{6}$$

9.
$$\sqrt{5}(4+2\sqrt{15})$$
$$= \sqrt{5}\cdot 4 + \sqrt{5}\cdot 2\sqrt{15}$$
$$= 4\sqrt{5} + 2\sqrt{75}$$
$$= 4\sqrt{5} + 2\sqrt{25\cdot 3}$$
$$= 4\sqrt{5} + 2\sqrt{25}\sqrt{3}$$
$$= 4\sqrt{5} + 2\cdot 5\sqrt{3}$$
$$= 4\sqrt{5} + 10\sqrt{3}$$

11.
$$2\sqrt{p}(\sqrt{2p} - p\sqrt{2})$$
$$= 2\sqrt{p}\cdot\sqrt{2p} - 2\sqrt{p}\cdot p\sqrt{2}$$
$$= 2\sqrt{2p^2} - 2p\sqrt{2p}$$
$$= 2\sqrt{p^2\cdot 2} - 2p\sqrt{2p}$$
$$= 2\sqrt{p^2}\sqrt{2} - 2p\sqrt{2p}$$
$$= 2p\sqrt{2} - 2p\sqrt{2p}$$

13. $(3 + \sqrt{2})(1 - \sqrt{2})$
$= 3 \cdot 1 - 3 \cdot \sqrt{2} + \sqrt{2} \cdot 1 - \sqrt{2} \cdot \sqrt{2}$
$= 3 - 3\sqrt{2} + \sqrt{2} - 2$
$= 1 - 2\sqrt{2}$

15. $(4 - \sqrt{a})(4 + \sqrt{a})$
$= (4)^2 - (\sqrt{a})^2$
$= 16 - a$

17. $(2 + \sqrt{3})^2$
$= (2)^2 + 2(2)(\sqrt{3}) + (\sqrt{3})^2$
$= 4 + 4\sqrt{3} + 3$
$= 7 + 4\sqrt{3}$

19. $(2\sqrt{w} + \sqrt{5})(\sqrt{w} - 2\sqrt{5})$
$= 2\sqrt{w} \cdot \sqrt{w} - 2\sqrt{w} \cdot 2\sqrt{5} + \sqrt{5} \cdot \sqrt{w} - \sqrt{5} \cdot 2\sqrt{5}$
$= 2\sqrt{w^2} - 4\sqrt{5w} + \sqrt{5w} - 2 \cdot 5$
$= 2w - 3\sqrt{5w} - 10$

21. $\dfrac{9 - 3\sqrt{5}}{3}$
$= \dfrac{\cancel{3}(3 - \sqrt{5})}{\cancel{3}}$
$= 3 - \sqrt{5}$

23. $\dfrac{-8 + \sqrt{8}}{4}$
$= \dfrac{-8 + \sqrt{4 \cdot 2}}{4}$
$= \dfrac{-8 + \sqrt{4}\sqrt{2}}{4}$
$= \dfrac{-8 + 2\sqrt{2}}{4}$
$= \dfrac{\cancel{2}(-4 + \sqrt{2})}{\cancel{2} \cdot 2}$
$= \dfrac{-4 + \sqrt{2}}{2}$

25. $\dfrac{6a - \sqrt{18}}{6a}$
$= \dfrac{6a - \sqrt{9 \cdot 2}}{6a}$
$= \dfrac{6a - \sqrt{9}\sqrt{2}}{6a}$
$= \dfrac{6a - 3\sqrt{2}}{6a}$
$= \dfrac{\cancel{3}(2a - \sqrt{2})}{2 \cdot \cancel{3} \cdot a}$
$= \dfrac{2a - \sqrt{2}}{2a}$

27. $(2x + 1)^2 = 8$ Extract square roots
$2x + 1 = \pm\sqrt{8}$ Simplify the radical
$2x + 1 = \pm\sqrt{4}\sqrt{2}$
$2x + 1 = \pm 2\sqrt{2}$ Subtract 1 from both sides
$2x = -1 \pm 2\sqrt{2}$ Divide both sides by 2
$x = \dfrac{-1 \pm 2\sqrt{2}}{2}$

29. $3(2x - 8)^2 = 60$ Divide both sides by 3
$(2x - 8)^2 = 20$ Extract square roots
$2x - 8 = \pm\sqrt{20}$ Simplify the radical
$2x - 8 = \pm\sqrt{4}\sqrt{5}$
$2x - 8 = \pm 2\sqrt{5}$ Add 8 to both sides
$2x = 8 \pm 2\sqrt{5}$ Divide both sides by 2
$x = \dfrac{8 \pm 2\sqrt{5}}{2}$ Factor numerator
$x = \dfrac{\cancel{2}(4 \pm \sqrt{5})}{\cancel{2}}$ Divide out common factor
$x = 4 \pm \sqrt{5}$

31. $\dfrac{4}{3}(x+3)^2 = 24$ Multiply both sides by 3

$4(x+3)^2 = 72$ Divide both sides by 4

$(x+3)^2 = 18$ Extract square roots

$x + 3 = \pm\sqrt{18}$ Simplify the radical

$x + 3 = \pm\sqrt{9}\sqrt{2}$

$x + 3 = \pm 3\sqrt{2}$ Subtract 3 from both sides

$x = -3 \pm 3\sqrt{2}$

33. $x^2 + a^2 = b^2$ Subtract a^2 from both sides

$x^2 + a^2 - a^2 = b^2 - a^2$

$x^2 = b^2 - a^2$ Extract square roots

$x = \pm\sqrt{b^2 - a^2}$

35. $\dfrac{x^2}{4} - y^2 = 1$ Add y^2 to both sides

$\dfrac{x^2}{4} - y^2 + y^2 = 1 + y^2$

$\dfrac{x^2}{4} = 1 + y^2$ Multiply both sides by 4

$4 \cdot \dfrac{x^2}{4} = 4(1 + y^2)$

$x^2 = 4(1 + y^2)$ Extract square roots

$x = \pm\sqrt{4(1 + y^2)}$ Simplify radical

$x = \pm\sqrt{4}\sqrt{1 + y^2}$

$x = \pm 2\sqrt{1 + y^2}$

37. $\dfrac{5}{4} + \dfrac{3\sqrt{2}}{2}$

$= \dfrac{5}{4} + \dfrac{3\sqrt{2} \cdot 2}{2 \cdot 2}$

$= \dfrac{5}{4} + \dfrac{6\sqrt{2}}{4}$

$= \dfrac{5 + 6\sqrt{2}}{4}$

39. $\dfrac{3}{2a} + \dfrac{\sqrt{3}}{6a}$

$= \dfrac{3 \cdot 3}{2a \cdot 3} + \dfrac{\sqrt{3}}{6a}$

$= \dfrac{9}{6a} + \dfrac{\sqrt{3}}{6a}$

$= \dfrac{9 + \sqrt{3}}{6a}$

41. $\dfrac{3\sqrt{3}}{2} + 3$

$= \dfrac{3\sqrt{3}}{2} + \dfrac{3 \cdot 2}{1 \cdot 2}$

$= \dfrac{3\sqrt{3}}{2} + \dfrac{6}{2}$

$= \dfrac{3\sqrt{3} + 6}{2}$

43. $\dfrac{3}{4} - 2\sqrt{y}$

$= \dfrac{3}{4} - \dfrac{2\sqrt{y} \cdot 4}{1 \cdot 4}$

$= \dfrac{3}{4} - \dfrac{8\sqrt{y}}{4}$

$= \dfrac{3 - 8\sqrt{y}}{4}$

45. Only like radicals can be added or subtracted. Since the radicands are different, $8\sqrt{7} - 2\sqrt{5}$ cannot be subtracted.

47. The product rule for radicals states that two numbers can be multiplied if both are under the radical. $6\sqrt{8} = \sqrt{36}\sqrt{8} = \sqrt{36 \cdot 8} = \sqrt{288}$

49. Since 5 is a term of the numerator, not a factor, it cannot be canceled. If we factor the numerator, then $\dfrac{5 - 10\sqrt{3}}{5} = \dfrac{5(1 - 2\sqrt{3})}{5} = 1 - 2\sqrt{3}$.

51. Check $t = \sqrt{3}$:
$$t^2 - 2\sqrt{3}t + 3 = 0$$
$$(\sqrt{3})^2 - 2\sqrt{3}(\sqrt{3}) + 3 \overset{?}{=} 0$$
$$3 - 2 \cdot 3 + 3 \overset{?}{=} 0$$
$$3 - 6 + 3 \overset{?}{=} 0$$
$$-3 + 3 \overset{?}{=} 0$$
$$0 = 0$$

53. Check $s = \sqrt{3} + 2$:
$$s^2 + 1 = 4s$$
$$(\sqrt{3} + 2)^2 + 1 \overset{?}{=} 4(\sqrt{3} + 2)$$
$$(\sqrt{3})^2 + 2(\sqrt{3})(2) + 2^2 + 1 \overset{?}{=} 4\sqrt{3} + 4 \cdot 2$$
$$3 + 4\sqrt{3} + 4 + 1 \overset{?}{=} 4\sqrt{3} + 8$$
$$8 + 4\sqrt{3} = 8 + 4\sqrt{3}$$

55. Check $x = \dfrac{-3 + \sqrt{5}}{2}$:
$$x^2 + 3x + 1 = 0$$
$$\left(\dfrac{-3 + \sqrt{5}}{2}\right)^2 + 3\left(\dfrac{-3 + \sqrt{5}}{2}\right) + 1 \overset{?}{=} 0$$
$$\dfrac{(-3)^2 + 2(-3)\sqrt{5} + (\sqrt{5})^2}{2^2} + \dfrac{3}{1} \cdot \dfrac{-3 + \sqrt{5}}{2} + 1 \overset{?}{=} 0$$
$$\dfrac{9 - 6\sqrt{5} + 5}{4} + \dfrac{-9 + 3\sqrt{5}}{2} + 1 \overset{?}{=} 0$$
$$\dfrac{14 - 6\sqrt{5}}{4} + \dfrac{-9 + 3\sqrt{5}}{2} + 1 \overset{?}{=} 0$$
$$\dfrac{2(7 - 3\sqrt{5})}{2 \cdot 2} + \dfrac{-9 + 3\sqrt{5}}{2} + 1 \overset{?}{=} 0$$
$$\dfrac{7 - 3\sqrt{5}}{2} + \dfrac{-9 + 3\sqrt{5}}{2} + 1 \overset{?}{=} 0$$
$$\dfrac{7 - 3\sqrt{5} - 9 + 3\sqrt{5}}{2} + 1 \overset{?}{=} 0$$
$$\dfrac{-2}{2} + 1 \overset{?}{=} 0$$
$$0 = 0$$

57. $\dfrac{\sqrt{18}}{\sqrt{2}} = \sqrt{\dfrac{18}{2}} = \sqrt{9} = 3$

59. $\dfrac{\sqrt{75x^3}}{\sqrt{3x}} = \sqrt{\dfrac{75x^3}{3x}} = \sqrt{25x^2} = 5x$

61. $\dfrac{\sqrt{48b}}{\sqrt{27b}} = \sqrt{\dfrac{48b}{27b}} = \sqrt{\dfrac{16}{9}} = \dfrac{4}{3}$

63. $\dfrac{5}{\sqrt{2}} = \dfrac{5 \cdot \sqrt{2}}{\sqrt{2} \cdot \sqrt{2}} = \dfrac{5\sqrt{2}}{2}$

65. $\dfrac{a\sqrt{2}}{\sqrt{a}}$

$= \dfrac{a\sqrt{2}\cdot\sqrt{a}}{\sqrt{a}\cdot\sqrt{a}}$

$= \dfrac{a\sqrt{2a}}{a}$

$= \sqrt{2a}$

67. $\dfrac{b\sqrt{21}}{\sqrt{3b}}$

$= \dfrac{b\sqrt{21}\cdot\sqrt{3b}}{\sqrt{3b}\cdot\sqrt{3b}}$

$= \dfrac{b\sqrt{63b}}{3b}$

$= \dfrac{b\sqrt{9\cdot 7b}}{3b}$

$= \dfrac{b\sqrt{9}\sqrt{7b}}{3b}$

$= \dfrac{b\cdot 3\sqrt{7b}}{3b}$

$= \dfrac{3b\sqrt{7b}}{3b}$

$= \sqrt{7b}$

69. $\sqrt{\dfrac{7x}{12}}$

$= \dfrac{\sqrt{7x}\cdot\sqrt{12}}{\sqrt{12}\cdot\sqrt{12}}$

$= \dfrac{\sqrt{84x}}{12}$

$= \dfrac{\sqrt{4\cdot 21x}}{12}$

$= \dfrac{\sqrt{4}\sqrt{21x}}{12}$

$= \dfrac{2\sqrt{21x}}{12}$

$= \dfrac{\sqrt{21x}}{6}$

71. $\text{LCD} = 3\sqrt{3}$

$\dfrac{\dfrac{\sqrt{5}}{3}}{\dfrac{1}{\sqrt{3}}} = \dfrac{3\sqrt{3}\left(\dfrac{\sqrt{5}}{3}\right)}{3\sqrt{3}\left(\dfrac{1}{\sqrt{3}}\right)} = \dfrac{\dfrac{3\sqrt{3}}{1}\left(\dfrac{\sqrt{5}}{3}\right)}{\dfrac{3\sqrt{3}}{1}\left(\dfrac{1}{\sqrt{3}}\right)} = \dfrac{\sqrt{15}}{3}$

73. $\text{LCD} = \sqrt{6}$

$\dfrac{1 - \dfrac{2}{\sqrt{6}}}{\dfrac{\sqrt{2}}{\sqrt{6}}} = \dfrac{\sqrt{6}\left(1 - \dfrac{2}{\sqrt{6}}\right)}{\sqrt{6}\left(\dfrac{\sqrt{2}}{\sqrt{6}}\right)} = \dfrac{\sqrt{6}(1) - \dfrac{\sqrt{6}}{1}\left(\dfrac{2}{\sqrt{6}}\right)}{\dfrac{\sqrt{6}}{1}\left(\dfrac{\sqrt{2}}{\sqrt{6}}\right)} = \dfrac{\sqrt{6} - 2}{\sqrt{2}}$

Rationalizing the denominator,

$\dfrac{(\sqrt{6} - 2)\sqrt{2}}{\sqrt{2}\cdot\sqrt{2}} = \dfrac{\sqrt{6}\sqrt{2} - 2\sqrt{2}}{2} = \dfrac{\sqrt{12} - 2\sqrt{2}}{2} = \dfrac{\sqrt{4}\sqrt{3} - 2\sqrt{2}}{2} = \dfrac{2\sqrt{3} - 2\sqrt{2}}{2}$

Factoring the numerator to reduce,

$\dfrac{2\sqrt{3} - 2\sqrt{2}}{2} = \dfrac{2(\sqrt{3} - \sqrt{2})}{2} = \sqrt{3} - \sqrt{2}$

75. $\text{LCD} = 2\sqrt{2}$

$\dfrac{1 + \dfrac{1}{\sqrt{2}}}{2\sqrt{3} - \dfrac{\sqrt{3}}{2}} = \dfrac{2\sqrt{2}\left(1 + \dfrac{1}{\sqrt{2}}\right)}{2\sqrt{2}\left(2\sqrt{3} - \dfrac{\sqrt{3}}{2}\right)} = \dfrac{2\sqrt{2}(1) + \dfrac{2\sqrt{2}}{1}\left(\dfrac{1}{\sqrt{2}}\right)}{2\sqrt{2}(2\sqrt{3}) - \dfrac{2\sqrt{2}}{1}\left(\dfrac{\sqrt{3}}{2}\right)} = \dfrac{2\sqrt{2} + 2}{4\sqrt{6} - \sqrt{6}} = \dfrac{2\sqrt{2} - 2}{3\sqrt{6}}$

Rationalizing the denominator,

$\dfrac{(2\sqrt{2} + 2)\sqrt{6}}{3\sqrt{6}\cdot\sqrt{6}} = \dfrac{2\sqrt{12} + 2\sqrt{6}}{3\cdot 6} = \dfrac{2\sqrt{4}\sqrt{3} + 2\sqrt{6}}{18} = \dfrac{2\cdot 2\sqrt{3} + 2\sqrt{6}}{18} = \dfrac{4\sqrt{3} + 2\sqrt{6}}{18}$

(Problem continued on next page)

Factoring the numerator to reduce,

$$\frac{4\sqrt{3}+2\sqrt{6}}{18} = \frac{2\left(2\sqrt{3}+\sqrt{6}\right)}{18} = \frac{2\sqrt{3}+\sqrt{6}}{9}$$

77a. In the right triangle, the hypotenuse is w, one leg is the height, h, and the other leg is half of the side, $\frac{w}{2}$.

$$a^2 + b^2 = c^2$$
$$h^2 + \left(\frac{w}{2}\right)^2 = w^2$$
$$h^2 + \frac{w^2}{4} = w^2$$
$$h^2 = w^2 - \frac{w^2}{4}$$
$$h^2 = \frac{4w^2}{4} - \frac{w^2}{4}$$
$$h^2 = \frac{3w^2}{4}$$
$$h = \sqrt{\frac{3w^2}{4}}$$
$$h = \frac{\sqrt{w^2}\sqrt{3}}{\sqrt{4}}$$
$$h = \frac{w\sqrt{3}}{2}$$

b.
$$A = \frac{1}{2}bh$$
$$= \frac{1}{2}(w)\left(\frac{w\sqrt{3}}{2}\right)$$
$$= \frac{w^2\sqrt{3}}{4}$$

79a. For the right triangle in the base of the pyramid, each leg is k.
$$a^2 + b^2 = c^2$$
$$k^2 + k^2 = c^2$$
$$2k^2 = c^2$$
$$\sqrt{2k^2} = c$$
$$\sqrt{k^2}\sqrt{2} = c$$
$$k\sqrt{2} = c$$

For the right triangle inside the pyramid, the hypotenuse is $2k$, one leg is the height, h, and the other leg is half of c, $\frac{k\sqrt{2}}{2}$.

$$a^2 + b^2 = c^2$$
$$h^2 + \left(\frac{k\sqrt{2}}{2}\right)^2 = (2k)^2$$
$$h^2 + \frac{k^2(\sqrt{2})^2}{2^2} = 4k^2$$
$$h^2 + \frac{k^2 \cdot 2}{4} = 4k^2$$
$$h^2 + \frac{k^2}{2} = 4k^2$$
$$h^2 = 4k^2 - \frac{k^2}{2}$$
$$h^2 = \frac{8k^2}{2} - \frac{k^2}{2}$$
$$h^2 = \frac{7k^2}{2}$$
$$h = \sqrt{\frac{7k^2}{2}}$$
$$h = \frac{\sqrt{k^2}\sqrt{7}}{\sqrt{2}}$$
$$h = \frac{k\sqrt{7} \cdot \sqrt{2}}{\sqrt{2} \cdot \sqrt{2}}$$
$$h = \frac{k\sqrt{14}}{2}$$

b.
$$V = \frac{1}{3}s^2h$$
$$= \frac{1}{3}(k)^2\left(\frac{k\sqrt{14}}{2}\right)$$
$$= \frac{k^3\sqrt{14}}{6}$$

Homework 6.6

1. $\sqrt{x+4} = 5$ Square both sides

$(\sqrt{x+4})^2 = 5^2$

$x + 4 = 25$ Subtract 4 from both sides

$x = 21$

Check $x = 21$:

$\sqrt{x+4} = 5$

$\sqrt{21+4} \overset{?}{=} 5$

$\sqrt{25} \overset{?}{=} 5$

$5 = 5$

3. $\sqrt{x} - 4 = 5$ Add 4 to both sides

$\sqrt{x} = 9$ Square both sides

$(\sqrt{x})^2 = 9^2$

$x = 81$

Check $x = 81$:

$\sqrt{x} - 4 = 5$

$\sqrt{81} - 4 \overset{?}{=} 5$

$9 - 4 \overset{?}{=} 5$

$5 = 5$

5. $6 - \sqrt{x} = 8$ Subtract 6 from both sides

$-\sqrt{x} = 2$ Divide both sides by -1

$\sqrt{x} = -2$ Square both sides

$(\sqrt{x})^2 = (-2)^2$

$x = 4$

Therefore, NO solution.

Check $x = 4$:

$6 - \sqrt{x} = 8$

$6 - \sqrt{4} \overset{?}{=} 8$

$6 - 2 \overset{?}{=} 8$

$4 \neq 8$

7. $2 + 3\sqrt{x-1} = 8$ Subtract 2 from both sides

$3\sqrt{x-1} = 6$ Divide both sides by 3

$\sqrt{x-1} = 2$ Square both sides

$(\sqrt{x-1})^2 = 2^2$

$x - 1 = 4$ Add 1 to both sides

$x = 5$

Check $x = 5$:

$2 + 3\sqrt{x-1} = 8$

$2 + 3\sqrt{5-1} \overset{?}{=} 8$

$2 + 3\sqrt{4} \overset{?}{=} 8$

$2 + 3 \cdot 2 \overset{?}{=} 8$

$2 + 6 \overset{?}{=} 8$

$8 = 8$

9. $2\sqrt{3x+1} - 3 = 5$ Add 3 to both sides

$2\sqrt{3x+1} = 8$ Divide both sides by 2

$\sqrt{3x+1} = 4$ Square both sides

$(\sqrt{3x+1})^2 = 4^2$

$3x + 1 = 16$ Subtract 1 from both sides

$3x = 15$ Divide both sides by 3

$x = 5$

Check $x = 5$:

$2\sqrt{3x+1} - 3 = 5$

$2\sqrt{3(5)+1} - 3 \overset{?}{=} 5$

$2\sqrt{15+1} - 3 \overset{?}{=} 5$

$2\sqrt{16} - 3 \overset{?}{=} 5$

$2 \cdot 4 - 3 \overset{?}{=} 5$

$8 - 3 \overset{?}{=} 5$

$5 = 5$

11a. $x = 16$

$$\sqrt{x} = 4$$
$$(\sqrt{x})^2 = 4^2$$
$$x = 16$$

b. $x \approx 6$

$$\sqrt{x} = 2.5$$
$$(\sqrt{x})^2 = 2.5^2$$
$$x = 6.25$$

c. No solution

$$\sqrt{x} = -2$$
$$(\sqrt{x})^2 = (-2)^2$$
$$x = 4$$

Check $x = 4$
$$\sqrt{x} = -2$$
$$\sqrt{4} \overset{?}{=} -2$$
$$2 \neq -2$$

d. $x \approx 28$

$$\sqrt{x} = 5.3$$
$$(\sqrt{x})^2 = 5.3^2$$
$$x = 28.09$$

13a.

x	4	5	6	10	16	19	24
y	0	1	1.4	2.4	3.5	3.9	4.5

b.
$$\sqrt{x-4} = 3$$
$$(\sqrt{x-4})^2 = 3^2$$
$$x - 4 = 9$$
$$x = 13$$

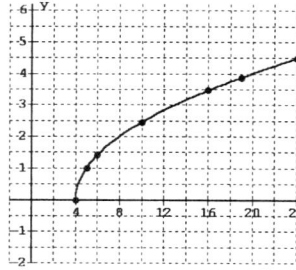

15a.

x	-3	-2	0	1	4	8	16
y	4	3	2.3	2	1.4	0.7	-0.4

b.
$$4 - \sqrt{x+3} = 1$$
$$-\sqrt{x+3} = -3$$
$$\sqrt{x+3} = 3$$
$$(\sqrt{x+3})^2 = 3^2$$
$$x + 3 = 9$$
$$x = 6$$

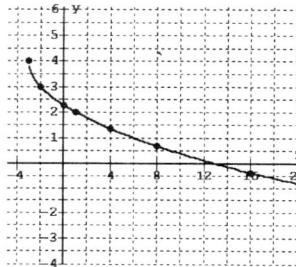

17.

$$\sqrt{x} = 3 - 2x \qquad \text{Square both sides}$$
$$(\sqrt{x})^2 = (3 - 2x)^2$$
$$x = (3)^2 - 2(3)(2x) + (2x)^2$$
$$x = 9 - 12x + 4x^2 \qquad \text{Write in standard form}$$
$$0 = 4x^2 - 13x + 9 \qquad \text{Factor}$$
$$0 = (4x - 9)(x - 1) \qquad \text{Set each factor equal to 0}$$
$$4x - 9 = 0 \quad \text{or} \quad x - 1 = 0$$
$$4x = 9 \qquad\qquad x = 1$$
$$x = \frac{9}{4}$$

Check $x = \frac{9}{4}$:
$$\sqrt{x} = 3 - 2x$$
$$\sqrt{\frac{9}{4}} \overset{?}{=} 3 - 2\left(\frac{9}{4}\right)$$
$$\frac{3}{2} \overset{?}{=} 3 - \frac{9}{2}$$
$$\frac{3}{2} \overset{?}{=} \frac{6}{2} - \frac{9}{2}$$
$$\frac{3}{2} \neq \frac{-3}{2}$$

Check $x = 1$:
$$\sqrt{x} = 3 - 2x$$
$$\sqrt{1} \overset{?}{=} 3 - 2(1)$$
$$1 \overset{?}{=} 3 - 2$$
$$1 = 1$$

Therefore, $x = 1$ is the only solution.

19.

$$\sqrt{x+4} + 2 = x \qquad \text{Subtract 2 from both sides}$$
$$\sqrt{x+4} = x - 2 \qquad \text{Square both sides}$$
$$(\sqrt{x+4})^2 = (x-2)^2$$
$$x + 4 = (x)^2 - 2(x)(2) + (2)^2$$
$$x + 4 = x^2 - 4x + 4 \qquad \text{Write in standard form}$$
$$0 = x^2 - 5x \qquad \text{Factor out } x$$
$$0 = x(x - 5) \qquad \text{Set each factor equal to 0}$$
$$x = 0 \quad \text{or} \quad x - 5 = 0$$
$$x = 5$$

Check $x = 0$:

$$\sqrt{x+4} + 2 = x$$
$$\sqrt{0+4} + 2 \overset{?}{=} 0$$
$$\sqrt{4} + 2 \overset{?}{=} 0$$
$$2 + 2 \overset{?}{=} 0$$
$$4 \neq 0$$

Check $x = 5$:

$$\sqrt{x+4} + 2 = x$$
$$\sqrt{5+4} + 2 \overset{?}{=} 5$$
$$\sqrt{9} + 2 \overset{?}{=} 5$$
$$3 + 2 \overset{?}{=} 5$$
$$5 = 5$$

Therefore, $x = 5$ is the only solution.

21.

$$x + \sqrt{2x+7} = -2 \qquad \text{Subtract } x \text{ from both sides}$$
$$\sqrt{2x+7} = -x - 2 \qquad \text{Square both sides}$$
$$(\sqrt{2x+7})^2 = (-x-2)^2$$
$$2x + 7 = (-x)^2 - 2(-x)(2) + (2)^2$$
$$2x + 7 = x^2 + 4x + 4 \qquad \text{Write in standard form}$$
$$0 = x^2 + 2x - 3 \qquad \text{Factor}$$
$$0 = (x + 3)(x - 1) \qquad \text{Set each factor equal to 0}$$
$$x + 3 = 0 \quad \text{or} \quad x - 1 = 0$$
$$x = -3 \qquad\qquad x = 1$$

Check $x = -3$:

$$x + \sqrt{2x+7} = -2$$
$$-3 + \sqrt{2(-3)+7} \overset{?}{=} -2$$
$$-3 + \sqrt{-6+7} \overset{?}{=} -2$$
$$-3 + \sqrt{1} \overset{?}{=} -2$$
$$-3 + 1 \overset{?}{=} -2$$
$$-2 = -2$$

Check $x = 1$:

$$x + \sqrt{2x+7} = -2$$
$$1 + \sqrt{2(1)+7} \overset{?}{=} -2$$
$$1 + \sqrt{2+7} \overset{?}{=} -2$$
$$1 + \sqrt{9} \overset{?}{=} -2$$
$$1 + 3 \overset{?}{=} -2$$
$$4 \neq -2$$

Therefore, $x = -3$ is the only solution.

23.

$$\sqrt{x+7} = 2x + 4 \qquad \text{Square both sides}$$
$$(\sqrt{x+7})^2 = (2x+4)^2$$
$$x + 7 = (2x)^2 + 2(2x)(4) + (4)^2$$
$$x + 7 = 4x^2 + 16x + 16 \qquad \text{Write in standard form}$$
$$0 = 4x^2 + 15x + 9 \qquad \text{Factor}$$
$$0 = (4x+3)(x+3) \qquad \text{Set each factor equal to 0}$$

$$4x + 3 = 0 \qquad \text{or} \qquad x + 3 = 0$$
$$4x = -3 \qquad\qquad\qquad x = -3$$
$$x = \frac{-3}{4}$$

Check $x = \frac{-3}{4}$:
$$\sqrt{x+7} = 2x + 4$$
$$\sqrt{\frac{-3}{4} + 7} \overset{?}{=} 2\left(\frac{-3}{4}\right) + 4$$
$$\sqrt{\frac{-3}{4} + \frac{28}{4}} \overset{?}{=} \frac{-3}{2} + 4$$
$$\sqrt{\frac{25}{4}} \overset{?}{=} \frac{-3}{2} + \frac{8}{2}$$
$$\frac{5}{2} = \frac{5}{2}$$

Check $x = -3$:
$$\sqrt{x+7} = 2x + 4$$
$$\sqrt{-3+7} \overset{?}{=} 2(-3) + 4$$
$$\sqrt{4} \overset{?}{=} -6 + 4$$
$$2 \neq -2$$

Therefore, $x = \frac{-3}{4}$ is the only solution.

25.

$$6 + \sqrt{5x - 4} - x = 4 \qquad \text{Subtract 6 and add } x \text{ to both sides}$$
$$\sqrt{5x - 4} = x - 2 \qquad \text{Square both sides}$$
$$(\sqrt{5x-4})^2 = (x-2)^2$$
$$5x - 4 = (x)^2 - 2(x)(2) + (2)^2$$
$$5x - 4 = x^2 - 4x + 4 \qquad \text{Write in standard form}$$
$$0 = x^2 - 9x + 8 \qquad \text{Factor}$$
$$0 = (x-8)(x-1) \qquad \text{Set each factor equal to 0}$$

$$x - 8 = 0 \qquad \text{or} \qquad x - 1 = 0$$
$$x = 8 \qquad\qquad\qquad x = 1$$

Check $x = 8$:
$$6 + \sqrt{5x - 4} - x = 4$$
$$6 + \sqrt{5(8) - 4} - 8 \overset{?}{=} 4$$
$$6 + \sqrt{40 - 4} - 8 \overset{?}{=} 4$$
$$6 + \sqrt{36} - 8 \overset{?}{=} 4$$
$$6 + 6 - 8 \overset{?}{=} 4$$
$$4 = 4$$

Check $x = 1$:
$$6 + \sqrt{5x - 4} - x = 4$$
$$6 + \sqrt{5(1) - 4} - 1 \overset{?}{=} 4$$
$$6 + \sqrt{5 - 4} - 1 \overset{?}{=} 4$$
$$6 + \sqrt{1} - 1 \overset{?}{=} 4$$
$$6 + 1 - 1 \overset{?}{=} 4$$
$$6 \neq 4$$

Therefore, $x = 8$ is the only solution.

Homework 6.6

13. a. Complete the table of values and graph $y = \sqrt{x - 4}$ on the grid.

x	4	5	6	10	16	19	24
y	0	1	1.4	2.4	3.5	3.9	4.5

b. Solve $\sqrt{x - 4} = 3$ graphically and algebraically. Do your answers agree?

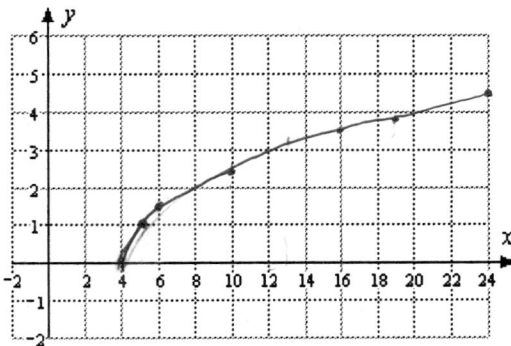

$$\left(\sqrt{x-4}\right)^2 = 3^2$$

$$x - 4 = 9$$

$$+4 \qquad +4$$

$$\boxed{x = 13}$$

Yes

14. a. Complete the table of values and graph $y = 2 - \sqrt{x}$ on the grid.

x	0	1	4	8	12	18	24
y	2	1	0	-.8	-1.5	-2.2	-2.9

b. Solve $2 - \sqrt{x} = -1$ graphically and algebraically. Do your answers agree?

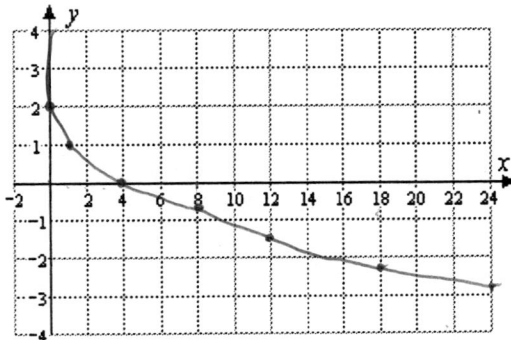

$$2 - \sqrt{x} = -1$$

$$-2 \qquad\qquad -2$$

$$\left(-\sqrt{x}\right)^2 = -3^2$$

$$\boxed{-x = 9}$$

15. a. Complete the table of values and graph $y = 4 - \sqrt{x + 3}$ on the grid.

x	-3	-2	0	1	4	8	16
y	4	3	2.3	2	1.4	0.7	-.04

b. Solve $4 - \sqrt{x + 3} = 1$ graphically and algebraically. Do your answers agree?

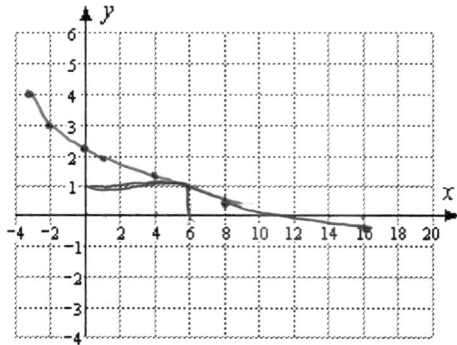

$$4 - \sqrt{x+3} = 1$$
$$\underline{-4 \qquad\qquad -4}$$
$$\left(-\sqrt{x+3}\right)^2 = (-3)^2$$
$$x + 3 = 9$$
$$\underline{-3 \qquad -3}$$
$$\boxed{x = 6}$$

16. a. Complete the table of values and graph $y = 2 + \sqrt{3x - 6}$ on the grid.

x	3	4	5	8	10	16	20
y	3.7	4.4	5	6.2	6.9	8.5	9.3

b. Solve $2 + \sqrt{3x - 6} = 8$ graphically and algebraically. Do your answers agree?

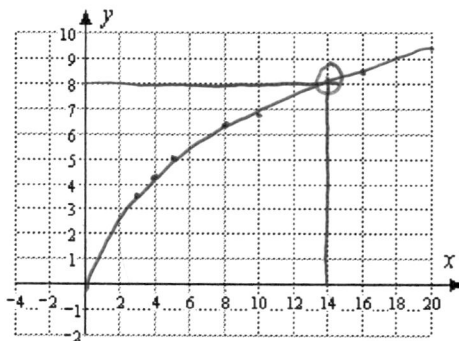

$$2 + \sqrt{3x-6} = 8 \qquad \boxed{Yes}$$
$$\underline{-2 \qquad\qquad -2}$$
$$\left(\sqrt{3x-6}\right)^2 = 6^2$$
$$3x - 6 = 36$$
$$\underline{+6 \qquad +6}$$
$$\frac{3x = 42}{3}$$
$$\boxed{x = 14}$$

27a. $d = \sqrt{12h}$
$d = \sqrt{12(1740)}$
$d = \sqrt{20{,}880}$
$d \approx 144.5$ km

b. $d = \sqrt{12h}$
$44 = \sqrt{12h}$
$(44)^2 = (\sqrt{12h})^2$
$1936 = 12h$
$\dfrac{1936}{12} = \dfrac{12h}{12}$
$161\frac{1}{3} = h$
It is at least $161\frac{1}{3}$ m.

29. $s = 3.9\sqrt{d}$
$615 = 3.9\sqrt{d}$
$\dfrac{615}{3.9} = \dfrac{3.9\sqrt{d}}{3.9}$
$\left(\dfrac{615}{3.9}\right)^2 = (\sqrt{d})^2$
$\dfrac{615^2}{3.9^2} = d$
$24{,}867 \approx d$
It is at least 24,867 ft.

31a. $r = \sqrt[3]{\dfrac{V}{12.57}}$
$r = \sqrt[3]{\dfrac{340}{12.57}}$
$r = \sqrt[3]{27.04852824\ldots}$
$r \approx 3$ in.

b. $r = \sqrt[3]{\dfrac{V}{12.57}}$
$5.5 = \sqrt[3]{\dfrac{V}{12.57}}$
$5.5^3 = \left(\sqrt[3]{\dfrac{V}{12.57}}\right)^3$
$166.375 = \dfrac{V}{12.57}$
$12.57(166.375) = 12.57\left(\dfrac{V}{12.57}\right)$
2091 in.$^3 \approx V$

33. $y = \sqrt{9 - x^2}$
$2 = \sqrt{9 - x^2}$ Square both sides
$2^2 = (\sqrt{9 - x^2})^2$
$4 = 9 - x^2$ Add x^2 and subtract 4 from both sides
$x^2 = 5$ Exract square roots
$x = \pm\sqrt{5}$
$x = \sqrt{5}$ or $-\sqrt{5}$

35. $2\sqrt[3]{x} + 15 = 5$ Subtract 15 from both sides
$2\sqrt[3]{x} = -10$ Divide both sides by 2
$\sqrt[3]{x} = -5$ Cube both sides
$(\sqrt[3]{x})^3 = (-5)^3$
$x = -125$

37. $\sqrt[3]{2x - 5} - 1 = 2$ Add 1 to both sides
$\sqrt[3]{2x - 5} = 3$ Cube both sides
$(\sqrt[3]{2x - 5})^3 = 3^3$
$2x - 5 = 27$ Add 5 to both sides
$2x = 32$ Divide both sides by 2
$x = 16$

39.

$$2 = 8 - 3\sqrt[3]{x^3 + 1}$$ Subtract 8 from both sides

$$-6 = -3\sqrt[3]{x^3 + 1}$$ Divide both sides by -3

$$2 = \sqrt[3]{x^3 + 1}$$ Cube both sides

$$2^3 = (\sqrt[3]{x^3 + 1})^3$$

$$8 = x^3 + 1$$ Subtract 1 from both sides

$$7 = x^3$$ Extract cube roots

$$\sqrt[3]{7} = x$$

Chapter 6 Summary and Review

1. <u>Third law of exponents:</u> $(a^m)^n = a^{m \cdot n}$
To raise a power to a power, multiply the exponents.

<u>First law of exponents:</u> $a^m \cdot a^n = a^{m+n}$
To multiply powers with the same base, add the exponents.

2. <u>Fourth law of exponents:</u> $(ab)^n = a^n b^n$

Example: $(2x)^3 = 2^3 x^3 = 8x^3$

<u>Fifth law of exponents:</u> $\left(\dfrac{a}{b}\right)^n = \dfrac{a^n}{b^n}$

Example: $\left(\dfrac{3}{x}\right)^2 = \dfrac{3^2}{x^2} = \dfrac{9}{x^2}$

3. <u>Negative exponent:</u> $a^{-n} = \dfrac{1}{a^n}$

Example: $5^{-2} = \dfrac{1}{5^2} = \dfrac{1}{25}$

4. <u>Second law of exponents:</u> $\dfrac{a^m}{a^n} = a^{m-n}$
To divide powers with the same base, subtract the exponents.

5. $1 = \dfrac{2^5}{2^5} = 2^{5-5} = 2^0$

6. A number is written in <u>scientific notation</u> if it is expressed as the product of a number between 1 and 10 times a power of 10.

7. <u>Product rule for radicals:</u> $\sqrt{ab} = \sqrt{a}\sqrt{b}$

<u>Quotient rule for radicals:</u> $\sqrt{\dfrac{a}{b}} = \dfrac{\sqrt{a}}{\sqrt{b}}$

8. $\sqrt{a+b} \neq \sqrt{a} + \sqrt{b};$

$\sqrt{a-b} \neq \sqrt{a} - \sqrt{b}$

9. <u>To simplify a square root:</u>
 1. Factor any perfect squares from the radicand.
 2. Use the product rule to write the radical as a product of two roots.
 3. Simplify the square root of the perfect squares.

10. We can simplify a sum or difference of square roots if they have identical radicands. To add or subtract the like radicals, add or subtract their coefficients and leave the radicand unchanged.

11. <u>To rationalize the denominator,</u> multiply the numerator and denominator by any radicals in the denominator that cannot be simplified.

12. <u>To solve a radical equation:</u>
 1. Isolate the radical.
 2. If the radical is a square root, square both sides. If the radical is a cube root, cube both sides.
  3. Solve the resulting equation.
 4. Check apparent solutions in the original equation. (The technique of squaring both sides may introduce extraneous solutions.)

13a. $a^4 \cdot a^6 = a^{4+6} = a^{10}$
  b. $(a^4)^6 = a^{4\cdot6} = a^{24}$

14a. $\dfrac{a^4}{a^6} = a^{4-6} = a^{-2} = \dfrac{1}{a^2}$
  b. $\dfrac{a^6}{a^4} = a^{6-4} = a^2$

15a. $(2a^2)^3 = 2^3(a^2)^3 = 8a^{2\cdot3} = 8a^6$
  b. $2a^2(a^2)^3 = 2a^2 \cdot a^{2\cdot3} = 2a^2 \cdot a^6 = 2a^{2+6} = 2a^8$

16a. $\left(\dfrac{-3u}{v^2}\right)^4 = \dfrac{(-3)^4 u^4}{(v^2)^4} = \dfrac{81u^4}{v^{2\cdot4}} = \dfrac{81u^4}{v^8}$
  b. $\dfrac{-3u^4}{v^2(v^4)} = \dfrac{-3u^4}{v^{2+4}} = \dfrac{-3u^4}{v^6}$

17. $-4x(-2x^2)^3$
   $= -4 \cdot x \cdot (-2)^3 \cdot (x^2)^3$
   $= -4 \cdot x \cdot (-8) \cdot x^6$
   $= 32x^7$

18. $-3w^2(-w^3)^2$
   $= -3 \cdot w^2 \cdot w^6$
   $= -3w^8$

19. $4t^2(t^2)^3 - (6t^4)^2$
   $= 4 \cdot t^2 \cdot t^6 - 6^2 \cdot (t^4)^2$
   $= 4 \cdot t^2 \cdot t^6 - 36 \cdot t^8$
   $= 4t^8 - 36t^8$
   $= -32t^8$

20. $(3v)^3(-v^3) - (2v)^2(-v^4)$
   $= 3^3 \cdot v^3 \cdot (-v^3) - 2^2 \cdot v^2 \cdot (-v^4)$
   $= 27 \cdot v^3 \cdot (-v^3) - 4 \cdot v^2 \cdot (-v^4)$
   $= -27v^6 + 4v^6$
   $= -23v^6$

21a. $3x^{-2} = 3 \cdot \dfrac{1}{x^2} = \dfrac{3}{x^2}$
  b. $(3x)^{-2} = \dfrac{1}{(3x)^2} = \dfrac{1}{3^2x^2} = \dfrac{1}{9x^2}$

22a. $(4y)^0 = 1$
  b. $4y^0 = 4 \cdot 1 = 4$

23a. $\left(\dfrac{5}{z}\right)^{-2} = \left(\dfrac{z}{5}\right)^2 = \dfrac{z^2}{5^2} = \dfrac{z^2}{25}$
  b. $\dfrac{5}{z^{-2}} = 5 \cdot z^2 = 5z^2$

24a. $\dfrac{16c^{-4}}{8c^{-8}} = \dfrac{16}{8} \cdot c^{-4-(-8)} = 2c^{-4+8} = 2c^4$
  b. $\dfrac{16c^{-4}}{-8c^8} = \dfrac{16}{-8} \cdot c^{-4-8} = -2c^{-12} = \dfrac{-2}{c^{12}}$

25. $3p^{-4}(2p^{-3})$
   $= (3 \cdot 2) \cdot p^{-4+(-3)}$
   $= 6 \cdot p^{-7}$
   $= \dfrac{6}{p^7}$

26. $2q^{-4}(2q)^{-3}$
   $= 2 \cdot q^{-4} \cdot 2^{-3} \cdot q^{-3}$
   $= 2^{1+(-3)} \cdot q^{-4+(-3)}$
   $= 2^{-2} \cdot q^{-7}$
   $= \dfrac{1}{2^2 q^7}$
   $= \dfrac{1}{4q^7}$

27. $\dfrac{(4k^{-3})^2}{2k^{-5}}$
   $= \dfrac{4^2 \cdot (k^{-3})^2}{2 \cdot k^{-5}}$
   $= \dfrac{16 \cdot k^{-3(2)}}{2 \cdot k^{-5}}$
   $= \dfrac{16 \cdot k^{-6}}{2 \cdot k^{-5}}$
   $= \dfrac{16}{2} \cdot k^{-6-(-5)}$
   $= 8 \cdot k^{-6+5}$
   $= 8 \cdot k^{-1}$
   $= \dfrac{8}{k}$

28.
$$\frac{6h^{-4}(2h^{-2})}{3h^{-3}}$$
$$= \frac{(6 \cdot 2) \cdot h^{-4+(-2)}}{3 \cdot h^{-3}}$$
$$= \frac{12 \cdot h^{-6}}{3 \cdot h^{-3}}$$
$$= \frac{12}{3} \cdot h^{-6-(-3)}$$
$$= 4 \cdot h^{-6+3}$$
$$= 4 \cdot h^{-3}$$
$$= \frac{4}{h^3}$$

29.
$$5g^{-6}(g^{-3})^{-2}$$
$$= 5 \cdot g^{-6} \cdot g^{-3(-2)}$$
$$= 5 \cdot g^{-6} \cdot g^{6}$$
$$= 5 \cdot g^{-6+6}$$
$$= 5 \cdot g^{0}$$
$$= 5 \cdot 1$$
$$= 5$$

30.
$$(8n)^{-2}(n^{-3})^{-4}$$
$$= 8^{-2} \cdot n^{-2} \cdot n^{-3(-4)}$$
$$= 8^{-2} \cdot n^{-2} \cdot n^{12}$$
$$= 8^{-2} \cdot n^{-2+12}$$
$$= 8^{-2} \cdot n^{10}$$
$$= \frac{n^{10}}{8^2}$$
$$= \frac{n^{10}}{64}$$

31.
$$\frac{2}{x^{-1}+y^{-1}} = \frac{2}{\dfrac{1}{x}+\dfrac{1}{y}} = \frac{xy(2)}{xy\left(\dfrac{1}{x}+\dfrac{1}{y}\right)} = \frac{xy(2)}{\dfrac{xy}{1}\left(\dfrac{1}{x}\right)+\dfrac{xy}{1}\left(\dfrac{1}{y}\right)} = \frac{2xy}{y+x}$$

32.
$$\frac{a^{-2}-b^{-2}}{a-b} = \frac{\dfrac{1}{a^2}-\dfrac{1}{b^2}}{a-b} = \frac{a^2b^2\left(\dfrac{1}{a^2}-\dfrac{1}{b^2}\right)}{a^2b^2(a-b)} = \frac{\dfrac{a^2b^2}{1}\left(\dfrac{1}{a^2}\right)-\dfrac{a^2b^2}{1}\left(\dfrac{1}{b^2}\right)}{a^2b^2(a)-a^2b^2(b)} = \frac{b^2-a^2}{a^3b^2-a^2b^3}$$

Factoring the numerator and the denominator to reduce,
$$= \frac{(b+a)(b-a)}{a^2b^2(a-b)} = \frac{(b+a)(-1)(a-b)}{a^2b^2(a-b)} = \frac{-(b+a)}{a^2b^2}$$

33. $586{,}000 = 5.86 \times 10^5$

34. $12{,}400{,}000 = 1.24 \times 10^7$

35. $0.0007 = 7 \times 10^{-4}$

36. $0.000\,009 = 9 \times 10^{-6}$

37.
$$483 \times 10^3$$
$$= (4.83 \times 10^2) \times 10^3$$
$$= 4.83 \times (10^2 \times 10^3)$$
$$= 4.83 \times 10^5$$

38.
$$0.0035 \times 10^2$$
$$= (3.5 \times 10^{-3}) \times 10^2$$
$$= 3.5 \times (10^{-3} \times 10^2)$$
$$= 3.5 \times 10^{-1}$$

39.
$$(48{,}000{,}000)(380{,}000{,}000)$$
$$= (4.8 \times 10^7) \times (3.8 \times 10^8)$$
$$= (4.8 \times 3.8) \times (10^7 \times 10^8)$$
$$= 18.24 \times 10^{15}$$
$$= 18{,}240{,}000{,}000{,}000{,}000$$

40.
$$(0.000\,002\,4)(1{,}900{,}000{,}000)$$
$$= (2.4 \times 10^{-6}) \times (1.9 \times 10^9)$$
$$= (2.4 \times 1.9) \times (10^{-6} \times 10^9)$$
$$= 4.56 \times 10^3$$
$$= 4{,}560$$

41.
$$= \frac{0.000\,000\,005}{0.0002}$$
$$= \frac{5 \times 10^{-9}}{2 \times 10^{-4}}$$
$$= \frac{5}{2} \times \frac{10^{-9}}{10^{-4}}$$
$$= 2.5 \times 10^{-5}$$
$$= 0.000\,025$$

42.
$$= \frac{38,500,000}{(0.0008)(0.0017)}$$
$$= \frac{3.85 \times 10^7}{(8 \times 10^{-4})(1.7 \times 10^{-3})}$$
$$= \frac{3.85 \times 10^7}{(8 \times 1.7) \times (10^{-4} \times 10^{-3})}$$
$$= \frac{3.85 \times 10^7}{13.6 \times 10^{-7}}$$
$$= \frac{3.85}{13.6} \times \frac{10^7}{10^{-7}}$$
$$\approx 0.283 \times 10^{14}$$
$$\approx 28,300,000,000,000$$

43. $(1.66 \times 10^{-27}) \times (6.02 \times 10^{23})$
$$= 1.66 \boxed{\text{EE}} \ 27 \ \boxed{+/-} \ \boxed{\times} \ 6.02 \ \boxed{\text{EE}} \ 23 \ \boxed{=}$$
$$= \boxed{0.00099932}$$
$$\approx 9.99 \times 10^{-4} \text{ kg}$$

44. $(1.67 \times 10^{-27}) \div (9.11 \times 10^{-31})$
$$= 1.67 \ \boxed{\text{EE}} \ 27 \ \boxed{+/-} \ \boxed{\div} \ 9.11 \ \boxed{\text{EE}} \ 31 \ \boxed{+/-} \ \boxed{=}$$
$$= \boxed{1.833150384 \quad 03}$$
$$\approx 1.83 \times 10^3 \text{ electrons}$$

45a. $\sqrt{4x^6} = 2x^3$
b. $\sqrt{4 + x^6}$ cannot be simplified.
c. $\sqrt{(4 + x)^6} = (4 + x)^3$

46a. $\sqrt{1 - w^9}$ cannot be simplified.
b. $\sqrt{1 - w^8}$ cannot be simplified.
c. $\sqrt{-w^8}$ is the square root of a negative number which is undefined.

47. $-\sqrt{27m^5}$
$$= -\sqrt{9m^4 \cdot 3m}$$
$$= -\sqrt{9m^4}\sqrt{3m}$$
$$= -3m^2\sqrt{3m}$$

48. $\pm\sqrt{98q^{99}}$
$$= \pm\sqrt{49q^{98} \cdot 2q}$$
$$= \pm\sqrt{49q^{98}}\sqrt{2q}$$
$$= \pm 7q^{49}\sqrt{2q}$$

49. $\sqrt{\dfrac{a^3c}{16}}$
$$= \frac{\sqrt{a^3c}}{\sqrt{16}}$$
$$= \frac{\sqrt{a^2 \cdot ac}}{\sqrt{16}}$$
$$= \frac{\sqrt{a^2}\sqrt{ac}}{\sqrt{16}}$$
$$= \frac{a\sqrt{ac}}{4}$$

50. $\sqrt{\dfrac{50b^7}{2g^4}}$
$$= \sqrt{\frac{25b^7}{g^4}}$$
$$= \frac{\sqrt{25b^7}}{\sqrt{g^4}}$$
$$= \frac{\sqrt{25b^6 \cdot b}}{\sqrt{g^4}}$$
$$= \frac{\sqrt{25b^6}\sqrt{b}}{\sqrt{g^4}}$$
$$= \frac{5b^3\sqrt{b}}{g^2}$$

51. $\dfrac{2}{3}b\sqrt{12b^3}$

$= \dfrac{2b}{3} \cdot \dfrac{\sqrt{4b^2 \cdot 3b}}{1}$

$= \dfrac{2b}{3} \cdot \dfrac{\sqrt{4b^2}\sqrt{3b}}{1}$

$= \dfrac{2b}{3} \cdot \dfrac{2b\sqrt{3b}}{1}$

$= \dfrac{4b^2\sqrt{3b}}{3}$

52. $\dfrac{4}{3a^2}\sqrt{45a^3}$

$= \dfrac{4}{3a^2} \cdot \dfrac{\sqrt{9a^2 \cdot 5a}}{1}$

$= \dfrac{4}{3a^2} \cdot \dfrac{\sqrt{9a^2}\sqrt{5a}}{1}$

$= \dfrac{4}{\cancel{3} \cancel{a} a} \cdot \dfrac{\cancel{3}\cancel{a}\sqrt{5a}}{1}$

$= \dfrac{4\sqrt{5a}}{a}$

53. $3\sqrt{24} + 2\sqrt{18} - 5\sqrt{6}$

$= 3\sqrt{4 \cdot 6} + 2\sqrt{9 \cdot 2} - 5\sqrt{6}$

$= 3\sqrt{4}\sqrt{6} + 2\sqrt{9}\sqrt{2} - 5\sqrt{6}$

$= 3 \cdot 2\sqrt{6} + 2 \cdot 3\sqrt{2} - 5\sqrt{6}$

$= 6\sqrt{6} + 6\sqrt{2} - 5\sqrt{6}$

$= \sqrt{6} + 6\sqrt{2}$

54. $2x\sqrt{x} - 3\sqrt{x^3} - 6\sqrt{x}$

$= 2x\sqrt{x} - 3\sqrt{x^2 \cdot x} - 6\sqrt{x}$

$= 2x\sqrt{x} - 3\sqrt{x^2}\sqrt{x} - 6\sqrt{x}$

$= 2x\sqrt{x} - 3x\sqrt{x} - 6\sqrt{x}$

$= -x\sqrt{x} - 6\sqrt{x}$

or $(-x - 6)\sqrt{x}$

55. $\dfrac{\sqrt{54w^{12}}}{\sqrt{9w^6}}$

$= \sqrt{\dfrac{54w^{12}}{9w^6}}$

$= \sqrt{6w^6}$

$= \sqrt{w^6 \cdot 6}$

$= \sqrt{w^6}\sqrt{6}$

$= w^3\sqrt{6}$

56. $\dfrac{\sqrt{24n^3}}{\sqrt{6n^5}}$

$= \sqrt{\dfrac{24n^3}{6n^5}}$

$= \sqrt{\dfrac{4}{n^2}}$

$= \dfrac{\sqrt{4}}{\sqrt{n^2}}$

$= \dfrac{2}{n}$

57. $\dfrac{6 - 3\sqrt{12}}{3}$

$= \dfrac{6 - 3\sqrt{4 \cdot 3}}{3}$

$= \dfrac{6 - 3\sqrt{4}\sqrt{3}}{3}$

$= \dfrac{6 - 3 \cdot 2\sqrt{3}}{3}$

$= \dfrac{6 - 6\sqrt{3}}{3}$

$= \dfrac{\cancel{3}(2 - 2\sqrt{3})}{\cancel{3}}$

$= 2 - 2\sqrt{3}$

58. $\dfrac{\sqrt{8} - \sqrt{12}}{6}$

$= \dfrac{\sqrt{4 \cdot 2} - \sqrt{4 \cdot 3}}{6}$

$= \dfrac{\sqrt{4}\sqrt{2} - \sqrt{4}\sqrt{3}}{6}$

$= \dfrac{2\sqrt{2} - 2\sqrt{3}}{6}$

$= \dfrac{\cancel{2}(\sqrt{2} - \sqrt{3})}{\cancel{2} \cdot 3}$

$= \dfrac{\sqrt{2} - \sqrt{3}}{3}$

59. $\dfrac{2}{3} - \dfrac{\sqrt{3}}{2}$

$= \dfrac{2 \cdot 2}{3 \cdot 2} - \dfrac{\sqrt{3} \cdot 3}{2 \cdot 3}$

$= \dfrac{4}{6} - \dfrac{3\sqrt{3}}{6}$

$= \dfrac{4 - 3\sqrt{3}}{6}$

60. $\dfrac{2\sqrt{3}}{5} - 1$

$= \dfrac{2\sqrt{3}}{5} - \dfrac{1 \cdot 5}{1 \cdot 5}$

$= \dfrac{2\sqrt{3}}{5} - \dfrac{5}{5}$

$= \dfrac{2\sqrt{3} - 5}{5}$

61.

$(3a - 2)^2 = 24$	Extract square roots
$3a - 2 = \pm\sqrt{24}$	Simplify the radical
$3a - 2 = \pm\sqrt{4}\sqrt{6}$	
$3a - 2 = \pm 2\sqrt{6}$	Add 2 to both sides
$3a = 2 \pm 2\sqrt{6}$	Divide both sides by 3
$a = \dfrac{2 \pm 2\sqrt{6}}{3}$	

62.

$$5(2d+1)^2 = 90 \qquad \text{Divide both sides by 5}$$
$$(2d+1)^2 = 18 \qquad \text{Extract square roots}$$
$$2d+1 = \pm\sqrt{18} \qquad \text{Simplify the radical}$$
$$2d+1 = \pm\sqrt{9}\sqrt{2}$$
$$2d+1 = \pm3\sqrt{2} \qquad \text{Subtract 1 from both sides}$$
$$2d = -1\pm3\sqrt{2} \qquad \text{Divide both sides by 2}$$
$$d = \frac{-1\pm3\sqrt{2}}{2}$$

63.

$$2a^2 + 4b^2 = c^2 \qquad \text{Subtract } 2a^2 \text{ from both sides}$$
$$2a^2 + 4b^2 - 2a^2 = c^2 - 2a^2$$
$$4b^2 = c^2 - 2a^2 \qquad \text{Divide both sides by 4}$$
$$\frac{4b^2}{4} = \frac{c^2 - 2a^2}{4}$$
$$b^2 = \frac{c^2 - 2a^2}{4} \qquad \text{Extract square roots}$$
$$b = \pm\sqrt{\frac{c^2 - 2a^2}{4}} \qquad \text{Quotient rule}$$
$$b = \frac{\pm\sqrt{c^2 - 2a^2}}{\sqrt{4}}$$
$$b = \frac{\pm\sqrt{c^2 - 2a^2}}{2}$$

64.

$$25w^2 - k = 16m \qquad \text{Add } k \text{ to both sides}$$
$$25w^2 - k + k = 16m + k$$
$$25w^2 = 16m + k \qquad \text{Divide both sides by 25}$$
$$\frac{25w^2}{25} = \frac{16m + k}{25}$$
$$w^2 = \frac{16m + k}{25} \qquad \text{Extract square roots}$$
$$w = \pm\sqrt{\frac{16m + k}{25}} \qquad \text{Quotient rule}$$
$$w = \frac{\pm\sqrt{16m + k}}{\sqrt{25}}$$
$$w = \frac{\pm\sqrt{16m + k}}{5}$$

65.
$$\sqrt{3}(\sqrt{2} - \sqrt{6})$$
$$= \sqrt{3}\cdot\sqrt{2} - \sqrt{3}\cdot\sqrt{6}$$
$$= \sqrt{6} - \sqrt{18}$$
$$= \sqrt{6} - \sqrt{9\cdot 2}$$
$$= \sqrt{6} - \sqrt{9}\sqrt{2}$$
$$= \sqrt{6} - 3\sqrt{2}$$

66.
$$3\sqrt{2}(8\sqrt{6} - \sqrt{12})$$
$$= 3\sqrt{2}\cdot 8\sqrt{6} - 3\sqrt{2}\cdot\sqrt{12}$$
$$= 24\sqrt{12} - 3\sqrt{24}$$
$$= 24\sqrt{4\cdot 3} - 3\sqrt{4\cdot 6}$$
$$= 24\sqrt{4}\sqrt{3} - 3\sqrt{4}\sqrt{6}$$
$$= 24\cdot 2\sqrt{3} - 3\cdot 2\sqrt{6}$$
$$= 48\sqrt{3} - 6\sqrt{6}$$

67. $(2 - \sqrt{d})(2 + \sqrt{d})$
$= (2)^2 - (\sqrt{d})^2$
$= 4 - d$

68. $(5 - 3\sqrt{2})(3 + \sqrt{2})$
$= 5 \cdot 3 + 5 \cdot \sqrt{2} - 3\sqrt{2} \cdot 3 - 3\sqrt{2} \cdot \sqrt{2}$
$= 15 + 5\sqrt{2} - 9\sqrt{2} - 3 \cdot 2$
$= 15 + 5\sqrt{2} - 9\sqrt{2} - 6$
$= 9 - 4\sqrt{2}$

69. $(\sqrt{7} + 3)^2$
$= (\sqrt{7})^2 + 2(\sqrt{7})(3) + (3)^2$
$= 7 + 6\sqrt{7} + 9$
$= 16 + 6\sqrt{7}$

70. $(3\sqrt{t} + 1)^2$
$= (3\sqrt{t})^2 + 2(3\sqrt{t})(1) + (1)^2$
$= 3^2(\sqrt{t})^2 + 6\sqrt{t} + 1$
$= 9t + 6\sqrt{t} + 1$

71. $\dfrac{2}{\sqrt{x}} = \dfrac{2 \cdot \sqrt{x}}{\sqrt{x} \cdot \sqrt{x}} = \dfrac{2\sqrt{x}}{x}$

72. $\sqrt{\dfrac{3a}{b}} = \dfrac{\sqrt{3a}}{\sqrt{b}} = \dfrac{\sqrt{3a} \cdot \sqrt{b}}{\sqrt{b} \cdot \sqrt{b}} = \dfrac{\sqrt{3ab}}{b}$

73. $\dfrac{2\sqrt{5}}{\sqrt{8}}$
$= \dfrac{2\sqrt{5} \cdot \sqrt{8}}{\sqrt{8} \cdot \sqrt{8}}$
$= \dfrac{2\sqrt{40}}{8}$
$= \dfrac{2\sqrt{4}\sqrt{10}}{8}$
$= \dfrac{2 \cdot 2\sqrt{10}}{8}$
$= \dfrac{4\sqrt{10}}{8}$
$= \dfrac{\sqrt{10}}{2}$

74. $\dfrac{a\sqrt{32}}{\sqrt{2a}}$
$= \dfrac{a\sqrt{32} \cdot \sqrt{2a}}{\sqrt{2a} \cdot \sqrt{2a}}$
$= \dfrac{a\sqrt{64a}}{2a}$
$= \dfrac{a\sqrt{64}\sqrt{a}}{2a}$
$= \dfrac{a \cdot 8\sqrt{a}}{2a}$
$= \dfrac{8a\sqrt{a}}{2a}$
$= 4\sqrt{a}$

75. $\dfrac{2}{\sqrt{7}} + \dfrac{3\sqrt{7}}{7}$
$= \dfrac{2 \cdot \sqrt{7}}{\sqrt{7} \cdot \sqrt{7}} + \dfrac{3\sqrt{7}}{7}$
$= \dfrac{2\sqrt{7}}{7} + \dfrac{3\sqrt{7}}{7}$
$= \dfrac{5\sqrt{7}}{7}$

76. $\dfrac{1}{2\sqrt{3}} - \dfrac{1}{3\sqrt{2}}$

$= \dfrac{1 \cdot 3\sqrt{2}}{2\sqrt{3} \cdot 3\sqrt{2}} - \dfrac{1 \cdot 2\sqrt{3}}{3\sqrt{2} \cdot 2\sqrt{3}}$

$= \dfrac{3\sqrt{2}}{6\sqrt{6}} - \dfrac{2\sqrt{3}}{6\sqrt{6}}$

$= \dfrac{3\sqrt{2} - 2\sqrt{3}}{6\sqrt{6}}$

$= \dfrac{(3\sqrt{2} - 2\sqrt{3}) \cdot \sqrt{6}}{6\sqrt{6} \cdot \sqrt{6}}$

$= \dfrac{3\sqrt{2} \cdot \sqrt{6} - 2\sqrt{3} \cdot \sqrt{6}}{6 \cdot 6}$

$= \dfrac{3\sqrt{12} - 2\sqrt{18}}{36}$

$= \dfrac{3\sqrt{4}\sqrt{3} - 2 \cdot \sqrt{9}\sqrt{2}}{36}$

$= \dfrac{3 \cdot 2\sqrt{3} - 2 \cdot 3\sqrt{2}}{36}$

$= \dfrac{6\sqrt{3} - 6\sqrt{2}}{36}$

$= \dfrac{6(\sqrt{3} - \sqrt{2})}{36}$

$= \dfrac{\sqrt{3} - \sqrt{2}}{6}$

77. Check $x = \dfrac{1+\sqrt{7}}{2}$:

$$2x^2 - 2x - 3 = 0$$

$$2\left(\dfrac{1+\sqrt{7}}{2}\right)^2 - 2\left(\dfrac{1+\sqrt{7}}{2}\right) - 3 \overset{?}{=} 0$$

$$\dfrac{2}{1} \cdot \dfrac{(1)^2 + 2(1)\sqrt{7} + (\sqrt{7})^2}{2^2} - \dfrac{2}{1} \cdot \dfrac{1+\sqrt{7}}{2} - 3 \overset{?}{=} 0$$

$$\dfrac{\cancel{2}}{1} \cdot \dfrac{1 + 2\sqrt{7} + 7}{\cancel{2} \cdot 2} - \dfrac{2 + 2\sqrt{7}}{2} - 3 \overset{?}{=} 0$$

$$\dfrac{8 + 2\sqrt{7}}{2} - \dfrac{2 + 2\sqrt{7}}{2} - 3 \overset{?}{=} 0$$

$$\dfrac{(8 + 2\sqrt{7}) - (2 + 2\sqrt{7})}{2} - 3 \overset{?}{=} 0$$

$$\dfrac{8 + 2\sqrt{7} - 2 - 2\sqrt{7}}{2} - 3 \overset{?}{=} 0$$

$$\dfrac{6}{2} - 3 \overset{?}{=} 0$$

$$0 = 0$$

78. Check $x = -2 - \sqrt{5}$:

$$x^2 + 4x - 1 = 0$$

$$(-2 - \sqrt{5})^2 + 4(-2 - \sqrt{5}) - 1 \overset{?}{=} 0$$

$$(-2)^2 - 2(-2)(\sqrt{5}) + (\sqrt{5})^2 - 8 - 4\sqrt{5} - 1 \overset{?}{=} 0$$

$$4 + 4\sqrt{5} + 5 - 8 - 4\sqrt{5} - 1 \overset{?}{=} 0$$

$$0 = 0$$

79.
$$a^2 + b^2 = c^2$$
$$(2\sqrt{a})^2 + (6\sqrt{a})^2 = c^2$$
$$2^2(\sqrt{a})^2 + 6^2(\sqrt{a})^2 = c^2$$
$$4a + 36a = c^2$$
$$40a = c^2$$
$$\sqrt{40a} = c$$
$$\sqrt{4}\sqrt{10a} = c$$
$$2\sqrt{10a} = c$$

80.
$$a^2 + b^2 = c^2$$
$$a^2 + (2b)^2 = (4b)^2$$
$$a^2 + 4b^2 = 16b^2 \qquad \text{Subtract } 4b^2 \text{ from both sides}$$
$$a^2 = 12b^2 \qquad \text{Extract square roots}$$
$$a = \sqrt{12b^2} \qquad \text{Simplify the radical}$$
$$a = \sqrt{4b^2}\sqrt{3}$$
$$a = 2b\sqrt{3}$$

81. $3\sqrt{x+2}-4=5$ Add 4 to both sides

$3\sqrt{x+2}=9$ Divide both sides by 3

$\sqrt{x+2}=3$ Square both sides

$(\sqrt{x+2})^2=3^2$

$x+2=9$ Subtract 2 from both sides

$x=7$

Check $x=7$:

$3\sqrt{x+2}-4=5$

$3\sqrt{7+2}-4\stackrel{?}{=}5$

$3\sqrt{9}-4\stackrel{?}{=}5$

$3\cdot 3-4\stackrel{?}{=}5$

$9-4\stackrel{?}{=}5$

$5=5$

Therefore, $x=7$ is the solution.

82. $\sqrt{x-3}+4=2$ Subtract 4 from both sides

$\sqrt{x-3}=-2$ Square both sides

$(\sqrt{x-3})^2=(-2)^2$

$x-3=4$ Add 3 to both sides

$x=7$

Check $x=7$:

$\sqrt{x-3}+4=2$

$\sqrt{7-3}+4\stackrel{?}{=}2$

$\sqrt{4}+4\stackrel{?}{=}2$

$2+4\stackrel{?}{=}2$

$6\neq 2$

Therefore, NO solution.

83. $\sqrt{2x+1}=x-7$ Square both sides

$(\sqrt{2x+1})^2=(x-7)^2$

$2x+1=(x)^2-2(x)(7)+(7)^2$

$2x+1=x^2-14x+49$ Write in standard form

$0=x^2-16x+48$ Factor

$0=(x-12)(x-4)$ Set each factor equal to 0

$x-12=0$ or $x-4=0$

$x=12$ $x=4$

Check $x=12$: $\sqrt{2x+1}=x-7$

$\sqrt{2(12)+1}\stackrel{?}{=}12-7$

$\sqrt{24+1}\stackrel{?}{=}5$

$\sqrt{25}\stackrel{?}{=}5$

$5=5$

Check $x=4$: $\sqrt{2x+1}=x-7$

$\sqrt{2(4)+1}\stackrel{?}{=}4-7$

$\sqrt{8+1}\stackrel{?}{=}-3$

$\sqrt{9}\stackrel{?}{=}-3$

$3\neq -3$

Therefore, $x=12$ is the only solution.

84. $4\sqrt{4x+1}=5x+2$ Square both sides

$(4\sqrt{4x+1})^2=(5x+2)^2$

$4^2(\sqrt{4x+1})^2=(5x)^2+2(5x)(2)+(2)^2$

$16(4x+1)=25x^2+20x+4$ Distribute 16 on the left side

$64x+16=25x^2+20x+4$ Write in standard form

$0=25x^2-44x-12$ Factor

$0=(25x+6)(x-2)$ Set each factor equal to 0

(Problem continued on next page)

218

$$25x + 6 = 0 \quad \text{or} \quad x - 2 = 0$$
$$25x = -6 \qquad\qquad x = 2$$
$$x = \frac{-6}{25}$$

Check $x = \frac{-6}{25}$: $\qquad 4\sqrt{4x+1} = 5x + 2$

$$4\sqrt{4\left(\frac{-6}{25}\right)+1} \stackrel{?}{=} 5\left(\frac{-6}{25}\right) + 2$$

$$4\sqrt{\frac{-24}{25} + \frac{25}{25}} \stackrel{?}{=} \frac{-6}{5} + \frac{10}{5}$$

$$4\sqrt{\frac{1}{25}} \stackrel{?}{=} \frac{4}{5}$$

$$4\left(\frac{1}{5}\right) \stackrel{?}{=} \frac{4}{5}$$

$$\frac{4}{5} = \frac{4}{5}$$

Check $x = 2$: $\qquad 4\sqrt{4x+1} = 5x + 2$

$$4\sqrt{4(2)+1} \stackrel{?}{=} 5(2) + 2$$

$$4\sqrt{8+1} \stackrel{?}{=} 10 + 2$$

$$4\sqrt{9} \stackrel{?}{=} 12$$

$$4(3) \stackrel{?}{=} 12$$

$$12 = 12$$

Therefore, $x = \frac{-6}{25}$ and $x = 2$ are solutions.

85. $\quad \sqrt[3]{3x+2} - 4 = 1 \qquad$ Add 4 to both sides

$\qquad \sqrt[3]{3x+2} = 5 \qquad$ Cube both sides

$\quad (\sqrt[3]{3x+2})^3 = 5^3$

$\qquad\quad 3x + 2 = 125 \quad$ Subtract 2 from both sides

$\qquad\qquad 3x = 123 \quad$ Divide both sides by 3

$\qquad\qquad\; x = 41$

86. $\quad 9 - 4\sqrt[3]{1-2x} = 17 \qquad$ Subtract 9 from both sides

$\qquad -4\sqrt[3]{1-2x} = 8 \qquad$ Divide both sides by -4

$\qquad\quad \sqrt[3]{1-2x} = -2 \qquad$ Cube both sides

$\quad (\sqrt[3]{1-2x})^3 = (-2)^3$

$\qquad\qquad 1 - 2x = -8 \qquad$ Subtract 1 from both sides

$\qquad\qquad\quad -2x = -9 \qquad$ Divide both sides by -2

$$x = \frac{9}{2}$$

87.

$$T = 2\pi\sqrt{\dfrac{L}{32}}$$

$$10.54 = 2\pi\sqrt{\dfrac{L}{32}} \qquad \text{Divide both sides by } 2\pi$$

$$\dfrac{10.54}{2\pi} = \dfrac{2\pi}{2\pi}\sqrt{\dfrac{L}{32}}$$

$$\dfrac{10.54}{2\pi} = \sqrt{\dfrac{L}{32}} \qquad \text{Square both sides}$$

$$\left(\dfrac{10.54}{2\pi}\right)^2 = \left(\sqrt{\dfrac{L}{32}}\right)^2$$

$$\left(\dfrac{10.54}{2\pi}\right)^2 = \dfrac{L}{32} \qquad \text{Multiply both sides by } 32$$

$$32\left(\dfrac{10.54}{2\pi}\right)^2 = 32\cdot\dfrac{L}{32}$$

$$90 \text{ ft} \approx L$$

88.

$$v = \sqrt{\dfrac{1.24\times 10^{12}}{R+h}}$$

$$17{,}187 = \sqrt{\dfrac{1.24\times 10^{12}}{3960+h}} \qquad \text{Square both sides}$$

$$17{,}187^2 = \left(\sqrt{\dfrac{1.24\times 10^{12}}{3960+h}}\right)^2$$

$$17{,}187^2 = \dfrac{1.24\times 10^{12}}{3960+h} \qquad \text{Multiply both sides by } (3960+h)$$

$$17{,}187^2(3960+h) = \dfrac{1.24\times 10^{12}}{3960+h}(3960+h)$$

$$17{,}187^2(3960) + 17{,}187^2 h = 1.24\times 10^{12} \qquad \text{Subtract } 17{,}187^2(3960) \text{ from both sides}$$

$$17{,}187^2 h = 1.24\times 10^{12} - 17{,}187^2(3960)$$

$$\dfrac{17{,}187^2 h}{17{,}187^2} = \dfrac{1.24\times 10^{12} - 17{,}187^2(3960)}{17{,}187^2} \qquad \text{Divide both sides by } 17{,}187^2$$

$$h \approx 238 \text{ mi}$$